HEAT PUM

SECOND EDITION

Pergamon Titles of Related Interest

CARTER & De VILLIERS
Principles of Passive Solar Building Design

DUNN & REAY
Heat Pipes, 3rd Edition

HORLOCK
Cogeneration: Combined Heat & Power

McVEIGH
Energy Around the World

McVEIGH
Sun Power, 2nd Edition

REAY
Industrial Energy Efficiency

REAY
Innovation for Energy Efficiency

REAY
Advances in Heat Pipe Technology

SODHA *et al.*
Solar Passive Building

Pergamon Related Journals
(free sample copy gladly sent on request)

Energy

Energy Conversion & Management

International Communications in Heat & Mass Transfer

International Journal of Heat & Mass Transfer

Journal of Heat Recovery Systems & CHP

Progress in Energy & Combustion Science

Solar & Wind Technology

Solar Energy

HEAT PUMPS

SECOND EDITION

D. A. REAY
David Reay and Associates, Whitley Bay, UK

and

D. B. A. MACMICHAEL
PA Technology, Royston, UK

PERGAMON PRESS
OXFORD · NEW YORK · BEIJING · FRANKFURT
SÃO PAULO · SYDNEY · TOKYO · TORONTO

U.K.	Pergamon Press, Headington Hill Hall, Oxford OX3 0BW, England
U.S.A.	Pergamon Press, Maxwell House, Fairview Park, Elmsford, New York 10523, U.S.A.
PEOPLE'S REPUBLIC OF CHINA	Pergamon Press, Room 4037, Qianmen Hotel, Beijing, People's Republic of China
FEDERAL REPUBLIC OF GERMANY	Pergamon Press, Hammerweg 6, D-6242 Kronberg, Federal Republic of Germany
BRAZIL	Pergamon Editora, Rua Eça de Queiros, 346, CEP 04011, Paraiso, São Paulo, Brazil
AUSTRALIA	Pergamon Press Australia, P.O. Box 544, Potts Point, N.S.W. 2011, Australia
JAPAN	Pergamon Press, 8th Floor, Matsuoka Central Building, 1-7-1 Nishishinjuku, Shinjuku-ku, Tokyo 160, Japan
CANADA	Pergamon Press Canada, Suite No. 271, 253 College Street, Toronto, Ontario, Canada M5T 1R5

First edition 1979
Second edition 1988

Library of Congress Cataloging in Publication Data

Reay, D. A. (David Anthony)
Heat pumps.
Includes index.
1. Heat pumps. I. Macmichael, D. B. A.
II. Title.
TJ262.R4 1987 621.402'5 87–22067

British Library Cataloguing in Publication Data

Reay, David A.
Heat pumps. — 2nd ed.
1. Heat pumps
I. Title II. Macmichael, D. B. A.
621.402'5 TJ262

ISBN 0–08–033463–6 Hardcover
ISBN 0–08–033462–8 Flexicover

Printed in Great Britain by A. Wheaton & Co. Ltd., Exeter

Contents

	Preface to the Second Edition	vii
	Preface to the First Edition	viii
	Acknowledgements	ix
	Introduction	1
Chapter 1	History of the Heat Pump	3
Chapter 2	Heat Pump Theory	14
Chapter 3	Practical Design	49
Chapter 4	Heat Pump Applications - Domestic	105
Chapter 5	Heat Pumps in Commercial and Municipal Buildings	160
Chapter 6	Heat Pump Applications in Industry	202
Appendix 1	Refrigerant Properties	289
Appendix 2	Bibliography	299
Appendix 3	Heat Pump Manufacturers	308
Appendix 4	Conversion Factors	329
Appendix 5	Nomenclature	331
	Index	333

Preface to the Second Edition

Since the first edition of this book was published in 1979, heat pumps have continued to play an important role in energy conservation, although this role varies in intensity, depending upon both geographical location and application. The use of small heat pumps for heating-only duties in single-family houses has not met with the success originally envisaged. However, large heat pumps in district heating schemes, particularly in Sweden, have continued to demonstrate strong market penetration. Where both heating and cooling can be combined, heat pumps have achieved viability in many application areas. Industrial applications of heat pumps are of considerable interest, but except for a comparatively few examples, this interest has not been reflected in the number of installations.

The last six years have seen an increase in development activity in the field of absorption cycle heat pumps, and this is reflected in several chapters. Sections on design have been updated, and some aspects of the theory of heat pumps have been expanded. The chapters dealing with heat pump applications have been extensively revised, and a number of new case studies have been incorporated.

The appendices have been updated, and the bibliography has been extended.

We hope that this new edition proves of interest and value to the reader.

June 1987

D. A. Reay
D. B. A. Macmichael

Preface to the First Edition

The need to conserve energy has become an everyday feature of our lives, at home, in offices and in factories. Energy saving is an area of activity which draws people together, as is shown by the activities of the International Energy Agency and the European Economic Community in funding joint energy projects.

One device which can make a significant contribution to energy conservation is the heat pump. By raising low grade (or temperature) heat to a more useful temperature, 'new' sources of heat become available, for example ambient air, and also, sources of waste heat which were previously considered unsuitable for recovering, because of their low temperature. The heat pump therefore can vastly increase the potential use of low grade energy, albeit with the expenditure of a proportion of high grade energy to achieve this.

We hope that this book, reviewing in some detail the current state of the art in heat pump design and application, will interest both those intimately familiar with one or more special aspects of this technology, and those who are on the periphery - for example plant managers who could benefit from low grade heat recovery in their processes, architects who are designing low energy housing schemes, or the local authority contemplating a new swimming pool and leisure complex.

D. A. Reay
D. B. A. Macmichael

Acknowledgements

The authors would like to acknowledge the assistance of several organisations in providing material for inclusion in this book. Reference is made in some cases in the text, but the list below gives the source of additional material.

Table 3.1 and the refrigerant nomenclature list in Appendix 1 are reprinted with permission from the 1977 Fundamentals Volume, ASHRAE HANDBOOK and Product Directory.

Figs. 3.4, 3.11 and 3.12 are reprinted with permission from the 1963 Equipment Volume, ASHRAE HANDBOOK and Product Directory.

Figs. 3.20 and 3.27 are reprinted with permission from the 1976 Systems Volume, ASHRAE HANDBOOK and Product Directory.

Fig. 3.7 is reprinted with the permission of James Howden Ltd, Glasgow.

The authors would also like to acknowledge the assistance of David Hodgett of the Electricity Council, for permission to use data reproduced in Chapter 6 on heat pumps in drying, and Mr J.L. Bowen, Marketing Manager of Dunham-Bush Ltd, for providing photos and data on chiller units and condensers (Figs. 3.23, 3.24, 5.5-5.7). American Air Filter (AAF) Ltd provided data on the Enercon system for Chapter 5, and Temperature Ltd are also thanked for providing data on the Versa Temp units. Thanks are also due to James Howden, Grasso, Westair Systems, Westinghouse, Glynwed, Sulzer, Carrier, Carlyle and Philips for the provision of data. Alan Deakin of ETSU, Harwell Laboratory, kindly provided photos of some heat pump installations. The European Commission (DG XII-D-3) is acknowledged as a source of R & D data and DG XVII is acknowledged as a source of case histories on demonstrations.

Last, but not least, we would like to thank our typist, Joan Tulip, for preparing the camera-ready copy.

June 1987

D. A. Reay
D. B. A. Macmichael

Introduction

The heat pump, the invention of one form of which is generally attributed to Lord Kelvin, is thermodynamically identical to the refrigerator. Most homes, foodstores, and large commercial buildings in the industrial societies have refrigerators or air conditioning plant. The principal difference between the heat pump and the refrigerator is in the role they play as far as the user is concerned. On the one hand refrigerators and air conditioners provide useful cooling, whereas the heat pump provides useful heat.

As will become evident later, there are quite a few books on heat pumps, some of which were written several decades ago. The literature on thermodynamics, air conditioning, and even industrial processes is full of references to heat pumps, and the number of installed units world-wide runs into 7 figures. There is, however, still a certain amount of mystique surrounding the operation of the heat pump, and one of the purposes of this book is to overcome this barrier to their understanding and application.

Heat pumps are available in many shapes, sizes and types, of which those operating on the vapour compression cycle are the most common. Other types include absorption cycle units (operating on a similar principle to the 'Electrolux' refrigerator) and thermoelectric devices. Heat pump sizes vary from a few Watts to several megawatts output, with, in the case of the various compression cycles, a multiplicity of compressor drives available, ranging from electric motors to internal and external combustion engines of all types. Compressors too can be one of a variety of configurations.

One vexed question concerning heat pumps is their reliability. Experiences in the late 1950's and early 1960's in the United States and Europe, involving component failures and high costs, led to a certain amount of disillusionment with the heat pump which has now largely been overcome - reliability and after-sales service is a major selling point of many domestic systems. Because of the importance of reliability, a number of case studies in the book examine this area.

Since the 1973 'energy crisis', the momentum of heat pump development has increased considerably, and this has motivated, and been motivated by, the need to apply heat pump systems in practical situations. Heat pump applications, and associated energy savings, are the 'raison d'etre' for heat pump development, and applications in the home, commercial and municipal buildings, and industry are many and varied. Encouragement for heat pump development is being given by many national governments and international organisations and in some instances programmes have been persued for over a decade.

As a result of some of these long-term R&D programmes, it is suggested by the International Energy Agency that much more competitive heat pump products will reach the market-place during the next few years. It is claimed that systems for heating detached houses, having an electricity demand of only 1.5-2.5kW will be available, costing typically £500/750 at the factory gate. Organisations promoting the interests of heat pump manufacturers have, however, to find ways to keep the heat pump in certain markets until such products are commercially available.

The first Chapter of this book briefly describes some of the historical landmarks in the development and application of heat pumps. This is followed by a detailed description of the various heat pump cycles, and their associated theory, concentrating mainly on the ubiquitous vapour compression cycle. Chapter 3 examines the components of a heat pump system - drive, compressor, heat exchangers etc - in some detail, including the more practical considerations to be taken into account in their selection.

Major sections of the book are devoted to considering the applications of the heat pump. The first major application area presented is the use of heat pumps in the home, for space heating, heat recovery from domestic appliances, and provision of hot water services. The various external and internal sources of heat which can be used are introduced in detail. In this and subsequent Chapters emphasis is put on commercially available systems, but important current development programmes are also described. Chapters 5 and 6, which deal respectively with commercial buildings and industrial applications of heat pumps, follow in a similar vein to Chapter 4, but both commence with an examination of the concept of heat reclaim and heat recovery from refrigeration plant, showing in one way the inevitable relationship between the refrigerator and the heat pump.

A number of appendices are included which, it is hoped, will be of considerable value to readers of varied backgrounds. The designer will be interested in the data on heat pump working fluids, and the potential user in the extensive list of manufacturers and agents worldwide. A comprehensive bibliography, including an historical section, is included, and the final Appendix gives common conversion factors relating British and SI units, as well as data on fuel calorific values and energy equivalents.

CHAPTER 1
History of the Heat Pump

The heat pump for many years remained an item of thermodynamic mystique, of interest to professors and research workers but seldom, it seems, appreciated in the 'real world'. To draw attention to those areas where development has taken place, this chapter traces the history of Lord Kelvin's 'heat multiplier'.

1.1 The Nineteenth Century

As described in Chapter 2, the basic principle of the vapour compression cycle heat pump derives from the work of Carnot in the early 19th Century, and his thesis on the Carnot cycle published in 1924. It was William Thomson (later to become Lord Kelvin) who first proposed a practical heat pump system, or 'heat multiplier' as it was then known, indicating that a refrigerating machine could also be used effectively for heating. In putting forward arguments for the development of such a system, Thomson was anticipating the fact that conventional energy reserves would not permit the continuing direct combustion of fuel in a furnace for heating - his 'heat multiplier' would use less fuel than a conventional furnace (ref.1.1). Illustrated in Fig.1.1, the heat pump proposed by Thomson used air as the working fluid. The ambient air was drawn in to a cylinder where it was expanded, thus reducing both its temperature and pressure. The air was then passed through an air-to-air heat exchanger, located outside, where the cooled air was able to pick up heat from the ambient air. Prior to being expelled into the building to be heated, the air was compressed back to atmospheric pressure, resulting in an increase in temperature above that of the ambient level. It is believed that a successful version of this machine was constructed in Switzerland (ref. 1.2). Thomson claimed that his heat pump was able to produce heat using only 3 per cent of the energy which would be needed for direct heating.

Absorption cycle machines have an even longer history, refrigerators based on the use of sulphuric acid and water dating from 1777 (ref.1.3). Systems using this fluid combination, improved and modified by Edmond Carre´, were used extensively in Paris cafes in the late 1800's. Slightly earlier than this, in 1859, a patent was filed by Ferdinand Carre´, the brother of Edmond, for the working fluid pair of ammonia and water. Again this system was used, in both continuous and discontinuous operating forms, for cooling and ice-making. Further background data on these and systems subsequently developed within the category of chemical heat pumps are given by Hodgett (ref.1.4).

1.2 The Twentieth Century

While one may dwell upon the history of refrigeration plant, particularly with respect to the latter part of the 19th Century, concentration on heat pumps necessitates a fairly rapid movement to the 1920's and 1930's, when one comes across the first heat pump installed in the United Kingdom (ref.1.5). Haldane reported in 1930 his work on the installation of and tests on a domestic heat pump* in Scotland in 1927. This unit provides heat for hot water and space heating, using outside air as the heat source.

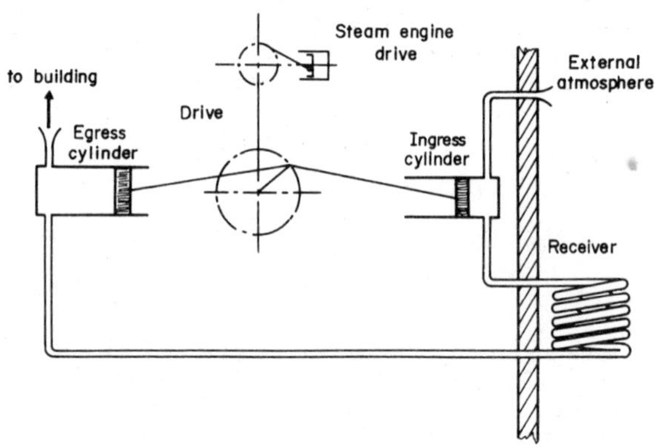

Fig.1.1 Layout of Thomson's 'Heat Multiplier'

Following on from this work, interest in the United States led toa number of design studies being initiated, some of which led to demonstrations (ref.1.6). 'Customized' systems for comfort heating were designed, but the number of projects reaching the demonstration stage was comparatively few because they were all privately financed.

The first major heat pump installation in Europe was commissioned during the period 1938-39 in Zurich. This unit, which used river water as the heat source, utilised a rotary compressor

* All heat pumps in this Chapter operate on the vapour compression cycle, unless otherwise stated.

with R12 as the working fluid (ref.1.7). Used to heat the Town Hall, the output of the Zurich heat pump was 175 kW, delivering water at a temperature of 60°C for space heating. A thermal storage system was incorporated in the circuit, in the form of a calorifier which could be boosted by electric heating at periods of peak demand. The system was also arranged so that it could provide cooling in summer months. This and other early heat pumps installed in Switzerland are listed in Table 1.1, (ref.1.8). The motivation behind early heat pump development in Switzerland was directed at reducing the high coal consumption in that country. Some of the installations have now been operating successfully for more than 30 years.

1.2.1 Early UK Heat Pumps The first heat pump installed in the United Kingdom which successfully demonstrated that a large building could be heated using this technique was located in Norwich (ref.1.9). The original schematic of the system is illustrated in Fig.1.2. The unit was installed in the offices of the Norwich Corporation Electricity Department, a building with a volume of 14,200 cubic metres. The heat source used was a river and the heat sink was circulating hot water, delivered at a temperature of 49°C. Sulphur dioxide was used as the refrigerant, and 'coefficients of performance' (defined in Chapter 2) of the order of 3 were achieved. A second-hand compressor, dating from 1926, was used, and it was driven using a belt drive from a DC electric motor. A power input of between 40 and 80 kW was required, depending upon the ambient and internal conditions.

The second heat pump installation most frequently cited in historical papers in the United Kingdom on the topic was the unit at the Royal Festival Hall, on the banks of the Thames in London, (ref.1.10). Again an experimental unit, it was designed to heat the Hall in winter and provide cooling in summer, the design heating load being of the order of 2.7 MW. (This proved to be an over-estimate of the actual peak heating demand). The heat source for the heat pump evaporator was the river Thames, and the condenser water outlet temperature was 71°C. When cooling, chilled water at 4°C was produced. The heat pump itself was unique in being constructed using detuned Rolls-Royce Merlin engines, converted to run on town gas and rated at 522 kW. The superchargers were used as the basis for the centrifugal compressors, and condenser heat output was supplemented by a waste heat boiler. Refrigerant R12 was used as the working fluid, the compressor coefficient of performance being 5.1, and the effective PER (Primary Energy Ratio - see Chapter 2) of the unit approached 1.5.

The system proved to be uneconomic, partly because of maintenance costs, but largely due to the overdesign. The capital cost as installed was £103,200 (at 1953 rates), but the designers estimated that a correctly sizes unit would have cost only £52,500. (Details of this and other early UK heat pumps are summarized in Table 1.2, ref.1.11).

TABLE 1.1 Early Swiss Heat Pump Installations

Year of Construction	Heat Source	Location	Output (kW)	Delivery Temp (°C)	Application
1938	River water	Zurich	175	70	Space heating
1939	Air	Zurich	58	30-40	Air conditioning
1941	River and waste water	Zurich	1500	23-45	Swimming pool heating
1941	Lake	Skeckborn	1950	70	Process heat in artificial silk mill
1941	Air	Landquart	122	-	Felt drying in paper mill
1942	River water	Zurich	7000	70	Space heating
1943	River water	Zurich	1750	50	Space heating
1943	-	Schonenwerd	250	50	Air conditioning in shoe factory
1944	Fermentation cellar	Largenthal	140	45	Heating & cooling in brewery
1945	Lake	Lugano	-	-	Space heating

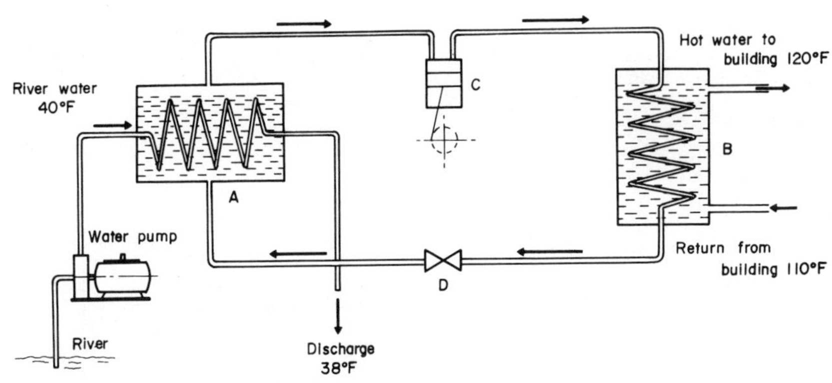

Fig.1.2 Layout of Sumner's heat pump for the Norwich Electricity Department

One of the earliest domestic heat pumps to operate successfully in the UK was installed in his own home by Sumner, the designer of the heat pump for the Norwich Electricity Building described above. Sumner's heat pump, (ref.1.12), was used in a single storey house, which was well-insulated, for whole house space heating. The heat was extracted from the outside air during the first few years of operation, but later a ground coil, extracting heat from below ground surface, at a depth of approximately 1 metre, was used. Heat dissipation into the house was via copper pipes located in the concrete floor. A coefficient of performance of 2.8 was achieved. The unit is still in operation today. A number of small heat pumps were produced on a commercial basis in the 1950's for domestic use (see also Chapter 5 - the Lucas unit). The most well-documented of these was the Ferranti 'fridge-heater' (ref.1.13). This system combined useful cooling and heating functions by extracting low grade heat from the larder, where food is stored, for upgrading for use in water heating. A 136 litre water storage tank was used as the heat sink, being supplied with 0.7 kW in the winter and 1.3 kW in warmer months. Power input to the compressor was 400 W, and the larder was cooled by an average 11°C. As a point of interest, the unit cost £141, but was classified as a luxury item which, at the time, attracted a purchase tax of 60 per cent, which deterred potential purchasers. However, those sold performed adequately.

The use of a heat pump at Nuffield College, Oxford, was first conceived in 1954, (ref.1.14). Illustrated in layout form in Fig.1.3, the system used low grade heat available in sewage as the heat source. The temperature of the sewage was between 16 and 24°C. The compressor was driven by a 31 kW diesel engine, giving an overall COP of about 4.

TABLE 1.2 Properties of Some Early Heat Pumps Used in the United Kingdom

Location	Date	Heat Source	Heat Sink	Power* Input	Power Output	Cost		Mean COP
						Capital	Running	
Norwich offices	1945	Water	Water	40–80kW	120–240kW	–	10.2p/kWh	3
Royal Festival Hall	1949	Water	Water	522kW (Gas)	2700kW	#103,200 (1953)	42p/GJ	1.5
Norwich House	1950	Concrete floor	Water	1.3kW	3.74kW	–	Same as conventional solid fuel system	2.8
BEAIRA	1951	Water	Water	3kW	7–15kW	–	–	2.2–5
ERA	1952	Soil	Water	7.5kW	25kW	#2.252 (1952)	#89p.a. (1955)	3
Various (Ferranti fridge-heater)	1954	Air	Water	0.4kW	0.7–1.3kW	#141 (1954)	–	3
Solar House (Denco Miller)	1956	Air	Air	–	6–12kW	#325 (1956)	#29 p.a. (1956)	–
Various (Brentford Electric)	1957	Air	Water	9kWh/day	–	–	40p/week (1957)	–
Nuffield College	1961	Sewage	Water	31kW (Diesel)	150kW	#9,310 (1962)	#896 p.a. (1962)	3.98

* Electricity unless otherwise stated.

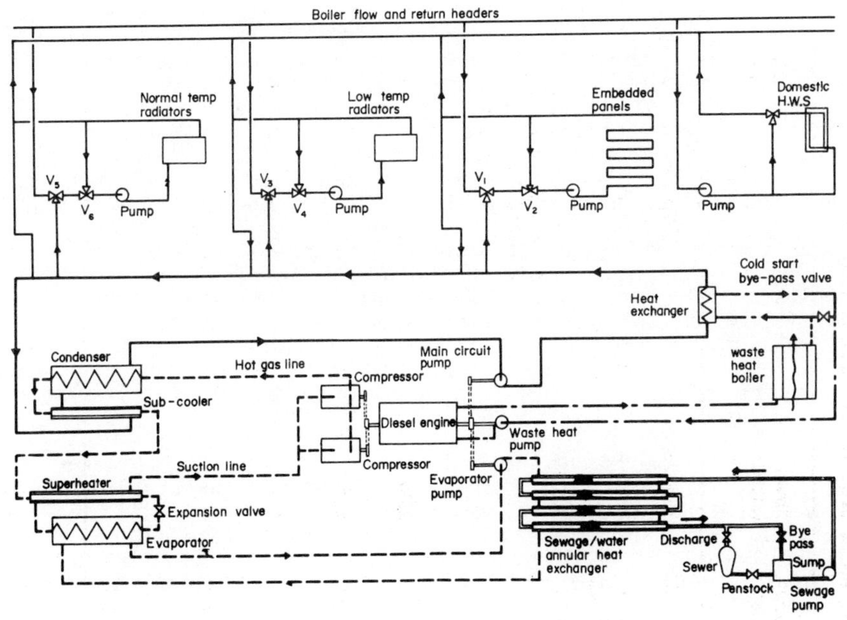

Fig.1.3 Overall Schematic Diagram of the First Nuffield College Installation at Oxford

A comparison was made of the running costs of this heat pump with those of a conventional oil-fired boiler and an electric heat pump. The actual operating cost of the diesel unit was 9.86 pence/therm, and that of a boiler 13.2 pence/therm. The predicted operating cost of an electric heat pump was 15.9 pence/therm, (all at 1963 costs, in 'old' currency). The cost of fuel oil at the time was 13.75 pence/gallon, and that of electricity 1.375 pence/kW.h. The capital cost of the installation was £9,310, or £73/kW of heat produced.

As illustrated in Fig.1.3, use was made of the exhaust heat from the engine to boost the temperature of water used to recover heat from the engine cooling jacket. The heat was distributed throughout the College via a hot water circuit at 49°C, providing 150 kW of the total design heating load of 450 kW.

While the original installation suffered because of lack of preventative maintenance, (which did not prevent it achieving almost 14,000 hours of running time), interest in the heat pump was sufficient to encourage in the early 1970's rehabilitation of the system. With support from the European Commission (ref. 1.15), a new packaged heat pump unit, this time using natural gas to fuel the prime mover, was commissioned in 1983.

1.2.2 Growth in the United States Reference has already been made to early projects in the USA. It was, however, in the late 1940's that the realization grew that heat pump development and marketing would best be met if 'unitary' systems were produced. These are factory engineered and assembled units which, like conventional domestic boilers, could be 'plumbed in' easily and cheaply by engineers in the home or small commercial premises. It was in 1952 that heat pumps developed along these lines were offered in quantity to the market, (ref.1.6).

In the first full year of production approximately 1000 units left the factories. This had doubled by 1954 and increased tenfold by 1957. In 1963 76,000 units were manufactured. The majority of these units were installed in southern areas of the USA, where the need for summer cooling existed, as well as the necessity for heating in winter, enabling the heat pump to compete effectively with heating-only conventional systems, based on boilers.

However, problems were encountered when these heat pumps were used in colder northern States, and under arduous winter conditions system durability was found to be seriously lacking. As a result the unitary heat pump achieved a growing reputation for unreliability. This led to a decline in the industry in the early 1960's, and it was not until 1971 that sales began to pick up again, as shown in Fig.1.4. The intervening period brought new, more reliable designs, and, more important, a major effort by manufacturers to make sure that installers and maintenance engineers were fully conversant with the product. Several studies of reliability were made (see Chapter 4), and the Air Conditioning and Refrigeration Institute in 1974 commenced a programme of heat pump certification, classifying products in terms of their heating and cooling duties. This is still in operation.

Although electricity costs in the USA fell during the 1960's, leading to a move away from heat pumps to all-electric direct heating, (in addition to the reduction in demand caused by unreliability), the 1973 'energy crisis' saw the start of a much stronger growth of interest in heat pumps, and by 1976 sales had topped 300,000 units per annum.

At the time the First Edition of this book was written, the total number of heat pumps installed in the USA exceeded 2 million, (based on 1977 data). A study carried out on behalf of the International Energy Agency (IEA) shows how the market has grown, up to 1981, and gives predictions for future growth, (ref.1.16). By 1981 annual heat pump shipments had reached 505,000 units, although some leveling of demand had been noticed. The total number of installations had reached 4.4 million, and air-air heat pump systems account for 80 per cent of annual sales. The breakdown of shipments for 1980, in terms of type, is given in Table 1.3.

TABLE 1.3 Heat Pump Shipments in the USA in 1980, (Ref.1.16)

Type	Number
Electric Air-Air	443,000
Electric Water-Air (warm water)	84,000
Electric Water-Air (cool water)	23,000
Gas Fired*	16,000
Commercial	11,000
Packaged Units	53,000
Total	630,000

* Air conditioners

The IEA study also included in its terms of reference projections for heat pump markets in the 1990's. Many market research organisations have been involved in such projections, including those acting on behalf of the IEA. The volatility of the market may be judged from the fact that annual sales of residential electric heat pumps are predicted to be between 154,000 and 1,700,000 in the USA in the 1990's, and sales of gas fired heat pumps for the same sector are projected to be 9,000 to 2,400,000.

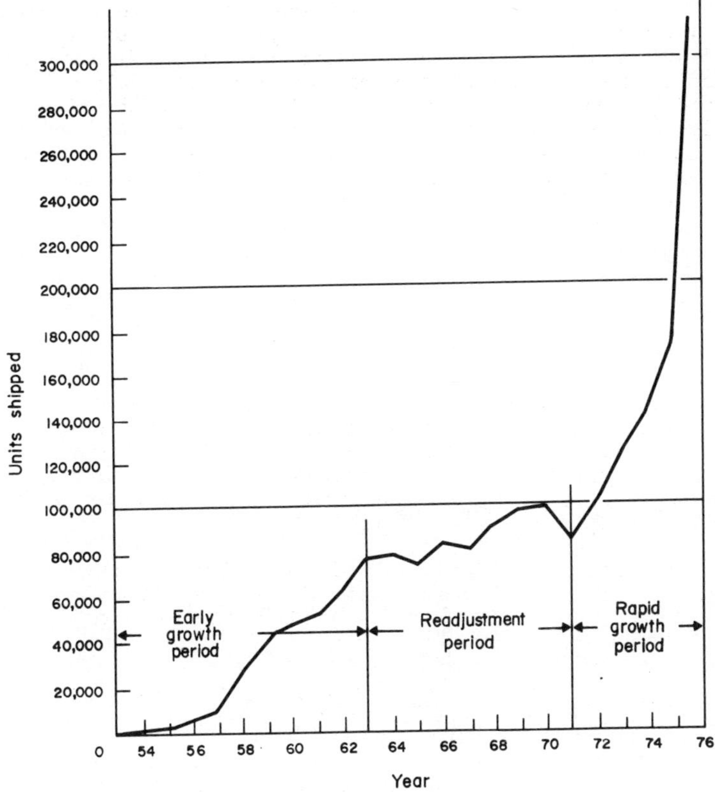

Fig.1.4 Manufacture of unitary heat pumps in the USA between 1954 and 1976

It should be pointed out that much of the study data arose during the 1979 oil price rise period. Subsequent events have in most instances made such projections look very optimistic.

This, of course, also applies to Europe. The number of heat pumps sold in West Germany, generally acknowledged to be the European market leader in the domestic field, has dropped significantly. From a figure of about 30,000 units in 1980, the annual sales have fallen to approximately 6000 units in 1985.

1.3 Current Status

Interest in the heat pump has never been as great as it is today, with manufacturers in Europe, Japan, and North America serving industrial, commercial buildings, and domestic applications. International organisations such as the International Energy Agency and the European Commission have major heat pump development programmes in hand, and demonstrations of new heat pump technology, or the use of existing technology in new applicatiion areas, are implemented. During the next few years a number of advanced heat pumps for use in the home will enter the market, using gas instead of electricity, and industrial applications will widen considerably, although on a much more selective basis than to date.

The heat pump enables us to use energy more effectively, and helps us to recover waste energy. These features should guarantee it a significant role in conserving our energy resources, in spite of short term influences of modest energy prices.

REFERENCES

1.1 Thomson, W. On the economy of the heating or cooling of buildings by means of currents of air. Proc. Glasgow Phil. Soc., Vol. III, pp 666-675, Dec. 1852.

1.2 Fearon, J. Heat from cold-energy recovery with heat pumps. Chartered Mechanical Engineer, pp 49-53, Sept. 1978.

1.3 Thevenot, R. A History of Refrigeration Throughout the World. International Institute of Refrigeration, Paris, 1979.

1.4 Hodgett, D.L. HTFS Design Report No.49: Heat Pumps. Part 8, Chemical Heat Pumps. Heat Transfer and Fluids Service, UKAEA Harwell, 1985.

1.5 Haldane, J.G.N. The heat pump - an economical method of producing low grade heat from electricity. I.E.E. Journal, Vol.68, pp 666-675, June 1930.

1.6 Pietsch, J.A. The unitary heat pump industry - 25 years of progress. ASHRAE Jnl., Vol.19, Pt.7, pp 15-18, July 1977.

1.7 Egle, M. The heating of the Zurich Town Hall by the heat pump. SEV Bulletin, Vol.29, pp 261-273, 27 May 1978.

1.8 Von Cube, H.L. and Steimle, F. Warmepumpen. Grundlagen und Praxis. VDI-Verlag GmbH, Dusseldorf, 1978.

1.9 Sumner, J.A. A summary of heat pump development and use in Great Britain. J.Inst. of Fuel, pp 318-321, Jan 1953.

1.10 Montagnon, P.E. and Ruckley, A.L. The Festival Hall heat pump. J. Inst. of Fuel, pp 1-17, Jan 1954.

1.11 Macadam, J.A. Heat pumps - the British experience. Building Research Establishment Note N117/74, Watford, Dec. 1974.

1.12 Sumner, J.A. Domestic Heat Pumps. Prism Press, Dorchester, 1976.

1.13 Butler, C. Ferranti 'fridge-heater'. Arch. J. Info. Sheet 28.J.1, May 31, 1956.

1.14 Kell, J.R. and Martin, P.L. The Nuffield College heat pump J.Inst. Heating & Ventil, Engnrs., pp 333-356, Jan. 1963.

1.15 Martin, P.L. and Oughton, D.R. The Nuffield College Phoenix. Proc. Int. Seminar on Energy Saving in Buildings, The Hague, 14-16 Nov. 1965. D. Reidel, Dordrecht, 1984.

1.16 Anon. Heat Pump Systems - A Technology Review. International Energy Agency/OECD, Paris, 1982.

CHAPTER 2
Heat Pump Theory

2.1 Introduction

The intention in this Chapter is to give the reader sufficient theoretical background to allow him to analyse and make comparison between different heat pump cycles using a variety of available working fluids. A good understanding of the theory will help the reader to appreciate the limitations of the heat pump, since these limitations are imposed not only by mechanical and engineering problems but also by the laws of nature.

It will be assumed that the reader understands the meaning of a thermodynamic 'state' and is familiar with the 'properties' which define the state: temperature, pressure, specific volume, enthalpy and entropy. These are the parameters which will be juggled in this Chapter and a full description of their meaning can be obtained from Rogers & Mayhew (ref.2.1).

A significant proportion of the Chapter will deal with the mechanical vapour compression heat pump cycle since this is by far the most common type of heat pump cycle. This will be followed by discussion on two heat-actuated cycles: the absorption cycle and a "Rankine/Rankine" cycle, the former being particularly important, and a brief description of a variety of other cycles which are receiving attention.

2.2 The Carnot Cycle

It was in 1824 that Carnot was the first to use a thermodynamic 'cycle' to describe a process, and the cycle which he then conceived remains the fundamental measure against which to judge heat pump efficiency.

The heat pump can be considered simply as a heat engine in reverse. The heat engine removes heat from a high temperature source and discharges heat to a low temperature and in so doing can deliver work. The heat pump requires a work input to remove heat from a low temperature and deliver it to a higher temperature as illustrated in Fig.2.1.

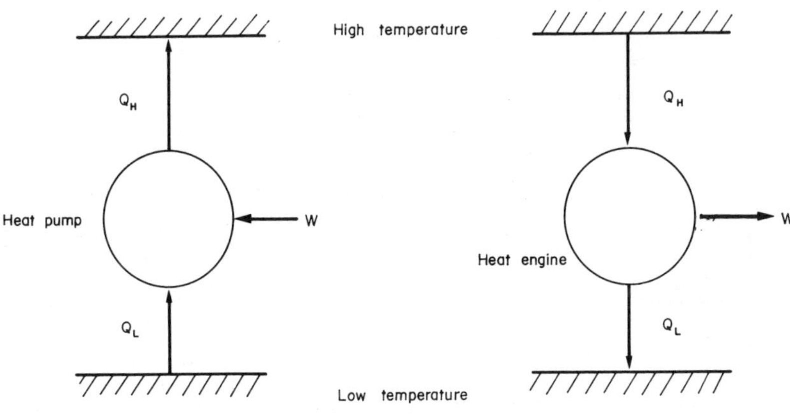

Fig.2.1 Thermodynamic model of heat pump and heat engine

It can be shown very simply that if these machines are both 'reversible' (a word used to describe thermodynamic processes in which there is no stray loss of heat or work) then there is a finite limit to how effective each one is. In effect the ratio Q_H/W is the same in both cases. If this were not the case, then a perpetual motion machine could be constructed by coupling the two together. This ratio is very important. In the case of a heat engine it is expressed as W/Q_H and called the thermal efficiency and in the case of a heat pump one may write Q_H/W and call it the coefficient of performance or COP. The reader should be wary of the refrigeration engineer who uses a different ratio, Q_L/W which we will refer to as COP (refrigeration) to make the distinction clear. Note that since $Q_H = W + Q_L$ then COP = COP (refrigeration) + 1.

The Carnot cycle in Fig.2.2 shows a very basic heat machine operating between two temperatures.

The cycle shown is for a heat pump. Heat is delivered isothermally at T_H and received isothermally at T_L . Expansion and compression are achieved isentropically, and the balance of work required is delivered by an external prime mover. By using the definition of entropy and the laws of thermodynamics it can be shown that the Carnot Coefficient of Performance is given by:

$$COP = \frac{T_L}{T_H - T_L} + 1$$

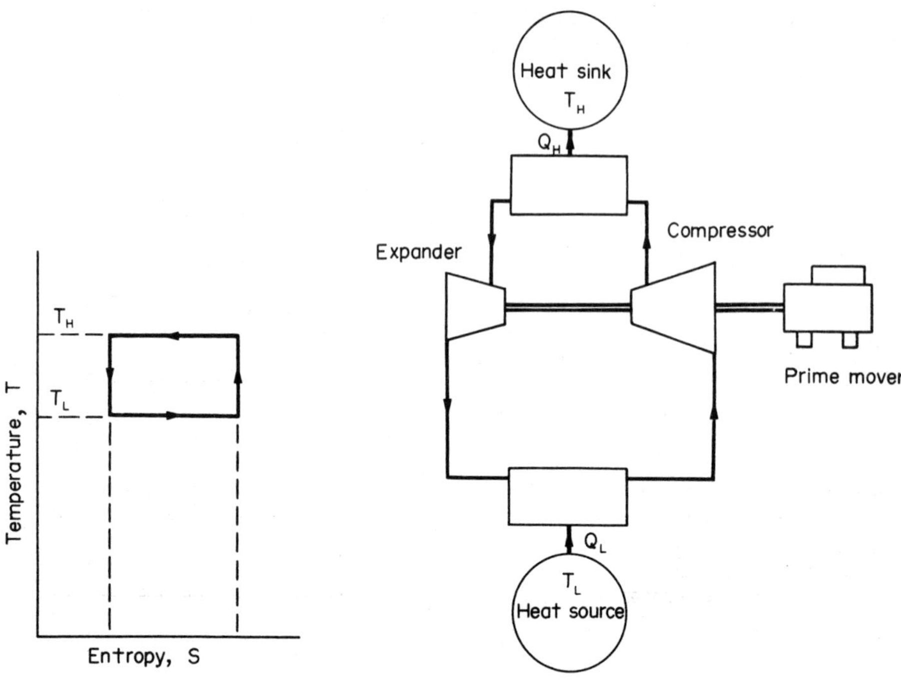

Fig.2.2 The Carnot ideal heat-pump cycle

No heat pump constructed within our simple Newtonian universe can have a better performance and all our practical cycles can do is struggle towards achieving this.

2.3 Mechanical Vapour Compression Cycle

In order to approach the simple Carnot cycle, indeed in order to make a useful heat pump, it is necessary to achieve the delivery and removal of heat under virtually isothermal conditions. To this end a working fluid is chosen which will change phase at useful temperatures and pressures. It will absorb heat by evaporation and it will deliver heat by condensation. These processes form the isothermal stages of the cycle. The cycle normally used employs compression of the dry vapour. This is because of mechanical limitations of the majority of compressor types used (see Chapter 3). The ingress of liquid refrigerant entrained in the vaporised refrigerant entering a compressor can damage the valves. The ingress of large volumes of liquid into a compressor would completely destroy it, if precautionary measures were not taken in the design (reciprocating piston compressors, for example, have spring-loaded cylinder heads to allow pressure relief).

The mechanical vapour compression cycle is shown in Fig.2.3 on a T-S (temperature-entropy) diagram.

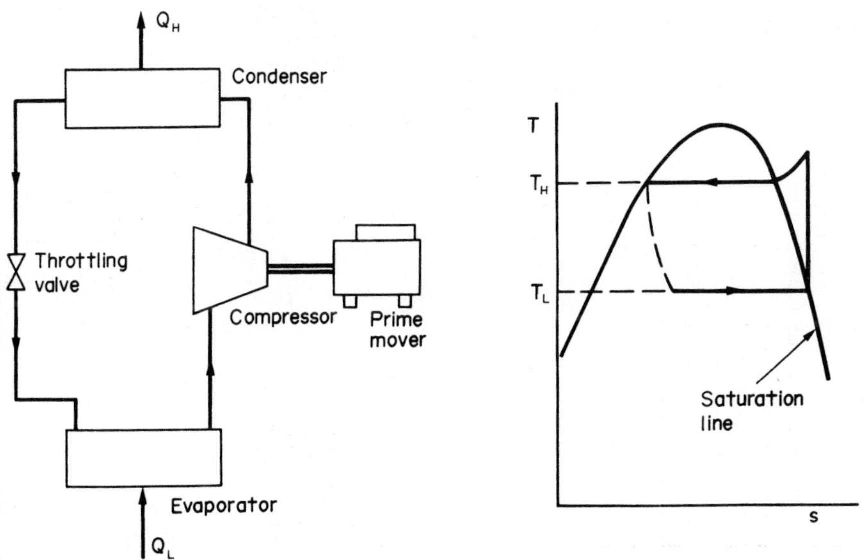

Fig.2.3 The mechanical vapour compression cycle

The discussion is restricted here to a dry-compression cycle using a throttling valve for expansion. This valve is either a variable nozzle or orifice, with external regulation, or else a fixed 'capillary' depending on the amount of control required. The fact that an expansion engine is not used means that a small amount of work is not available, which reduces the COP. In general it is found that the cost of recovering this work is not justified. The process of expansion through a nozzle is irreversible and for this reason has been shown dotted on the T-S diagram. It is generally regarded as 'adiabatic' for the purposes of analysis which means that there is no addition or removal of heat from the fluid in passing from the high pressure to the low pressure state.

It is at this stage that the engineer inside us cries 'enough' and demands a simpler demonstration of the work cycle. To this end it is universal practice to show the mechanical vapour compression cycle on a pressure-enthalpy (p-h) diagram. This is shown in Fig.2.4, and this type of diagram will be used for the remainder of this discussion.

This figure should now be studied carefully. Dense high pressure refrigerant is leaving the compressor at point 1. Because it is necessary to compress only dry vapour and because of the slope of the isentropic lines, this vapour is superheated, that is, it must be cooled at constant pressure before it begins to condense, at point 2. Between 2 and 3 condensation continues at constant temperatures until there is no vapour left. This illustrates that the condensing heat exchanger, the condenser, will almost always have to cope with a certain amount of superheat at the high temperature end.

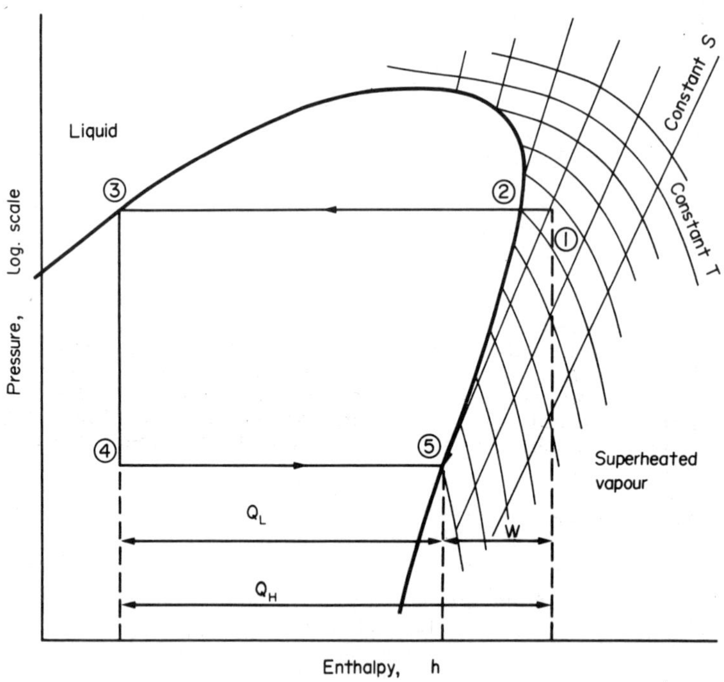

Fig.2.4 The ideal vapour compression cycle on a pressure-enthalpy diagram

The adiabatic expansion can be shown on a p-h diagram as a straight line from 3 to 4. This is one reason why this type of diagram is so convenient. In order to analyse a cycle it is necessary only to know the refrigerant state on entering and on leaving the compressor. The remainder is a matter of straight lines.

Evaporation takes place at constant pressure and temperature from point 4 to point 5. Note that the expansion has resulted in a two-phase mixture of liquid and vapour refrigerant. It is all at low temperature and pressure, but a certain fraction of it (sometimes as much as 50 per cent by weight) is not useful, because it is entering the evaporator as a vapour.

Between point 5 and point 1 on Fig.2.4 one has isentropic compression of the dry vapour. In practice this can not be achieved, but we are still looking at an ideal cycle, although it is unable to attain Carnot cycle efficiency because of the irreversible expansion nozzle. The second major advantage of the p-h diagram now becomes apparent. Because the horizontal axis is enthalpy, it gives a direct measure of Q_H, Q_L and W. The simple relationship $Q_H = Q_L + W$ can be clearly seen, and more importantly, the diagram gives a feeling for COP. To obtain a high COP, Q_H must be large and W, the work of compression must be small. It is therefore possible to rapidly assess a working fluid from a glance at the p-h diagram.

The mechanical vapour compression cycle just described is identical for either heat pump or refrigeration duty. It is often referred to as the reversed Rankine cycle or more loosely still, the Rankine cycle. The true Rankine cycle however is predominantly applied in the analysis of the steam turbine cycle used for electric power generation. On the T-S chart this is a clockwise cycle, with evaporation and condensation, but there are two differences between the Rankine cycle and mechanical vapour compression. The first is in direction, the Rankine cycle being a power cycle, delivering power by expansion of vapour in a turbine. This leads to the term 'reverse Rankine'. The second difference is that the Rankine cycle uses compression of 100 per cent liquid. A true reverse of this would be in an expansion engine, not in the irreversible expansion nozzle. In practice the difference is not very substantial so if the less rigorous usage is found in this text, it is hoped the reader will bear with it.

2.4 The Practical Cycle

The working cycle described in the previous Section is an ideal cycle. Although it takes into account practical limitations in the need for dry vapour compression and the absence of an expansion engine it still assumes components of 100 per cent efficiency. The ways in which real machines differ from the ideal will now be shown.

One component dominates design of the heat pump cycle and this is the compressor. It has already been explained that it is essential to compress only dry vapour and for this purpose, a certain amount of superheat is applied to the refrigerant before it enters the compressor. This is shown in Fig.2.5 where the refrigerant is now entering the compressor at stage 5' instead of state 5. This superheat essentially gives a safety margin, to reduce the risk of liquid droplets entering the compressor. One disadvantage of superheat is that the compressor must have an increased size, since it is dealing with a less dense vapour, at the same mass flow rate. A more important problem is that the compressor discharge temperature is increased and for many compressors, one operating limitation is the discharge temperature since a high temperature can damage the discharge valve.

This leads into the second major change to the working cycle which is due to the compressor efficiency. Because of heat transfer between the working fluid and the compressor, and because of irreversibilities in the flow through the compressor the compressor will increase the enthalpy by a greater amount

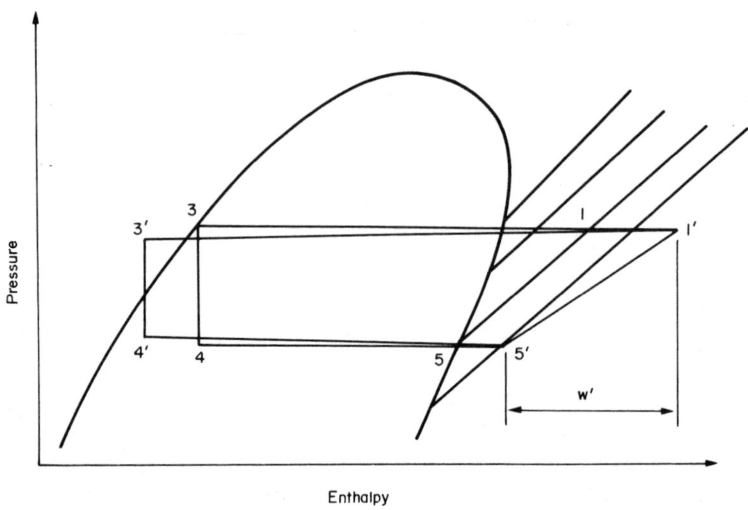

Fig.2.5 The practical vapour compression cycle

than is necessary, which means an increased discharge temperature. This is shown at 1' on Fig.2.5. The amount of increase is often expressed by the 'isentropic efficiency' of the compressor. The enthalpy added by a real compressor is given by W'. and the enthalpy required for isentropic compression is W. The 'isentropic efficiency' is given by W/W'. A practical reciprocating compressor might have an isentropic efficiency of 70 per cent. Note that isentropic compression involves the minimum amount of work required for an uncooled compressor. The work could be reduced further by cooling, but since one needs heat output at high temperature, this is not always advantageous or, indeed, possible.

Two other measures of compressor efficiency can be used. The 'mechanical efficiency' is a measure of how much of the work applied to the shaft is delivered to the working fluid.

$$\text{Mechanical efficiency} = \frac{\text{Power input to compressor}}{\text{Enthalpy increase x mass flow rate}}$$

A typical figure might be 95 per cent. Note that this is as equally important as the isentropic efficiency and these two must be taken into account when working out the COP of a real system.

Finally there is the 'volumetric efficiency': this does not influence the COP of the complete cycle but does influence the capital cost of equipment since it is used to select the size of the compressor.

$$\text{Volumetric efficiency} = \frac{\text{Mass throughput x Suction specific volume}}{\text{Compressor swept volume per unit time}}$$

95 per cent is a reasonably typical figure for volumetric efficiency.

Apart from the compressor, there are practical inefficiencies in other components of the working cycle. As the working fluid passes through the heat exchangers, there is a small pressure loss. This is because the velocity is maintained reasonably high to avoid isolated areas of oil collecting. The effect of these pressure drops is a divergance from the isothermal performance of the heat exchangers as shown in Fig.2.5. Because this can be a matter of a degree or less, this is slightly exaggerated to make it clear. The effect is apparent both in the evaporator and in the condenser.

The last deviation from the ideal cycle which we will deal with here is the question of 'subcooling'. In the ideal cycle, it was shown that throttling started from a point (point 3) on the liquid saturation line. Any pressure loss, however, in the pipe between the condenser and the throttling valve will cause the formation of some vapour, which will impair the valve's performance. It is therefore desirable to subcool, to point 3' for example. This subcooling will also reduce the proportion of vapour entering the evaporator. In order to achieve this subcooling however, one needs a source at a convenient lower temperature, and the normal condenser cooling water (or air) cannot cope with this duty since it is generally the aim of the heat pump to maintain this flow as hot as possible.

The necessity, which has already been explained, to superheat the working fluid as it leaves the evaporator provides a convenient and elegant solution. Heat removed from the condensate during subcooling at T_H can be used to superheat the suction vapour at T_L by incorporating a heat exchanger called a subcooler or an intercooler, as shown in Fig.2.6.

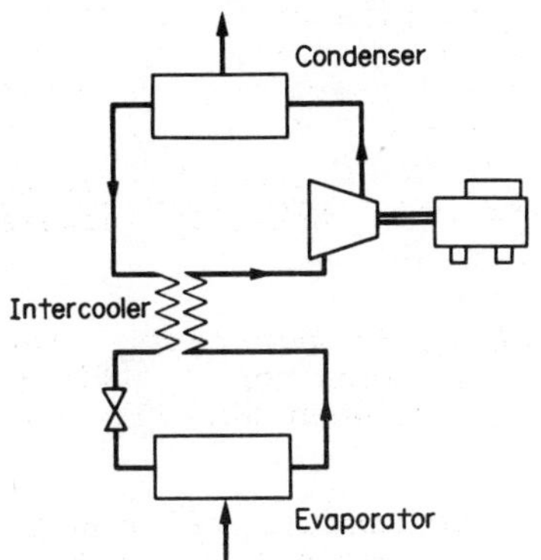

Fig.2.6 Vapour compression cycle with intercooler

Note that the intercooler does not directly affect the COP since the extra enthalpy delivered by the high temperature flow between 3 and 3' is not delivered externally, but is consumed between 5 and 5'. The intercooler does indirectly improve COP because it allows heat recovery at T_L to be more nearly isothermal.

2.5 Calculation of COP

This Section will consider in realistic terms an application where a heat pump may be considered for waste heat recovery. A comparison will be made between the performance which can be obtained and the performance of an ideal Carnot heat pump.

Let one suppose that one has a factory which is using water for cleaning various components. The water is maintained in large tanks at 65°C and after the cleaning, is sent to waste at 35°C. The intention is to install a heat pump to recover heat from the waste and to use this to maintain the temperature of the water tanks. For the time being the fact that the purpose might be better served by the use of a heat exchanger will be ignored.

Carnot states that the highest COP one can expect is given by:

$$\text{Carnot COP} = \frac{T_L}{T_H - T_L} + 1$$

where T_L and T_H are measured in absolute units. In this case:

$$\text{Carnot COP} = \frac{273 + 35}{65 - 35} + 1$$

$$= 11.3$$

This is a handsome COP to obtain, so one may now investigate what can be done in practice.

The first stage is to select the necessary evaporating and condensing temperatures. This depends on the size of heat exchangers one is prepared to buy, but for the time being it will be assumed that:

$$T_H = 75°C$$

$$T_L = 15°C$$

Note that a large Δt is needed in the evaporator, since the waste must be cooled from 35°C to, say, 20°C to give up useful heat.

Now either by prolonged study of the refrigerant tables in Appendix 1, or by asking someone who knows, one has arrived at the decision to use Refrigerant 12 (R-12) as the working fluid. The choice of working fluid is not always straight forward, and is discussed in Chapter 3, but in this case R-12 has been selected and one can now contemplate the p-h diagram in Fig.2.7.

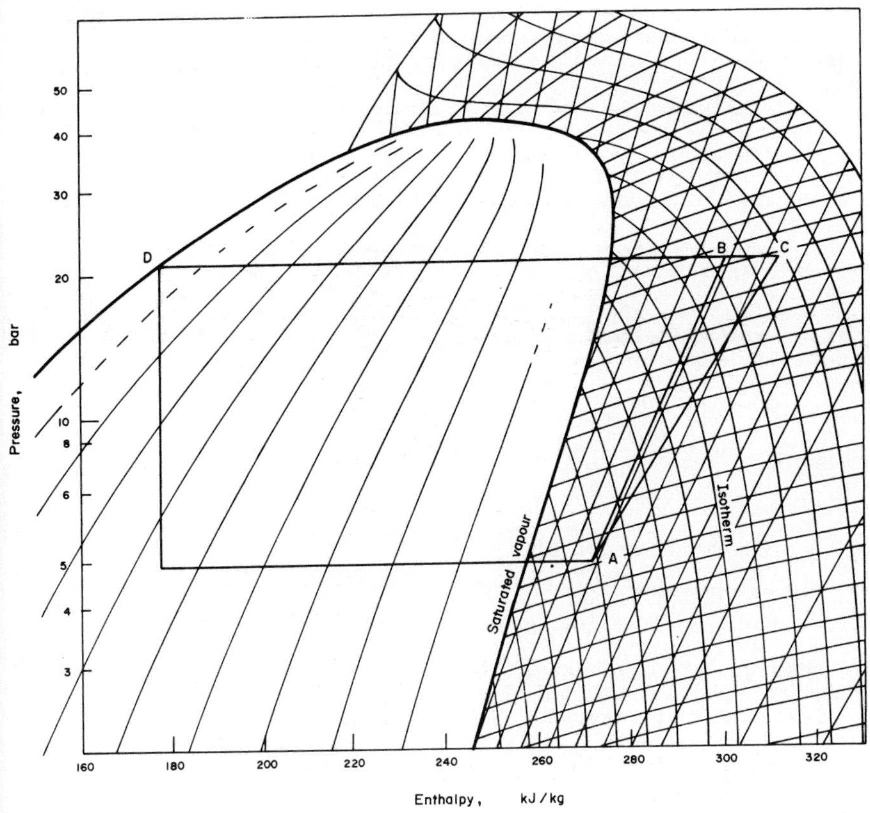

Fig.2.7 Heat pump cycle example using R-12

Plotting the working cycle always starts with consideration of the compressor conditions. In this case the compressor chosen requires a 20°C superheat to the suction vapour, and so the state of the suction gas is given by point 'A'. This is found by following the constant pressure line corresponding to 15°C evaporation (4.9 bar) until the 35°C isotherm is reached. Point A has a specific enthalpy of 271 kJ/kg. Following an isentropic line up to the pressure corresponding to the condensation temperature of 75°C (21 bar) gives the discharge state from an isentropic compressor. This has specific enthalpy of 300 kJ/kg, and is point B. The true discharge state (c) is calculated using the compressor isentropic efficiency:

$$\text{Isentropic efficiency} = \frac{h_B - h_A}{h_C - h_A}$$

from which it follows that with an efficiency of 70 per cent, h_C = 312 kJ/kg and this is plotted in Fig.2.7.

The enthalpy change across the condenser can be found by plotting point D, which corresponds to saturated liquid at the discharge pressure (neglecting heat exchanger pressure loss).

h_D = 177 kJ/kg.

$$\text{Practical cycle COP} = \frac{h_C - h_D}{h_C - h_A}$$

$$= \frac{312 - 177}{312 - 271} = 3.29$$

Additionally it must be remembered that the compressor has a limited mechanical efficiency, so that a small extra amount of work must be input.

$$\text{Overall COP} = 3.29 \times 0.95$$

$$= 3.13$$

To summarize:

Apparent Carnot COP	11.3
Carnot COP after selection of heat exchanger Δt	4.8
Cycle COP including thermodynamnic inefficiencies	3.3
Cycle COP including mechanical inefficiencies	3.1

The message of the above figures is plain. Design of heat exchangers should not be overlooked, since reduction in Δt can pay dividends in terms of COP. Applications where latent heat is exchanged on both sides of the heat exchangers are more suitable in this respect (e.g. distillation applications, see Chapter 6).

2.6 The Use of PER

The COP gives a measure of the usefulness of the heat pump unit, in producing large amounts of heat from a small amount of work. It does not, however express the fact that energy available as work is normally more valuable than energy available as heat. This becomes apparent when one tries to decide how best to drive the compressor. If one uses an electric motor, one has to use power which is inefficiently generated; if some form of heat engine is used, one is only able to harness some of the heat available in the fuel as work. Ideally a heat pump where free work is available should be contemplated e.g. wind or water power, but this is not always possible.

In order to assess different heat pump systems using compressor drives from different fuel or energy sources the PER or Primary Energy Ratio is applied.

The PER takes into account not only the heat pump COP but also the efficiency of conversion of the primary fuel (e.g. oil, gas, coal or solar heat) into the work which drives the pump. This is particularly relevant to the two 'heat-actuated' heat pumps which are discussed later in this Chapter since in these combined cycles it is not always possible to distinguish between the flows of heat and work.

The PER is defined as follows:

$$\text{PER} = \frac{\text{Useful heat delivered by heat pump}}{\text{Primary energy consumed}}$$

It is often possible to use an alternative definition, when a heat engine with thermal efficiency is used to drive a heat pump compressor. In this case:

$$\text{PER} = \eta \times \text{COP}$$

When using a heat pump for domestic heating or process heating or in any application where the only benefit is from the supply of heat, then the PER shows how much profit the heat pump will make as compared with a conventional boiler (for hot water or steam heating) or compared with directly fired heating. For example the heat pump described in the previous example could be driven by a diesel engine. A diesel engine is a fairly good type of heat engine, and at a suitable rating it could be stated that $\eta = 0.4$.

Therefore

$$\text{PER} = 0.4 \times 3.1 = 1.24$$

In other words, the heat pump would give 24 per cent more heat than by direct combustion of the fuel. If in addition one makes use of the 35 per cent of the primary energy which is recoverable waste heat from the diesel engine (via the hot exhaust and radiator) then the PER is correspondingly increased:

$$\text{PER} = 1.24 + 0.35 = 1.59$$

This can be directly compared with an indirectly-fired heating system, using a boiler for which the efficiency (or PER) may be only 0.7 or 0.8. The comparison shows that in this case the heat pump gives twice the heat output per unit of fuel consumed.

2.7 The Absorption Cycle

The absorption cycle manifests itself in refrigeration in many different shapes and forms, but its usefulness in heat pumping is still limited in many applications by high capital cost. Because of the complications of the cycle and the difficulty involved in making even a simple or approximate analysis, no apologies are made for any crude simplifications made in the interest of clarify. An introduction into the basic workings of the absorption cycle is followed by a description of perhaps the best known varient, the Platen-Munters or 'Electrolux' refrigeration cycle. The performance of a typical industrial absorption cycle heat pump is then illustrated using derived data.

Fig.2.8 shows the very basic absorption cycle, with the simple vapour-compression cycle alongside for comparison. It is clear that an absorption heat pump contains an evaporator and condenser which operate in exactly the same way as for the vapour compression cycle. Heat is taken in at the evaporator, causing the refrigerant to evaporator at low pressure and heat is released by condensation at high pressure. In the absorption cycle, however, there is a secondary circuit around which a liquid absorbent or solvent flows. The evaporated refrigerant vapour is absorbed into this at low pressure, and there is a net surfeit of heat for this process. The solvent, now diluted by refrigerant is raised to the high pressure by a liquid pump. High pressure refrigerant vapour is then produced by the addition of heat to the mixture, in the generator.

Table 2.1 Carnot COP of Absorption Heat Pump

	T_G = 150°C	40°C	60°C	80°C
	-10°C	2.37	1.80	1.48
T_E	0°C	2.77	1.97	1.56
	10°C	3.45	2.20	1.67

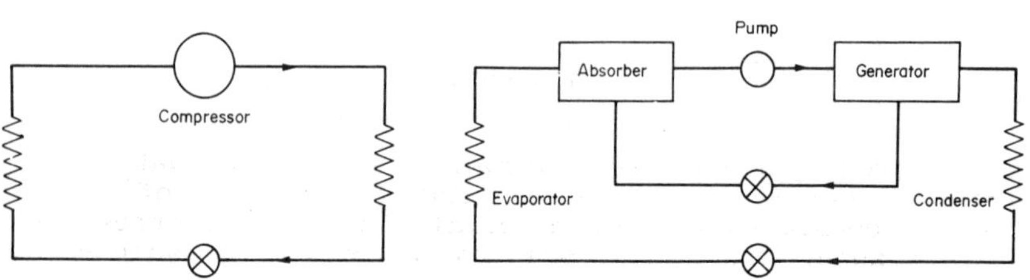

Fig.2.8 Simple absorption cycle and vapour compression cycle

Because the liquid solvent/refrigerant mixture is virtually incompressible the work of the pump is genuinely negligible, and the primary energy source is the heat required at the generator which is always the hottest part of the cycle. Heat liberated from the absorber can be added to the heat from the conddenser so that it can be made certain that as a heat pump the COP is always greater than unity.

As with other cycles, comparison can be made with the theoretical Carnot efficiency. The absorption heat pump is only a heat engine driving a heat pump, with the temperatures defined as shown in Fig. 2.9.

$$\text{Carnot COP} = 1 + \frac{T_E\ (T_G - T_A)}{T_G\ (T_C - T_E)}$$

Some values are given in Table 2.1 for what might be regarded as typical temperatures, for a home-heating application, using the convenient simplifying assumption that $T_A = T_C$.

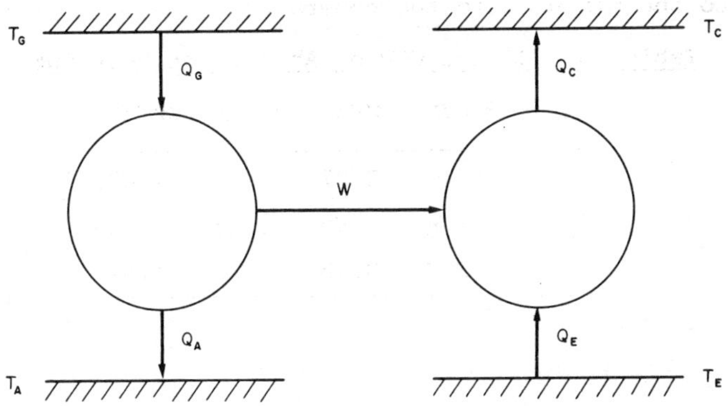

Fig.2.9 Carnot absorption cycle

These figures are not very large, considering that they are theoretical maxima. The generating temperature of 150°C is limited by considerations of material stability or stress levels due to pressure, and a first step to increasing COP values should be increasing T_G .

We can visualise the practical constraints which restrict the absorption cycle performance by studying a pressure-temperature-concentration or p-t-x chart. Fig.2.10 gives a crudely generalised chart to illustrate the basic points. In practice p-t-x charts for real refrigerant/absorbant pairs show a marked deviation from the linear due to strong affinity between the working fluids.

The two pressures of the system are shown in the two horizontal lines, the higher being the generator and condenser pressure and the lower the absorber and evaporator pressure. The condenser pressure is that corresponding to 100 per cent refrigerant and the evaporator temperature also corresponds to 100 per cent refrigerant, at the lower pressure. The two vertical lines correspond to the concentrations ideally attainable for the temperatures and pressures shown in the absorber and generator.

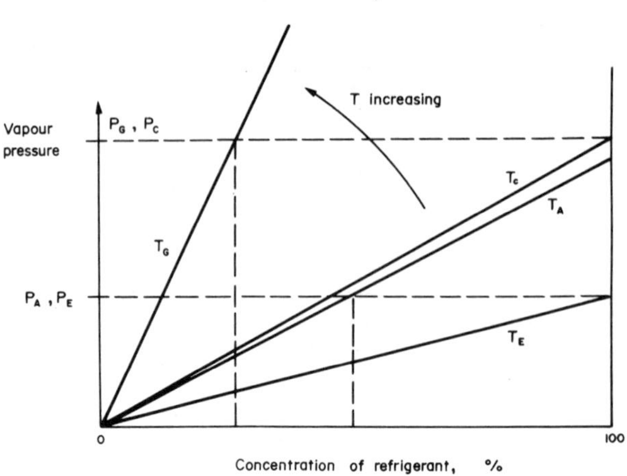

Fig.2.10 Generalised "p-t-x" chart

Now consider that the generator temperature is fixed. As the evaporator temperature is lowered, the two concentrations become more similar, which reduces the cycle efficiency because less refrigerant is produced for a given amount of circulating solvent. Alternatively, raising either the absorber or condenser temperature has the same effect. This explains in simple terms why this type of heat pump can only achieve low values of COP.

Another representation of the process is a plot of the log of the vapour pressure against the inverse of absolute temperature. Its convenience arises because the relationship for the absorbent, refrigerant and mixtures of the two appear as straight lines. An arbitrary cycle is illustrated in Fig.2.11. An increase in the inverse temperature is conventionally plotted from right to left as shown so as to have increasing temperature in the usual direction. The numbered points on Fig.2.11 correspond with those in Fig.2.12, a basic practical absorption output. The terminology rich and poor solutions refers to the high and low concentrations respectively of the refrigerant vapour in the absorbent. (This can be confusing if the refrigerant is water since the rich solution is the most dilute by the usual chemical definitions).

An important feature of the cycle which can be deduced from Fig.2.11 is the relatively small number of assumptions which are required to quantify the basic thermodynamics of the system. The evaporating and condensing temperatures are established by the intended application and the effectiveness of the condenser and evaporator. These temperatures determine the two operating pressures because of the unique pressure/temperature relationship of the pure refrigerant. It is reasonable to assume that the condenser and absorber release heat to the same sink and under similar cooling conditions; consequently they will have similar operating temperatures. Because the condenser temperature is established the absorber temperature is known and hence the position of the rich solution line. The generator

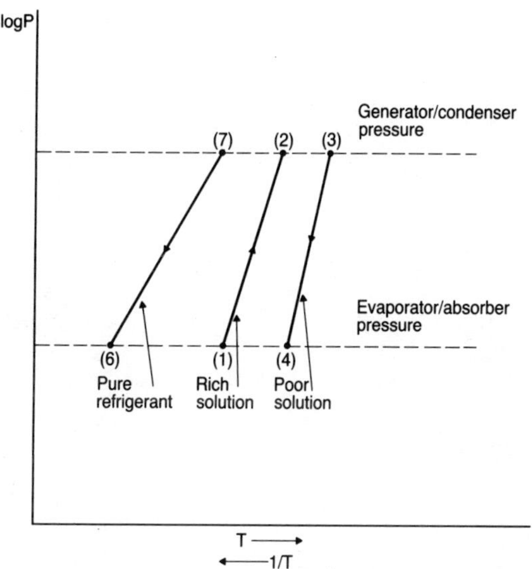

Fig.2.11 Absorption cycle operation shown as log of vapour pressure against reciprocal temperature

pressure is fixed by the operating conditions but there is some flexibility in the maximum temperature represented by position 3. It is partly influenced by practical considerations such as the temperature of the heat source. Variation in the temperature of position 3 affects the temperature of position 4 and an assumption must be made regarding the maximum generator temperature. Its value influences the absorber operation and it affects the efficiency of the system.

It can also be seen from this that there will be a combination of temperatures for which the absorption cycle will simply not work, because there is no difference in concentration between absorber and generator. As this limit is approached, the COP will drop to unity.

In practical terms the system differs in several respects from what has been described so far. Real pairs of refrigerant/solvent are selected for their strong mutual affinity and for other properties which are listed below and in Table 2.2

(1) Strong affinity to allow large concentration changes in solution

(2) High volatility ratio to aid generator efficiency

(3) Reasonable operating pressure, limited by refrigerant vapour pressure

(4) Chemical stability to reduce breakdown in generator.

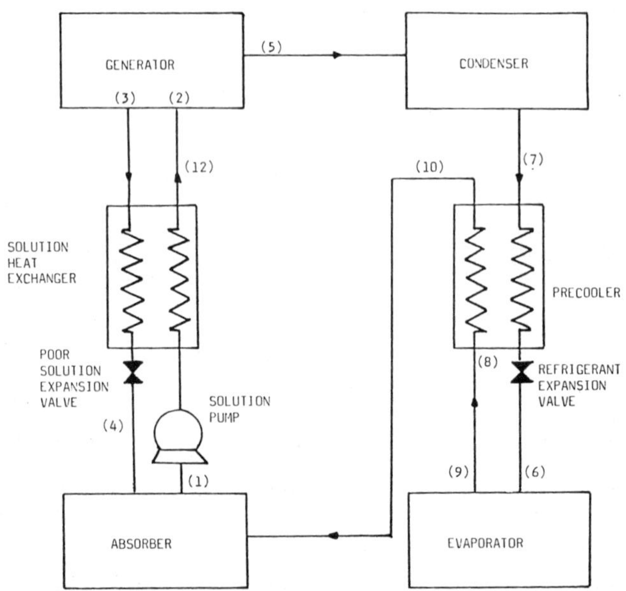

Fig.2.12 Basic Absorption Cycle

The absorption cycle has been widely used for air conditioning and refrigeration and in this field only two refrigerant/solvent pairs have come to real fruition, ammonia/water and water/lithium bromide, (see also Chapter 4). Both pairs have merits in different fields and fairly significant disadvantages when heat pumps are considered. These are listed below. (The relative merits achieved different degrees of importance, depending on the application - see Chapter 3).

Ammonia/Water

1. Because both ammonia and water exert vapour pressure in the generator, a reflux condenser is required to separate the fluids.

2. Ammonia exerts a fairly high vapour pressure, for example, 20 bar at 50°C.

3. Ammonia is a slightly toxic vapour and certain safety precautions are necessary.

4. Ammonia corrodes copper which cannot be used for heat exchangers, which increases production costs.

Water/Lithium Bromide

1. The onset of crystallisation limits the concentration attainable and therefore the cycle efficiency.

2. Water (the refrigerant) exerts a very low vapour pressure and therefore the whole cycle is at subatmospheric pressure.

3. Water freezes at 0°C and therefore the evaporator cannot be exposed below this temperature. This eliminates this type of heat pump from any air-source applications.

4. Lithium bromide is highly corrosive and necessitates careful selection of containment vessels.

Many other combinations have been, and are being, investigated and some predicted results indicate that combinations of conventional refrigerants with organic solvents may have potential. The disadvantage of using fluorocarbon refrigerants, however, is that they are relatively less stable at the high (generator) temperature. They are prone to progressive decomposition.

It is possible to associate specific physical properties with the requirements of absorption cycle fluid pairs and these are summarised in Table 2.2.

Most of the specifications in Table 2.2 are self explanatory with the possible exception of the requirement to have a large negative deviation from Raoult's Law. This law is concerned with the vapour pressure of solutions and it is expressed as a function of the vapour pressures of the pure components and their mole fractions. It is stated:

$$p = xP_R + (1-x)P_A$$

Where

P = solution vapour pressure
x = mole fraction of refrigerant
P_R = vapour pressure of pure refrigerant
P_A = vapour pressure of pure absorbent

Solutions which have a negative deviation from Raoult's Law have a lower solution vapour pressure for a given mole fraction of refrigerant than that predicted by the law. Conversely, for a given solution vapour pressure the mole fraction of refrigerant increases with increasing deviation from the law. This is advantageous since the volume of solution is reduced from a given refrigerant flow rate. In principle, the deviation effect could be enhanced by selecting an absorbent with a low molecular weight but this would involve the use of fluids with low boiling points and high vapour pressures and these are highly unsatisfactory features, (ref.2.6).

TABLE 2.2 Desirable Absorption Cycle Fluid Properties

Fluid	Desirable Properties	Reason
Refrigerant	High latent heat	Reduces mass flow
	Moderate pressure at condensing temperature	Reduces strength requirement for condenser and generator
	Relatively low triple point	Limit on evaporator temperature
	Relatively high critical point	Limit on condensing temperature
	Low liquid specific heat	Reduces effect of pre-cooler
	Vapour specific heat equal to liquid	Precooler effectiveness
	Low vapour specific volume	Ease of vapour transport
Absorbent	Negligible vapour pressure	No rectification requirement
	Easily absorbs refrigerant	Fundamental to process
Solution	Negative deviation from Raoult's Law	Reduces refrigerant flow rate
	Low specific heat	Reduces solution heat exchanger duty
	Low specific volume	Reduces pump work
All Fluids	High thermal conductivity	Increases heat transfer coefficients
	Low viscosity	Increases heat transfer coefficients and reduces pipe work losses
	Low surface tension	Improves absorber operation
	Low toxicity	Safety
	Chemically stable	System life

The chemical stability and toxicity of the fluids have a more obvious significance but it is important to include corrosion properties within the definition of stability. The toxicity of the fluids can restrict their use, particularly in domestic applications. UK regulations, for example, do not permit the domestic use of certain refrigerants (Group 3) and only allow certain systems for others (Group 2).

In the practical cycle shown in Fig.2.12 the precooler (or intercooler) plays the same role as for the vapour compression cycle and gives more nearly isothermal evaporation. The solution heat exchanger is very important for economy. Because the solvent is merely circulating to and fro between the generator and the absorber, it will just drain heat from the generator to the lower temperature level which will reduce the system PER. The purpose of the solution heat exchanger is simply to recover a reasonable proportion of this heat.

The generator is drawn here schematically. It may contain a rectifier, sometimes called a reflux condenser, or when constructed externally to the generator a 'dephlegmator' (this latter term applying specifically to the removal of water from ammonia). The purpose of the rectifier and the coil removing the liquid from the bottom of the generator is to maintain the temperature distribution within the column so that, in the case of an ammonia/water system, the ammonia liquor is successively evaporated then cooled as it passes up. Each stage progressively dehydrates the ammonia. A full description of this process will be found in Wood (Ref.2.7).

It is essential that the refrigerant leaving the generator is free of solvent because otherwise the evaporator will cease to function as designed. The evaporation will favour the refrig erant, and the liquid in the evaporator will contain an increasing fraction of solvent, which will progressively raise the evaporating temperature, or lower the pressure. Both effects are disastrous to system efficiency.

2.8 Comparison between Absorption Cycle and Vapour Compression Cycles

Before a comparison between the two heat pump systems described so far can be made it is important to establish the criteria which would be employed in the selection of some specific heating system. Clearly, the single most important factor is the ability of the system to provide the required quantity of heat at the desired temperature. Given this basic feature, the most appropriate device would be that which incurred the lowest total cost over its life. Capital, fuel and maintenance costs are the important factors which must be considered. Heat pumps are attractive because their fuel costs are lower than equivalent direct heating systems but there are several methods of assessing performance and it is important to appreciate their significance. The problem can be illustrated by considering the three most common types of heat pump which are:

a. Vapour compression driven by an electric motor
b. Vapour compression driven by a gas engine
c. Absorption cycle

The most direct measure of the performance of the electric vapour compression heat pumps is to consider the electrical energy supplied and the heat output, i.e. the COP. This measure would enable the user to assess the economic performance of a particular design. However, from a national or global standpoint, this assessment is inadequate because it does not account for losses in the electrical generation and distribution systems as discussed earlier.

A gas engine driven vapour compression system uses primary energy in a more direct manner than the electric drive but there are still complicating factors. The COP is normally calculated on the basis of the power supplied to the compressor and the heat pump output; this is directly comparable with that calculated for an electric drive. Its value ignores the efficiency of the gas engine and a second possible assessment is to use the energy supplied to the engine and the heat output from the pump. In practical installations, however, there is usually heat recovery from the engine which considerably boosts the total heat output. In a typical case approximately 30% of the total heat output is due to engine heat recovery, (ref.2.8). Thus, at the risk of repeating earlier statements, a further measure of performance is based on the energy supplied to the gas engine and the total heat output from the installation. This is usually defined as the Primary Energy Ratio for gas engined systems. It is assumed that most absorption cycle systems will be gas fired in which case the heat pump makes direct use of primary energy and the COP and PER are identical.

None of the various methods of assessing performance are incorrect; they indicate different attributes of the various systems. The large differences in apparent effectiveness due to the measure of performance used can be seen from the following simple example. The COP of an electric or gas engine driven vapour compression heat pump can typically be about 3, (process application), in some cases significantly higher. An absorption cycle COP is more likely to be in the range 1.25-1.50. If the efficiency of electrical generation and distribution is taken as 27% then the PER of an electric heat pump is 0.8. A similar efficiency can be expected for a gas engine but heat recovery increases the PER to the range 1.25-1.50. It can be seen that the absorption cycle and gas engine driven vapour compression cycle are more useful in the conservation of resources but this is of interest to the user only if the fuel price reflects the conservation potential. An electric heat pump will be cheaper to operate if the cost per Joule of electricity is no more than approximately twice the cost per Joule of gas.

The similarity of the performance of the absorption cycle and gas engined vapour compression cycle is illustrated in Fig.2.13. This compares the actual performance of the gas engine system

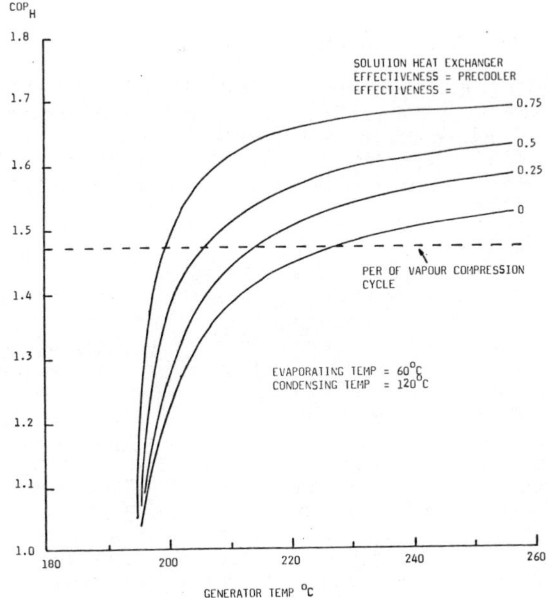

Fig.2.13 Comparison of Absorption Cycle and Gas Engined Vapour Compression Cycle Performance

developed at NEI International Research & Development Company* with the theoretical performance of an intended absorption cycle device. It is unlikely that in practice the absorption cycle will operate over these particular temperature ranges. Fig.2.13 indicates that the absorption cycle is more efficient over a large range of conditions but the curves assume that there are no losses in the generator system. In a real system generator losses are likely to be in the range 10%-20% and if these are taken into account the apparent advantage of the absorption cycle is removed.

The other features separate the vapour compression and absorption cycle systems. The absorption cycle is mechanically simple and should give a long life with minimum maintenance requirements. In contrast the gas engined designs are relatively complex and require all the servicing associated with internal combustion engines.

* See Chapter 6 for a detailed description.

2.9 Absorption Cycle Derivatives

Derivatives of the basic absorption cycle heat pump are numerous and are the subject of significant research and development work. The addition of further stages to a single stage absorption cycle heat pump can lead to improved COP and greater temperature lifts. Also of significance, and described in Chapter 6, is the heat transformer, which is marketed for industrial applications by a number of manufacturers in Japan and Europe.

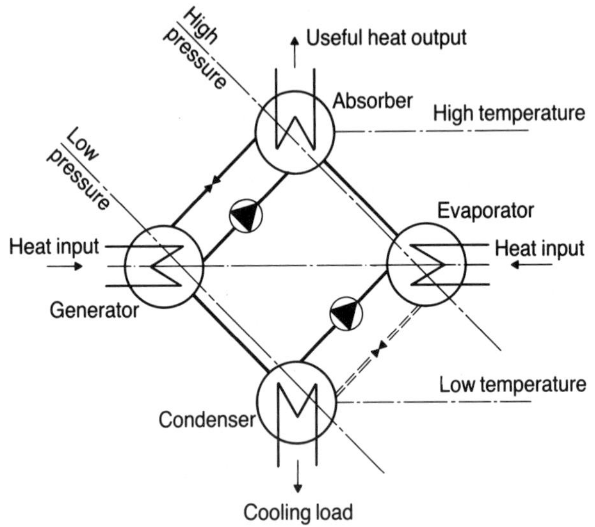

Fig.2.14 Operation of a Heat Transformer

The operation of a heat transformer, which is a reversed absorption cycle heat pump, may be explained with reference to Fig. 2.14. In a conventional system, high temperature heat is used in the generator, the evaporator taking in low grade heat and the absorber and condenser producing heat at an intermediate temperature. In a heat transformer, however, heat at an intermediate temperature, normally originating from a waste source, is supplied to the generator and evaporator, as shown in the figure. High grade heat - the useful heat output - is extracted from the absorber, while the condenser is cooled.

Unlike the heat pump, where the amount of useful heat generally exceeds that taken in at the evaporator, the heat transformer is only able to convert a fraction of the intermediate temperature energy into high grade heat. This proportion depends upon the system design and the working fluids used. In an example developed by MAN New Technology in West Germany, about 30% of the heat in a waste source at 100°C was transformed into useful heat at 135°C.

2.10 Other Types of Heat Pump

Apart from the heat pump cycles already described, there is a multitude of other devices which can be described as heat pumps and which range from the expensive to the ineffective. Some are novel, and some are old friends. Grouped together here are a representative sample of these devices and perhaps this will be a useful source of information and inspiration.

Development from the drawing board to hardware can be restricted by financial or technical constraints which must be overcome. In the case of some of the more exotic heat pumps prejudice against innovation will also be met.

2.9.1 The Stirling Cycle In 1816 an external combustion engine was patented using air as the working fluid. This was the Stirling engine, working on what is now called the Stirling cycle which involves isothermal heat exchange and constant specific volume pressure change. This latter is effected by internal heat exchange with a regenerator.

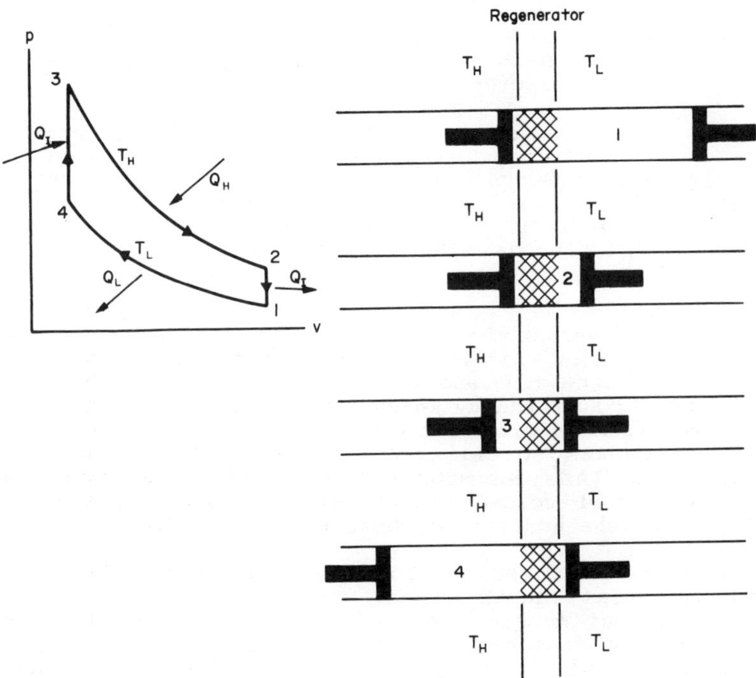

Fig.2.15 The Stirling Cycle

The heat flows shown on the pv diagram in Fig.2.15 give the key to understanding the cycle. As gas is compressed isothermally from 1 to 2, the gas is cooled externally. It is passed through the regenerator and heated internally. Between 3 and 4 the gas is externally heated and allowed to expand, in doing so it delivers useful work then finally the gas is passed back through the regenerator, which internally cools the gas back to state 1.

The Stirling cycle is interesting bcause all the processes described in the ideal cycle are reversible, and the external heat exchangers are isothermal which means that an ideal Stirling heat engine could equal the efficiency of the Carnot cycle. It is of interest because, in reverse, the same applies to a Stirling refrigeration machine or heat pump.

In practice there are several idealizations which make the Stirling machine difficult to design, and bulky. These are listed below.

(1) The pistons illustrated in the figure move intermittently, not sinusoidally.

(2) The regenerator is frictionless and 100 per cent effective.

(3) Ideal heat exchangers are assumed for external heat transfer.

Heat exchangers always pose problems with external combustion engines. In effect, an extra two temperature differences are being introduced between heat source and sink. Despite these obstacles, Stirling cycle machines have been used successfully in some low-temperature refrigeration applications and Fig. 2.16 shows a neat design which uses a 'vee' compressor to give the necessary piston timing.

In position 1 the working fluid, a gas, is being compressed and is simultaneously delivering heat isothermally to the heat sink, which is the hot end of the cycle. At 2 the gas has been passed through the regenerator at nearly constant volume and is then expanded with the addition of heat. The regenerator has cooled the gas so this is the low temperature end of the cycle. From 3-4 the gas returns through the regenerator, again at nearly constant volume but this time with the gas much less dense and with the addition of heat from the regenerator.

It is apparent that the separate constant volume and constant temperature steps of the ideal cycle have been compromised in this practical machine, but efficiencies are still respectable and a great deal of work continues in this field. A different type of unit is described by Wurm (ref.2.4) which is a free piston Stirling engine/Stirling heat pump cycle which he describes as a 'Duplex Stirling Cycle', illustrated in Fig. 2.17.

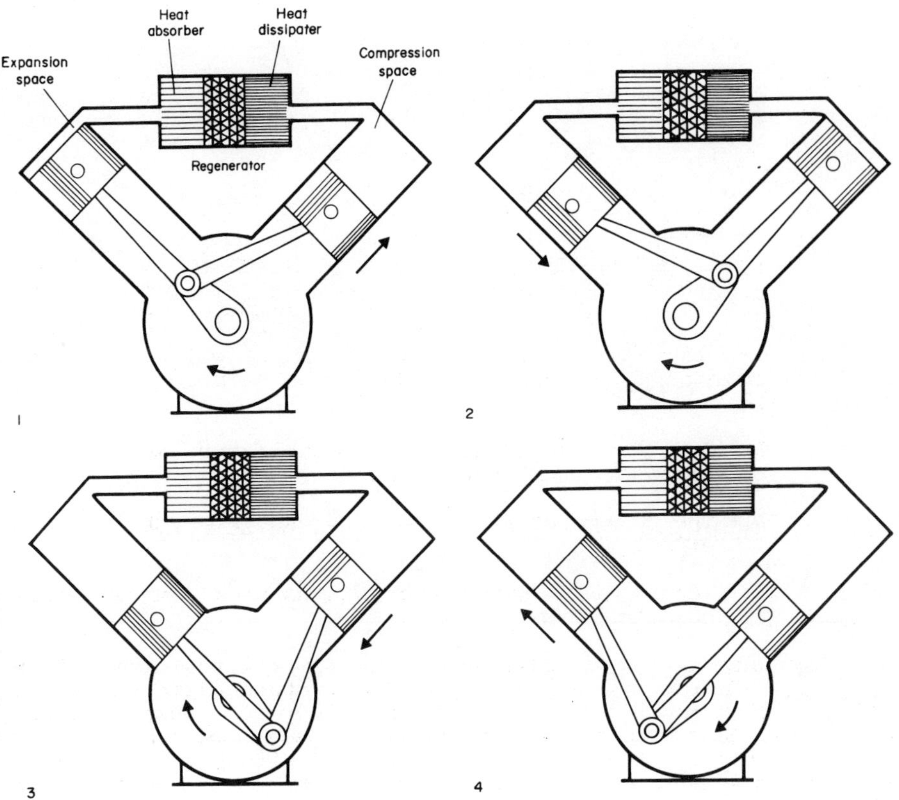

Fig.2.16 A Stirling Cycle Refrigerator

By assuming isothermal heat exchange Wurm writes that the heat pump is delivering:

$$Q = RM_{He} T_2 \ln(p_2/p_3)$$

By writing similar expressions for the engine and balancing the heat and work exchangers overall, Wurm calculated the system PER. The results are given in Table 2.3.

While these data appear reasonably favourable, a true picture of the system feasibility can only be achieved if working models are constructed and operated with all the appropriate heat exchangers.

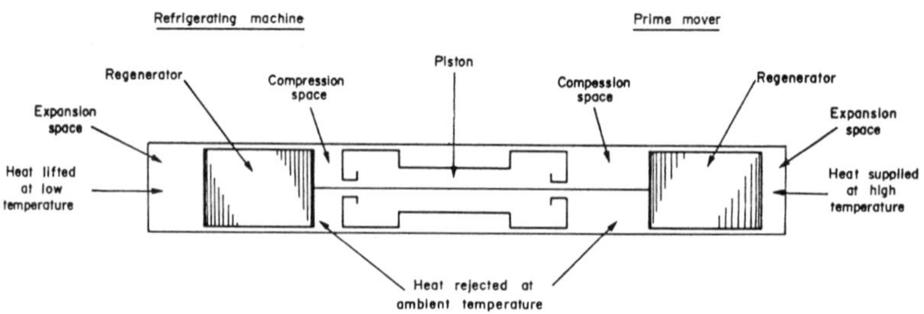

Fig.2.17 The 'Duplex' Stirling Cycle

Table 2.3 PER Values for a Stirling/Stirling Heat Pump

Engine Efficiency %	Ambient Temperature °C	Minimum Cycle Temperature °C	Maximum Cycle Temperature °C	PER
	-29	-45	61	1.39
	-18	-34	62	1.47
25	-7	-23	64	1.57
	4	-12	65	1.70
	-29	-45	64	1.49
	-18	-34	65	1.58
30	-7	-23	66	1.70
	4	-12	67	1.85
	-29	-45	66	1.59
	-18	-34	67	1.70
35	-7	-23	68	1.83
	4	-12	69	1.99

2.10.2 The Brayton Cycle This cycle is thought of most commonly in connection with the gas turbine engine, although this of course uses an open cycle, drawing air from the atmosphere and exhausting to it, in effect using it as an infinite heat sink. When the closed power cycle is considered the 'atmospheric' leg is shown in Fig. 2.18 as cooling at constant pressure.

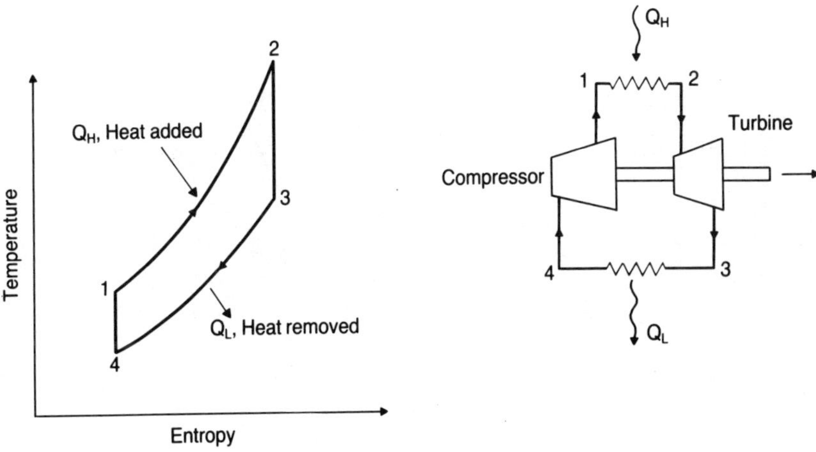

Fig.2.18 Brayton Power Cycle

The principal use of the Brayton cycle to the heat pump designer is that it gives the opportunity of an external combustion prime mover with reasonable efficiency. If rotary machines can be made to handle the temperatures and pressure ratios involved, then they would be ideal heat pump prime movers. All the heat rejected at the low temperature end of the cycle can be added to the heat output and provided that a high temperature can be attained at point 2 on the cycle, then system PER can become very respectable.

In an industrial heat pump application where there are perhaps tens of megawatts to be recovered then one can envisage a gas turbine being used without stepping beyond current technology. Ideally the gas turbine would be coupled to a centrifugal compressor, (which would run at a similar speed) which would drive a Rankine heat pump cycle. All the waste heat from the gas turbine appears in the exhaust, and an exhaust gas heat exchanger would recover much of this to add to the recovered heat. The first cost, and the running cost of this type of installation would be very great, so despite large potential savings in fuel it may be sometime before one sees such machines running.

The cycle now is simply the reverse of the power cycle, shown inFig.2.19 as closed. There are two basic ways in which it can be usefully applied in heat pumping (although Wurm classifies seventeen variations) by using an open cycle. These are shown below. In the first case ambient air is drawn in at point 1, it is heated by compression to 2 and between 2 and 3 is used to deliver heat where it is needed (via a heat exchanger or two) and after doing work by expansion the air is released to the atmosphere. Note that it is difficult to achieve a substantial temperature difference between T_2(ambient) and T3 (internal).

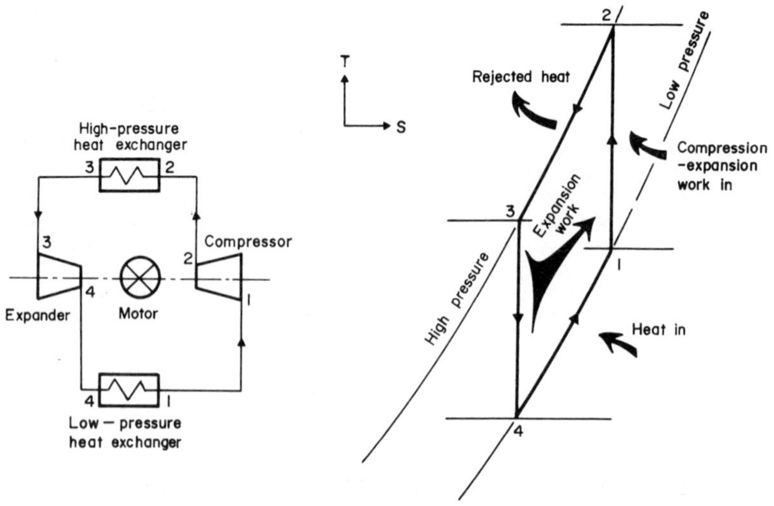

Fig.2.19 Brayton Heat Pump Cycle

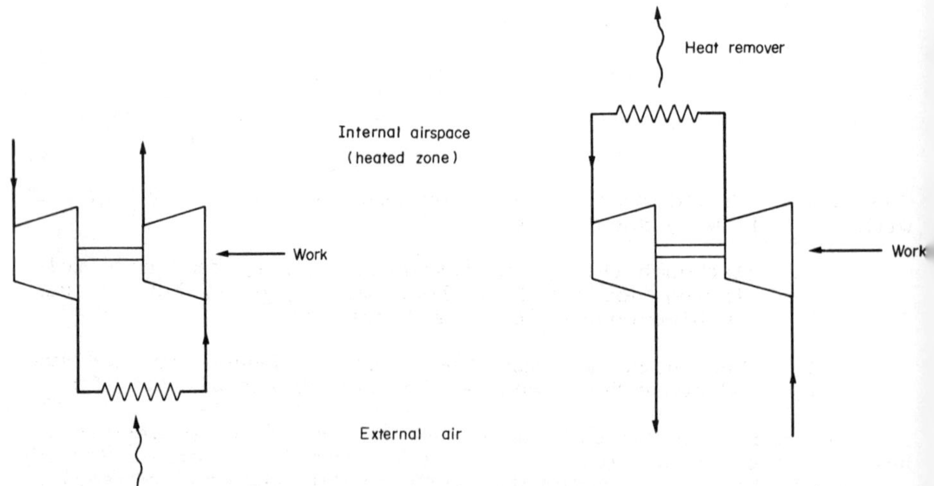

Fig.2.20 Open Cycle Brayton Heat Pumps

The basic alternative shown in Fig.2.20 is to extract internal air at point 3, to expand it to subatmospheric pressure then warm it by an ambient air heat exchanger and recompress it before release at point 2. Apart from the difficulties of subatmospheric pressure machinery, this configuration also has the frosting problems of an external heat exchanger.

The major use of Brayton cycle cooling has been in air conditioning passenger aircraft. Because a small supply of compressed air is available from the engines, and because this can be readily cooled to ambient temperature using outside air, it is simplicity itself to expand the air to cabin pressure, thereby providing pressurization as well as a supply of fresh cool air. Systems are also used for road vehicle air conditioning (ref.2.5).

Wurm (ref.2.4) has made a theoretical analysis of a Brayton/Brayton heat pump. The results of this are given in Table 2.4.

Table 2.4 PER of Brayton/Brayton Heat Pump

Ambient Temperature °C	Internal Temperature (min) °C	Max Cycle Temperature °C	PER	Prime Mover efficiency
-29	21	56.7	1.50)	
-7		59.4	1.76)	25
-15.6			1.94)	
-29		59.4	1.63)	
-18			1.75)	30
-7		62.2	1.91)	
-15.6			2.1)	
-29		61.7	1.74)	35
-7		64.4	2.06)	

The reader should temper his enthusiasm for these PER figures with the following sobering facts.

1. Although this is a Brayton/Brayton system, no simplifying designs have been made and we have a gas turbine coupled to an air compressor.

2. The high maximum temperatures needed to achieve adequate heat transfer to the ambient air at 21°C.

Brayton cycle heat pumps have the potential to achieve high heat delivery temperatures, as, of course, the working fluid is not limited by considerations common to vapour compression and some absorption cycle systems. Work at CEM CERCEM in France (ref.2.8) on a sub-atmospheric open cycle Brayton heat pump was recently carried out, directed at applications in spray drying. The system is illustrated schematically in Fig.2.21.

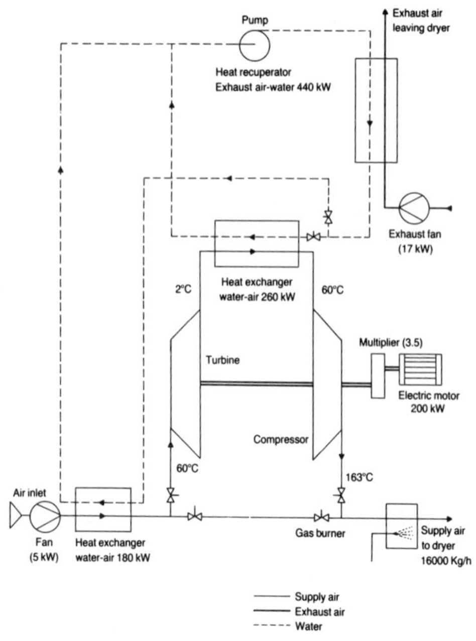

Fig.2.21 Proposed Use of a Brayton Cycle Heat Pump for Industrial Dryers

In this example, air at 1 bar pressure is heated using reject heat to 60°C and then expanded through the turbine to 0.5 bar and 2°C. The air is then reheated to 60°C, again using reject heat, and compressed to 1 bar at 165°C. The high grade heat is then used in a spray dryer.

The turbo-compressor was designed to deliver 4 kg/s of hot air at 165°C, consuming 235 kW. An electric motor drive was selected to drive the turbo-compressor, and the COP was calculated to be 2.75.

2.10.3 Thermoelectric heat pumps Peltier or thermoelectric types are reasonably common and have been used in one or two applications where they are well suited to the requirements.

Very simply, thermoelectric current is generated at junctions between dissimilar materials and is a function of the temperature of the junction. In reverse, passing current through such junctions will cause heat to flow. A basic thermoelectric heat pump can be made from a junction between 'n' type and 'p' type semiconductor as shown in Fig.2.22.

In this case the effect of current flowing in the direction shown is to remove heat from the n/p junction towards the current electrodes.

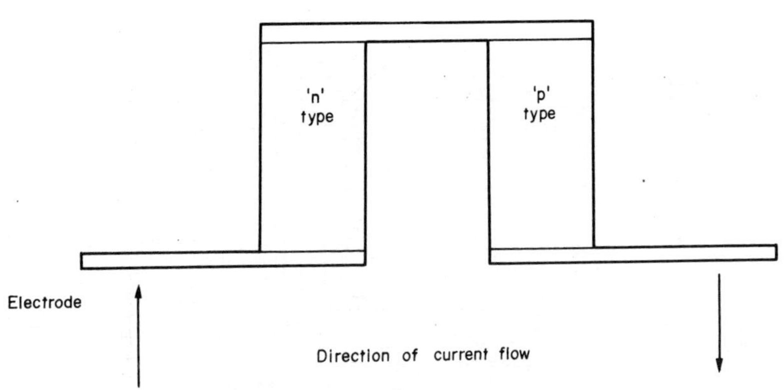

Fig.2.22 Schematic Thermoelectric Junction

The quantity of heat pumped is releated to the Seebeck coefficient (S). A figure of merit commonly used for thermoelectric materials is $S^2/k\ell$ where k is the conductivity and ℓ the electtrical resistivity. This is a function of temperature and therefore the material must be chosen with the temperature range taken into consideration.

The heat balance for one leg of the heat pump shown in Fig. 2.22 is given by the following equation.

$$Q = S \times T_j \times I - \frac{I^2 \ell (L/A)}{2} - \frac{k\Delta T}{L/A} - I^2 R$$

The four terms of the equation relate respectively to the thermoelectric heat pumping effect, the heat lost because of bulk resistive heating, the heat lost because of conduction from the warm end and resistive heating at the junction contact. Note that if the last term is neglected then for a given I and L/A then Q remains constant irrespective of size. This leads to a certain amount of miniaturization down to typical sizes of a few millimetres.

Practical heat pumps are built from numbers of elements in series electrically, and in series and parallel thermally.

This leads to problems of achieving thermal contact and electric isolation.

Despite the problems of finding cheap effective semiconducting materials and assembling them into a neat package, these devices have been used in some interesting applications. Values of COP similar to those achieved by the absorption cycle are claimed.

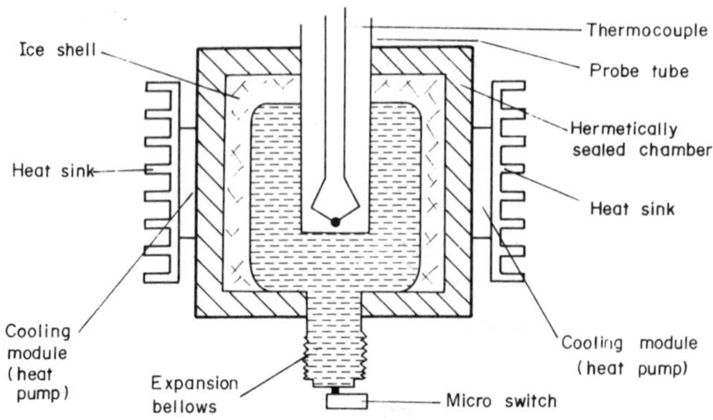

Fig.2.23 Zeref Thermoelectric Ice Point Control

One application described by Huck (ref.2.9) has been to cool the mercury in fluorescent lighting tube to maintain the optimum pressure. This is attractive because the unit is small and an electricity supply is already connected to the installation. The small extra consumption was rewarded by dramatically improved light output.

A more common application is in the maintenance of the cold junction of a thermocouple at constant temperature (e.g. 0°C - See Fig.2.23). The thermoelectric system is ideal since it is reversible and thus gives completely proportional control and it is easy to miniaturize, unlike any other refrigeration system.

An example of a thermoelectric heat pump developed for domestic applications is the 3.8 kW unit developed by Laboratoires de Marcoussis (ref.2.10). The system was designed as an air-water heat pump, extracting heat from ambient air to heat water to 35°C for underfloor heating. The P-type materials used are tellurium, bismuth and antimony, while the N-type materials comprise tellurium, bismuth and selenium.

The performance of the system is illustrated in Fig.2.24. COP's of between 1.5 and 2.0 were achieved, and the capital cost of the system was similar to that of electric-drive vapour compression cycle units.

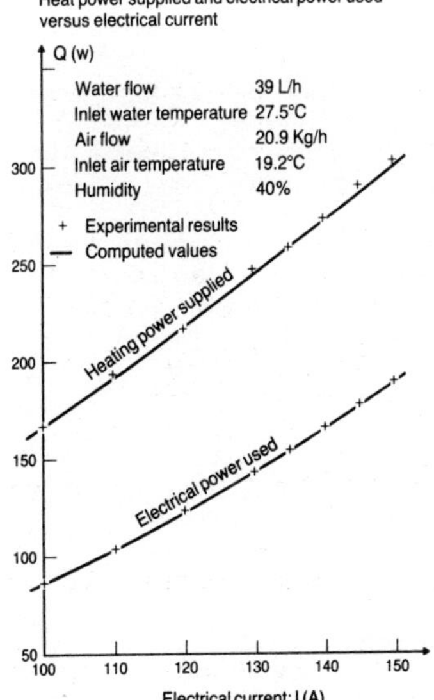

Fig.2.24 Performance of a Domestic Thermoelectric Heat Pump

REFERENCES

2.1 Rogers, G.F.C. and Mayhew, Y.R. Engineering Thermodynamics, Work and Heat Transfer. Longmans, London, 1964.

2.2 Fischer, R.D. et al. Solar powered heat pump utilizing pivoting-tip rotating equipment. Phase 1 Summary Report, 1 July 1974 -1 September 1975, NSF/RANN/SE/C-876/FR /75/3, 1975.

2.3 Anon. Systems de chaufage et climatisation domestique utilisant une pompe a chaleur actionnee thermiquement. Report of Battelle, Geneva, 190-4100 JPB/YT/fsch, 25 March 1977.

2.4 Wurm, J. An assessment of selected heat pump systems. American Gas Association Report on Project HC-4-20, Feb. 1977.

2.5 Anon. Rovac patent. US Patent 3686 893.

2.6 Heppenstall, T. A theoretical and experimental investigation of absorption cycle heat pumps for industrial process. Contractors Report, CEC, Brussels, 1986.

2.7 Wood, B.D. Applications of Thermodynamics. Addison-Wesley, Reading, Mass, 1969.

2.8 Flaux, J.P. Study and realization of an industrial Brayton cycle heat pump for high temperatures (150-300°C). Commission of the European Communities Report No. EUR 9849FR, Brussels, 1985. (In French).

2.9 Huck, W. Thermoelectric heat pumping. Proc. ASHRAE Annual Meeting, Vancouver, B.C., June 13-15, 1960.

2.10 Dubois, P. Thermoelectric heat pump for domestic heating Commission of the European Communities Report No. EUR 9308 FR, Brussels, 1985 (In French).

CHAPTER 3
Practical Design

3.1 Introduction

Having found an application and chosen a thermodynamic cycle to suit it, it is necessary to study and select the hardware required to construct a heat pump. This Chapter is concerned with the basic areas of design, the choice and type of major components and the control systems needed to make the unit operate efficiently.

As described in Chapter 2, a considerable number of heat pump cycles are available. Thhis Chapter however relates principally to the vapour compression cycle, with a variety of drives, with some brief comments on simple absorption cycle systems.

3.2 Working Fluid Selection

The range of working fluids is quite literally unlimited. Any fluid which can be made to evaporate between 1 bar and 20 bar at useful temperatures is interesting, and there may be more "dark horses" outside this range.

Following conventional practice, refrigerants are numbered in accordance with the ASHRAE Standard List of Refrigerants. These are tabulated in Appendix 1. They are classified as halocarbons, cyclic organic compounds, azeotropes, hydrocarbons, oxygen compounds, nitrogen compounds, inorganic compounds and unsaturated organic compounds. This discussion will be concerned primarily with the halocarbon compounds which are commonly known under the trade names "Freon" (Du Pont) "Arcton" (ICI) "Genetron" (Allied Chemical) etc. We shall simply refer to them by their ASHRAE designation e.g. refrigerant 12, CCl_2F_2 is R-12.

It might be thought that the choice of working fluid will have a direct bearing on the COP. In practice COP remains remarkably constant over a range of refrigerants with dramatically differing pressures and densities, for the same evaporating and condensing temperatures. This is illustrated in Table 3.1.

In this table the COP of different refrigerants has been calculated for an evaporating temperature of -13°C and a condensing temperature of 30°C. The method used was broadly as described in the previous Chapter. Saturated suction vapour has been assumed, except for the cases of R-113, R-114 and R-115 for which the discharge gas is assumed to be on the saturation line with the suction being slightly superheated. This is necessary because isentropic compression of these two vapours brings them back towards the saturation line on the p-h diagram. Isentropic compression has been assumed throughout.

Note that the COP is almost invariably within the range 4.8+ 10 per cent. The important exceptions to this are those fluids for which the condensing temperature (30°C) is close to the critical temperature. The reason for this is clear if reference is made to Fig.2.4 of Chapter 2. As the condensing temperature approaches the critical temperature, the amount of latent heat available rapidly reduces. Heat is still available from the superheated vapour but the COP is drastically reduced. This is most obvious for R-170 (COP = 2.41, critical temperature = 32°C) and R-744 (COP = 2.56, critical temperature = 31°C).

It would appear from the foregoing that apart from systematic comparison of refrigerants, the best way to ensure a reasonable performance is to work with a fluid well below its critical pressure. In practice, however, this tends to lead to vapours with unacceptably low densities which can lead to much more serious limitations than mere loss of efficiency.

Referring back to Table 3.1, consider the column headed "compressor displacement". The swept volume of the compressor required to serve this one duty varies by a factor of greater than 500:1 for comparatively minor changes in input power. Obviously the price and efficiency of the different compressors needed for different refrigerants is of much greater significance than the theoretical value of COP. This consideration tends to suggest operating somewhat closer to the critical temperature that might seem advisable from efficiency considerations. Studying the figures for compression ratio also leads to the same conclusion: operating at low vapour pressures will lead to disproportionately large compressors.

This leaves us now with a band of refrigerants between R-13B1 and R-40, methyl chloride, from which to select a suitable working fluid. All these fluids fall within a reasonably narrow band in respect of compression ratio and swept volume and so the final choice can be made on the basis of safety, efficiency and cost.

On safety grounds, flammable fluids such as propylene, propane and methylchloride could be excluded from consideration. This depends of course on the application and the safety precautions appropriate. In general it is advisable to avoid flammable fluids when there is not some degree of technical supervision e.g. in a domestic application where a heat pump may not be regularly maintained, or it may be maintained on a do-it-yourself basis.

TABLE 3.1 Comparative Refrigerant Properties

No.	Name	Evaporator Pressure (Absolute) bar	Condenser Pressure (Absolute) bar	Compression Ratio	Refrigerating Effect kg/kg	Refrigerant Circulated g/sec	Suction Specific Volume m^3/kg	Compressor displacement $m^3/sec \times 10^{-3}$	Shaft Power kW	COP	Discharge Temp. °C	Critical Temp. °C
170	Ethane	16.3	46.6	2.86	136	25.8	0.033	0.85	1.46	2.41	50	32
744A	Nitrous Oxide	21.3	64.6	3.03	198	17.8	0.017	0.31	0.98	3.60		37
744	Carbon Dioxide	22.9	72.1	3.15	129	27.4	0.017	0.45	1.37	2.56	66	31
13B1	Bromotrifluoromethane (BTM)	5.38	18.0	3.36	68	51.9	0.024	1.24	0.77	4.25	51	67
1270	Propylene	3.56	12.5	3.51	402	8.3	0.163	1.43	0.78	4.51	42	92
290	Propane	2.89	10.7	3.70	281	12.5	0.155	1.93	0.77	4.58	36	97
502	22/115 Azeotrope	3.49	13.1	3.75	106	33.1	0.051	1.70	0.80	4.37	37	82
22	Chlorodifluoromethane	2.95	11.9	4.03	163	21.6	0.077	1.68	0.75	4.66	53	96
115	Chloropentafluoroethane	2.66	10.4	3.89	68	52.0	0.048	2.50	0.87	4.02	30	80
717	Ammonia	2.36	11.7	4.94	1103	3.2	0.509	1.62	0.74	4.76	99	133
500	12/152a Azeotrope	2.14	8.79	4.12	141	24.9	0.094	2.34	0.75	4.65	41	106
12	Dichlordifluoromethane	1.82	7.44	4.08	116	30.2	0.091	2.75	0.75	4.70	38	112
40	Methyl Chloride	1.46	6.53	4.48	349	10.0	0.28	2.81	0.72	4.90	78	143
600a	Isobutane	0.90	4.10	4.54	259	13.5	0.40	5.43	0.81	4.36	27	135
764	Sulphur Dioxide	0.81	4.58	5.63	329	10.7	0.40	4.29	0.72	4.87	88	158
630	Methylamine	0.68	4.24	6.13	707	5.0	0.97	4.83	0.73	4.81		175
600	Butane	0.57	2.85	5.07	299	11.8	0.62	7.32	0.71	4.95	31	152
114	Dichlorotetrafluoroethane	0.47	2.53	5.42	100	35.1	0.27	9.50	0.78	4.49	30	146
160	Ethyl Chloride	0.32	1.86	5.83	331	11.0	1.07	11.7	0.68	5.21	41	187
631	Ethylamine	0.23	1.70	7.40	525	6.7	2.02	18.3	0.64	5.52		183
11	Trichlorofluoromethane	0.21	1.25	6.19	155	22.6	0.76	17.2	0.70	5.03	44	198
661	Methyl Formate	0.13	0.96	7.74	440	8.0	3.01	24.1				214
610	Ethyl Ether	0.10	0.85	8.20	294	11.9	2.18	26.1	0.61	5.74		194
30	Methylene Chloride	0.081	0.69	8.60	313	11.3	3.12	35.1	0.72	4.90	96	237
113	Trichlorotrifluoroethane	0.071	0.54	8.02	125	28.2	1.71	48.2	0.73	4.84	30	214
1130	Dichloroethylene	0.058	0.48	8.42	266	13.2	3.97	52.5	0.73	4.83		243
1120	Trichloroethylene	0.014	0.13	11.65	213	16.5	14.32	237	0.73	3.82		271

Note. R21 has been omitted from this list because of its toxicity.

Ammonia is an unusual case because it is toxic, being exceeded only by sulphur dioxide in its Underwriter Laboratories toxicity rating. Despite this it has been widely accepted in domestic applications in absorption cycle refrigerators and codes of practice are well established. Nonetheless, as can be seen from Table 3.1, acceptable alternatives in the form of non-flammable, non toxic halocarbons are available and it would be advisable to select one of these. (As discussed in Chapter 4, ammonia is important for the domestic absorption cycle units).

It would be rash here to select and recommend one particular fluid, but the figures make it clear why R-22 and R-12 are so widely used.* The rather marginal advantages of the azeotropes and of the recent innovations 12B1 and 13B1 are outweighed by their much greater cost. A summary of approximate relative cost is given in Table 3.2. 12B1 (Bromochlorodifluoromethane, BCF) and 13B1 (Bromotrifluoromethane, BTM) have found widespread acceptance as fire extinguisher fluids, where inertness is a great merit.

TABLE 3.2 Relative Costs of Refrigerants (Courtesy of ICI Limited, Mond Division)

R-11	1.0
R-12	1.14
R-13 B1	6.4
R-22	2.0
R-114	2.0
R-502	2.95

The question of chemical stability is of great importance to the heat pump designer. Working fluids have been used for many years at refrigerating and air conditioning temperatures but in heat pumps they must be run at temperatures which are sometimes substantially higher. Breakdown of the refrigerant occurs mainly at the compressor discharge port. The discharge temperature is the hottest part of the working cycle. At this temperature the refrigerant, with a small quantity of oil, is being forced past various metal surfaces at high velocity, and the metal surface can catalyse the decomposition reaction. It is difficult to simulate these conditions in sealed tube tests or any form of life tests and one must rely on direct operating experience, but nontheless, refrigerants can be ranked in order of thermal stability, and Ref.3.1 lists maximum continuous exposure temperatures for refrigerants, as shown in Table 3.3.

* At the time of writing, indications are that a number of working fluids may be banned, or at least limited in their availability, due to suggestions that they have a detrimental effect on the ozone layer surrounding the Earth. The fluids most affected are R-11, R-12 and R-114. A number of alternative fluids are being proposed, including R-124, chlorotetrafluoroethane.

TABLE 3.3 Maximum Continuous Exposure Temperatures

Refrigerant	Maximum Temperature for Continuous Exposure in the Presence of Oil, Steel and Copper, °C
R-11	107
R-113	107
R-12	121
R-500	121
R-114	121
R-22	135 - 150
R-502	135 - 150
R-13	$>$ 150

The products of refrigerant breakdown are generally acidic and are harmful to the mechanical components and have a direct bearing on the lifetime of the machine. An example is given in Chapter 6 of the selection procedure for refrigerants for a high temperature heat pump.

3.3 Compressors

This section contains an introductory description of the various types of compressors appropriate to vapour compression cycle units, after which there is a short summary of those compressor requirements by which the heat pump differs from the refrigerator. The discussion commences with the smallest compressors and works through to the largest.

Compressors are broadly divided into two types which are referred to as 'wet' and 'dry'. For the purposes of this discussion a wet compressor has greater than 15 ppm of oil in the vapour being compressed, in some cases as much as 20 per cent by volume. A dry compressor is one in which the oil serves no useful purpose in the compressor and has only got there by leakage from bearing seals and is generally less than 5 ppm.

3.3.1 Rotary vane compressors These are small compressors operating normally with low compression ratios and low pressures, also called sliding vane compressors. A unit is shown schematically in Fig.3.1.

They have characteristically a high volumetric efficiency and are well suited to high rotational speeds.

Compression takes place between the sliding vanes which means that the strength of the vanes, and their sealing problems are of paramount importance. The vanes can be of cast iron, but are more commonly made from reinforced pheolic resin. There may be only two blades or any larger number, as shown in the figure. Increasing the number of blades will increase frictional losses but improves volumetric efficiency, so the number finally selected is a compromise.

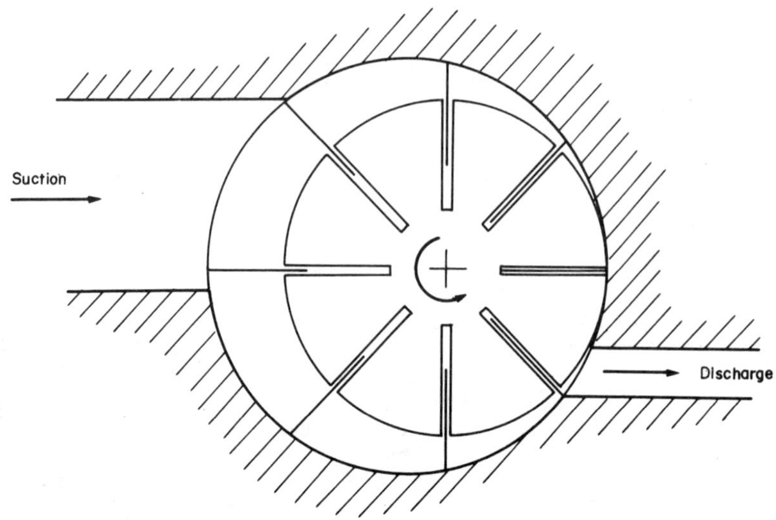

Fig.3.1 Schematic Rotary Vane Compressor

Sealing of the blades to the cylinder is by hydrodynamic lubrication. In other words a film of oil is maintained inside the cylinder, and the vanes slide round on this. The force which this film must support depends on rotational speed, and pressure difference across the machine. The blades are also held out by the underblade pressure, which is fed from the discharge pressure.

Unlike reciprocating compressors, the oil sump is usually maintained at the discharge pressure which obviates the need for any oil pump. Oil feed problems are critical because any breakdown of the film can lead to rapid degradation of the sealing surfaces.

Rankine-Rankine cycle units, where a turbine drives the compressor, have used rotary vane systems for both expander and compressor. In an attempt to achieve high efficiencies, a compressor/expansion engine has been operated with hydrodynamic vanes lubricated only by the working fluid. Each vane has a pivoting tip or 'shoe' which is able to rotate slightly so that the outer surface can conform to the cylinder wall. Obviously a successful development along these lines would obviate many lubrication problems. Such a unit is shown in Fig.3.2.

Because the inlet flow is continuous, no valve is necessary. Discharge valves are sometimes used however, depending on the number of blades and the pressure requirements.

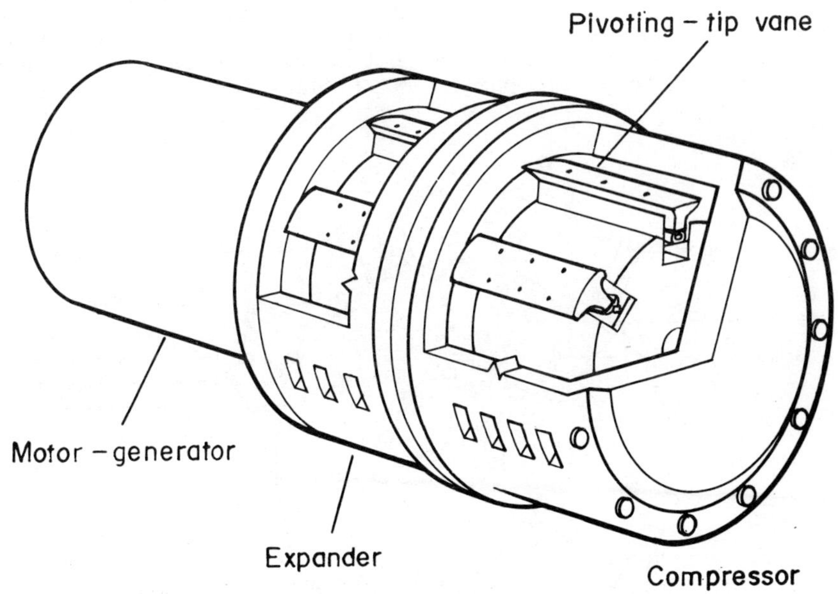

Fig.3.2 Pivoting-tip Rotary Vane Compressor/Expander

Rotary vane compressors have to date usually been found in applications of less than 5 kW drive, generally a great deal less. Discharge pressures seldom exceed 10 bar and pressure differences are limited to about 4 bar. Although these limits are being constantly eroded by development they may prevent the rotary vane compressor from performing high temperature heat pump duties.

3.3.2 Reciprocating compressors Reciprocating compressors are the commonest type of compressor in every day use. For air compression as well as for refrigerating or heat pump duties the range of sizes is truly impressive. Very small fractional horse power units compete with rotary vane compressors whilst at the other end of the scale, 100-150 kW drive compressors can be used. The reason for their extensive use is the simplicity of the basic design and the reasonable efficiencies which can be achieved.

Compressors of this type (and others) are referred to as 'open' or 'hermetic', or sometimes 'semi-hermetic'. An open compressor is one in which the input drive shaft emerges from the casing which contains the refrigerant via a seal. Drive can then be from an electric motor, an internal combustion engine, steam turbine or any other source of work. With hermetic construction however, the seal is eliminated and an electric motor is contained within the casing, which is welded up. This has the two advantages of eliminating leakage around the seal and providing motor cooling, but can lead to chemical problems if the motor should overheat. Semi-hermetic construction describes a casing of the hermetic type, but with a bolted rather than welded casing, to permit easier servicing.

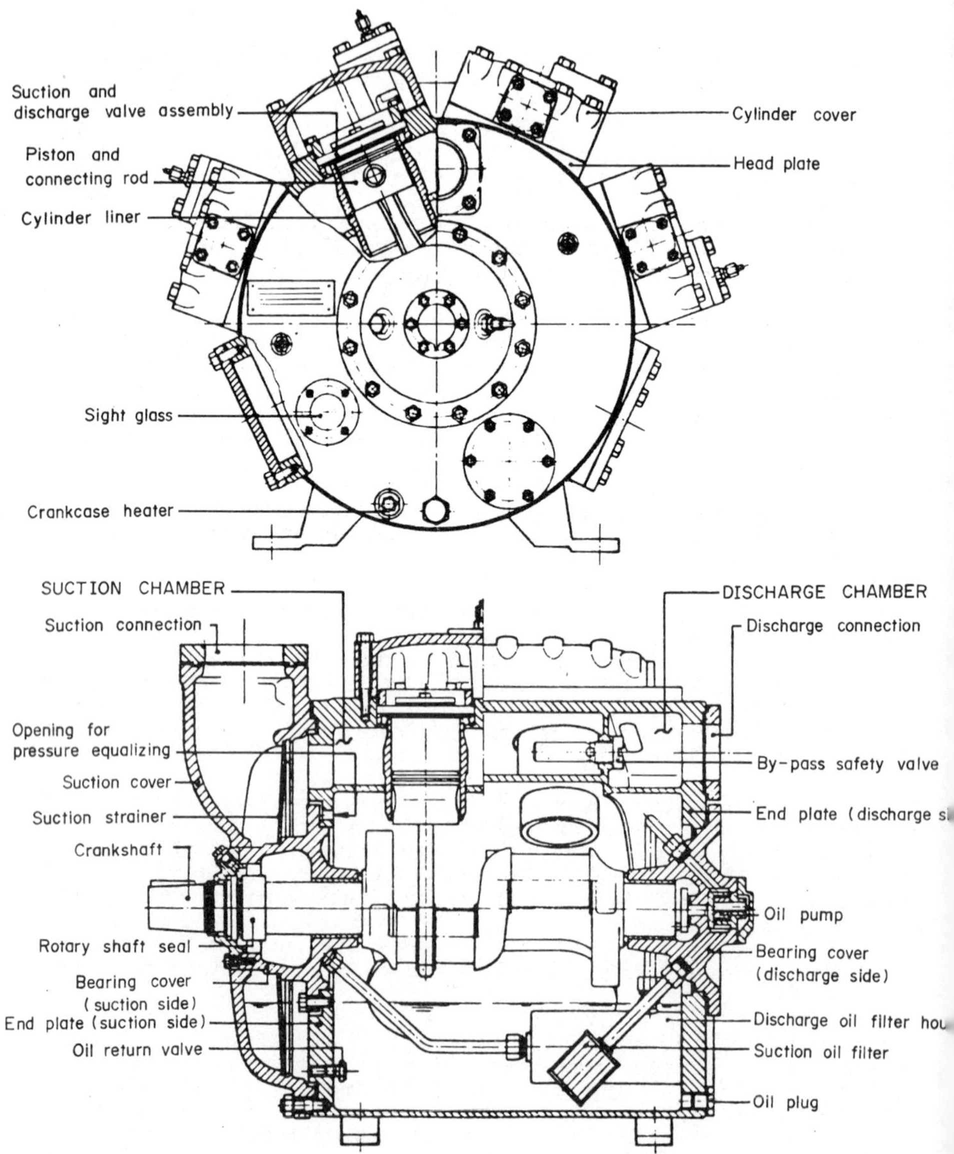

Fig.3.3 Grasso Reciprocating Compressor

Note in Fig.3.3 that the arrangement of the pistons allows a similar crankshaft to be used for 2, 4 or 6 pistons which permits standardisation of components. Pistons are made either from aluminium with the use of piston rings or from cast iron, which can eliminate the need for rings in very small compressors. The size of the piston, connecting rod and crankshaft are important because their strength will determine the maximum pressure difference which the compressor can handle.

Oil is fed from the sump into the crankshaft in the same way as in an automotive engine. Lubricant for the cylinders is either fed from the crankshaft or simply is oil splashed up from the sump. In either case a film of oil is maintained on the cylinder wall, and a proportion of this will pass out of the compressor and into the working circuit. On the suction side, when this oil returns with the returning refrigerant, it is hoped that it falls out of suspension and returns to the sump. To allow this to happen an equalising port is provided which maintains the sump at the suction pressure.

When the refrigerant enters the compressor it is directed onto the cylinder walls to cool them as far as possible. This is important thermodynamically since heating the gas in the cylinder reduces the isentropic efficiency and increases the discharge temperature. This cooling helps to reduce the rate of oil breakdown and consequently the rate of wear of cylinder walls and discharge valves.

One of the worst conditions to which the compressor will be exposed is known as 'slugging'. Under an adverse combination of circumstances, the suction side refrigerant can contain significant quantities of liquid. When these are small droplets they can cause damage to the valves, where velocities are high. More importantly, when these are large slugs of liquid substantial mechanical damage can result unless precautions are taken. The cylinder heads of most reciprocating compressors are now spring-loaded to permit them to lift under slugging conditions. This is preferable to bending pistons and connecting rods or breaking crankshafts.

The valves are generally of simple construction, although the design is critical to compressor performance. They can be reeds, either free or clamped or else rings or plates with a light backing spring. The purpose of the valve is to give unrestricted flow in one direction, with a gas tight seal against return flow.

The compressor is normally protected by a battery of electronic devices which switch off in the event of difficulty. Commonly discharge pressure and temperature are monitored, as is suction pressure and oil pressure. Under certain circumstances differential pressure can also become critical and must be included if there is a danger of exceeding the maximum. Some compressors also have separate cylinder head temperature sensors in case one cylinder only is over-heating. A typical compressor specification is included in Table 3.4.

TABLE 3.4 Limiting Conditions for a Typical Reciprocating Compressor

Max. discharge pressure	21 bar
Max. discharge temperature	140°C
Max. pressure difference	18.5 bar
Max. suction pressure	5 bar
Max. pressure ratio	10:1

The performance of a compressor is normally expressed by refrigerating capacity and power input curves, for given conditions. For heat pump applications it is important to have heat rejection curves in addition. An example of these is given in Fig. 3.4.

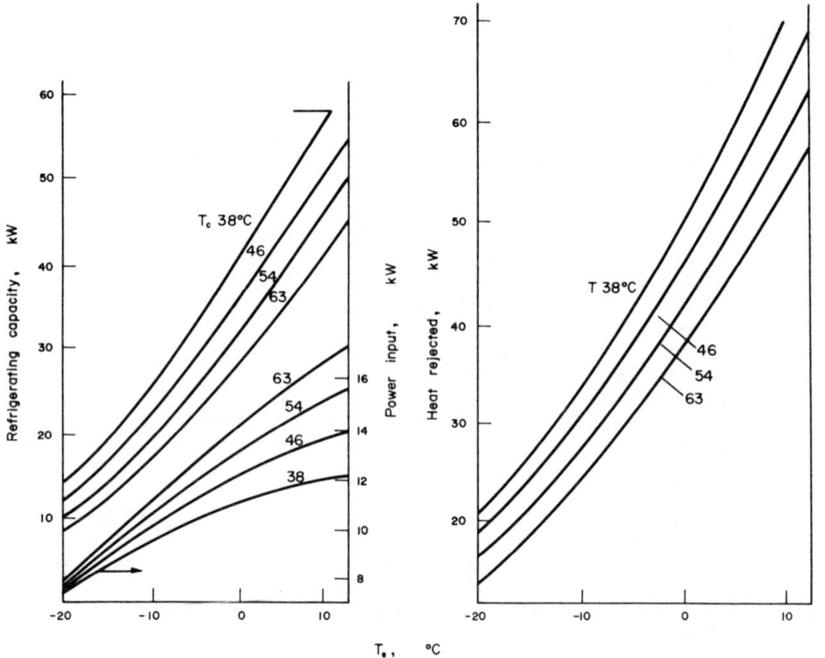

Fig.3.4 Performance Curves for a Reciprocating Compressor with R-22 (Source: ASHRAE Handbook)

With this information and the use of the R-22 p-h chart Appendix 1 it is possible to calculate the mechanical, volumetric and isentropic efficiencies. In practice the design engineer does not need to know these figures and he can complete his design with knowledge of only the heat added/rejected and the shaft power. The reciprocating compressor is not necessarily always wet. Lubrication of the cylinder walls is normally needed with the 'trunk piston' construction described so far, because the piston carries lateral loads from the connecting rods. A more advanced (and more expensive), design shown in Fig.3.5 uses a 'crosshead' construction which isolates the piston from the crankshaft and sump.

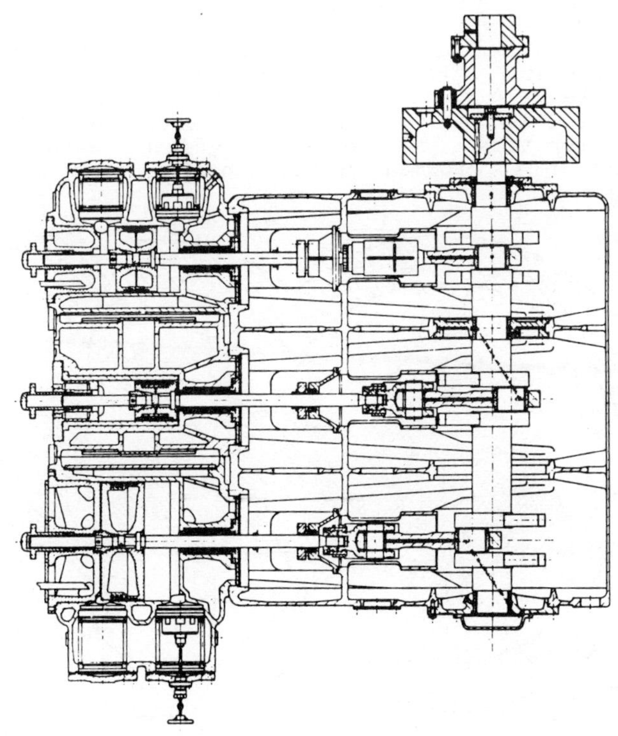

Fig.3.5 Cross Section of a Linde 'oil-free' Compressor

With this type of compressor the piston does not actually touch the cylinder, once it has been run in. The seal is hydrodynamically formed on a thin film of vapour, aided by labirynth grooves in the piston, or possibly PTFE piston rings. A more conventional seal is required on the crosshead bearing to contain the oil in the sump.

Horizontal air compressors of this type manufactured by one company (Ingersol Rand) go up to 4500 kW in capacity! In general however they are not suitable for heat pumps because of their size, the size of the foundations necessary and the cost, except in specific cases such as district heating, where existing hard standings could be used (e.g. an old power station).

3.3.3 Screw compressors The commonest type of screw compressor is often called the Lysholm compressor, after its Swedish inventor. It is manufactured in the UK under licence by James Howden, Glasgow. The principal of operation is easy to follow in general terms, but becomes more complex if the details are examined. The description here is restricted to general principles.

The compressor consists of two rotors running together within a sealing sleeve. The rotors are identified as 'male' and 'female'. The male rotor has a number of lobes on it (usually 4) which are semicircular in section and formed in a helix along the rotor body. The female rotor has a number of flutes and channels in it, formed in the opposite helix to the male, as shown in Fig.3.6.

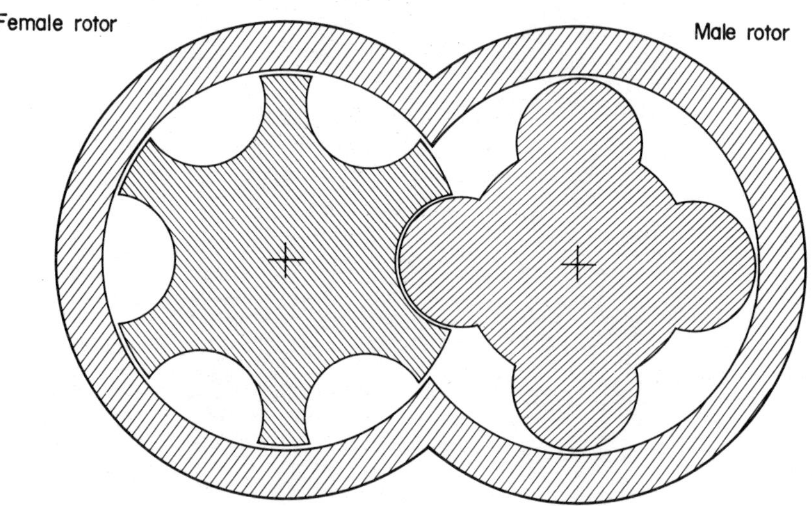

Fig.3.6 Cross Section of Lysholm Compressor Rotors

Compression is achieved in a volume within one channel of the female rotor, sealed by the discharge end face of the casing, the cylindrical inside of the casing, and the interface between the male and female rotors. Discharge occurs through a port in the end of the casing, and all the vapour is discharged, so there is no re-expansion loss. Such a compressor is shown in Fig.3.7.

When a screw compressor is designed to run dry, the it is an oil-free compressor offering high throughputs at reasonable efficiencies, although discharge temperatures tend to be high, there are sealing problems and pressures are limited. The oil-injected screw compressor is in comparison ideally suited to many refrigeration and heat pump applications.

Fig.3.7 General View of Lysholm Screw Compressor

The principle of oil injection is that oil is continuously pumped into the suction volume of the compressor, to about 1 per cent by volume. Because the ratio by weight of oil to refrigerant fluid is much greater, the oil removes the heat of compression directly from the refrigerant, permitting it to be discharged at remarkably low temperatures. In addition, the oil provides a close seal between the rotors and the casing and permits close juxtaposition of the bearings and the rotors. The main disadvantage of oil injection is that the oil must be separated after discharge, and because the oil and refrigerant are mixed at high pressure, oil dilution can be a problem.

Oil injected screw compressors can effectively meet any combination of pressures which a heat pump would reasonably demand. They are suited to medium and large powers. The principal disadvantages are the cost relative to reciprocating machines and the noise level. Current (1986) commercially available oil injected screw compressors can accommodate suction pressures of up to 9 bar(abs) and discharge pressures of 26 bar(abs).

There is a second type of screw compressor (the 'Zimmern' screw) which is currently available and is illustrated in Fig.3.8.

With this type there is only one compression screw. The gas is trapped within its helix by two secondary wheels which rotate in synchronization. This type of compressor has symmetric pressure loadings on the main rotor which is an advantage and it is also claimed to be cheaper to manufacture. It will be interesting to see whether this compressor can achieve the widespread market acceptance which the Lysholm compressor deserves but has not yet attained.

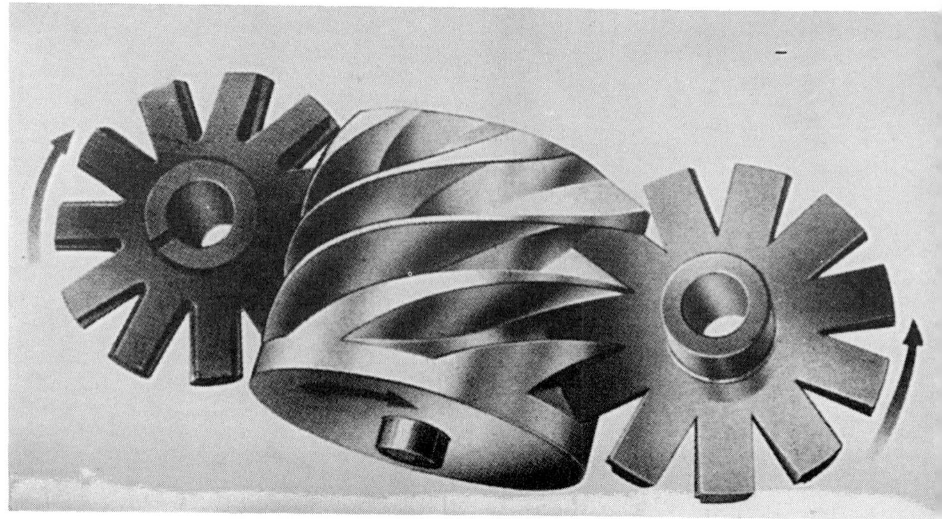

Fig.3.8 Principal Components of Zimmern Screw Compressor

3.3.4 Centrifugal compressors The centrifugal compressor is a well developed machine because of its use in the compression stage of small gas turbines. It also is used in many high-flow applications in the chemical and fuel industries. Unlike the other types described so far, it is exclusively used as a dry compressor, since the presence of oil in the gas or vapour has no beneficial effects.

Compressors are available in single stage or two stagte form (Fig.3.9) with either open or hermetic drive. Pressure ratios of 3-4 are considered about optimum. Hermetic units are popular because the high rotational speeds lead to seal problems, unless there is an integral step-up gearbox. Multi-stage compressors, an example of which is shown in Fig.3.10, are made either by linking stages via external pipework (which has the advantage of permitting inter-stage cooling, but is more expensive) or by combining a number of stages within the barrel of a single machine.

Because these compressors are suited to large capacities due to the high volumetric throughputs, and to small temperature differences due to the low pressure ratios, they are primarily used in water/water heat recovery systems, with large shell and tube heat exchangers. This probably means that in heat pump guise, apart from air conditioning large buildings, their most important role has been in large industrial process units.

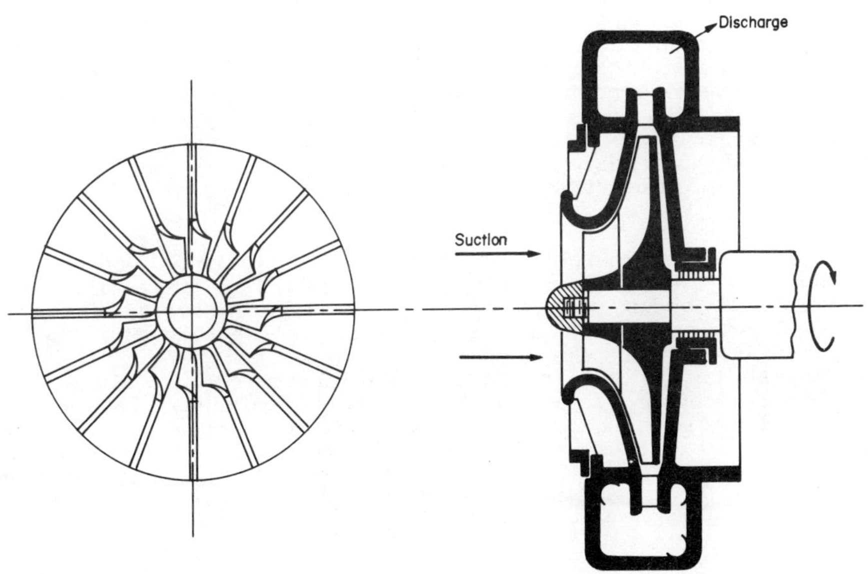

Fig.3.9 Simple Centrifugal Compressor

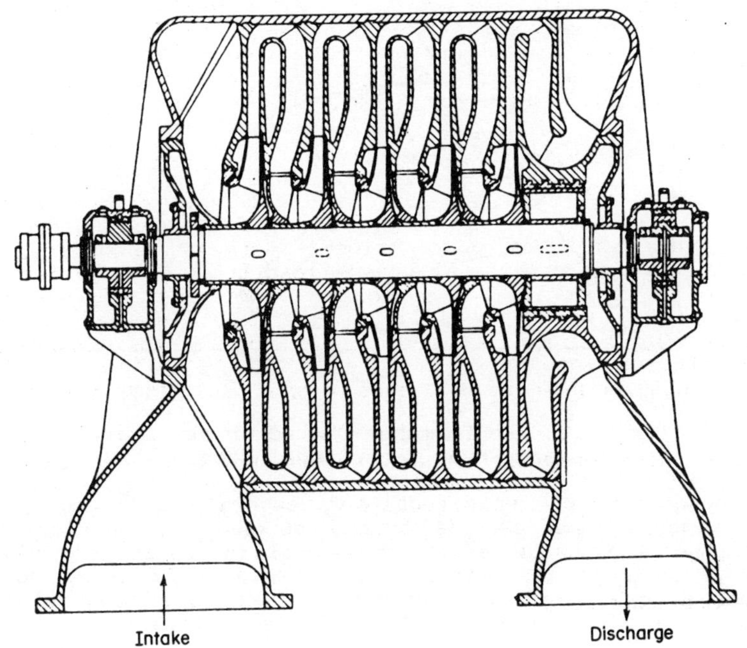

Fig.3.10 Multi-stage Centrifugal Compressor

Refrigerant selection for centrifugal compressors is a complex subject. The speed of sound in the refrigerant vapour is critical because the peripheral velocity of the compressor is directly related to it, for a given compression ratio. Specific volume is very important because capacity is not as easily controlled as on a reciprocating compressor, particularly if constant-speed drive is used. Sometimes a refrigerant may be selected to give a deliberately low capacity for a given compressor in order that the efficiency is high at the design condition.

The performance characteristic of a centrifugal compressor is normally shown on a pressure rise Vs flow rate diagram, and Fig.3.11 shows such a diagram for a variable speed unit. The most important feature is the line delineating the region of potential surge. Surge results from the compressor stalling causing an on/off cycling effect with a period of only a few seconds. Although it is not immediately detrimental to efficiency or damaging, it is noisy and also will accelerate deterioration of the compressor and the refrigerant. While surge is a major headache for part-load applications such as air conditioning, for continuous-drive heat pump duties it is possible to avoid it.

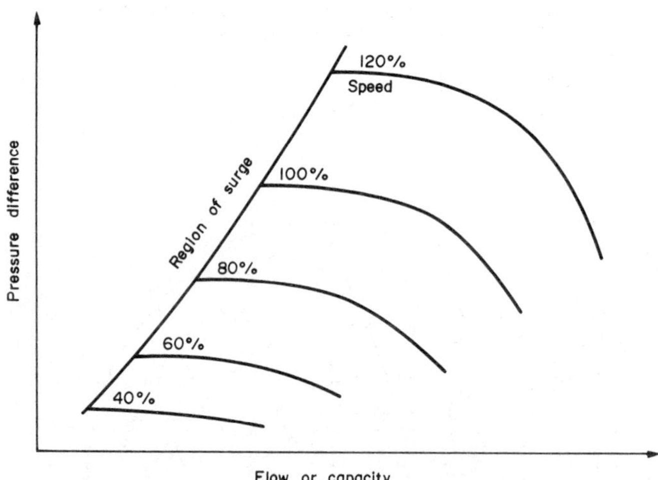

Fig.3.11 Performance Characteristics of a Variable Speed Centrifugal Compressor

Capacity control of centrifugal compressors is achieved by inlet guide vanes, which can be turned to set the suction vapour into either a favourable or unfavourable direction of rotation. Although this is less efficient than variable speed control, it does allow operation over a larger range of conditions and is therefore almost universally used, (see Fig.3.12).

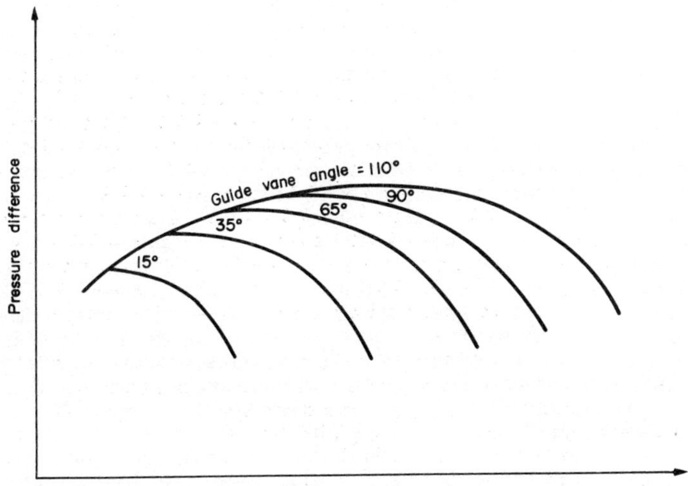

Fig.3.12 Performance Characteristics of a Centrifugal Compressor with Variable Inlet Guide Vanes

A typical isentropic efficiency characteristic of a centrifugal compressor is shown in Fig.3.13.

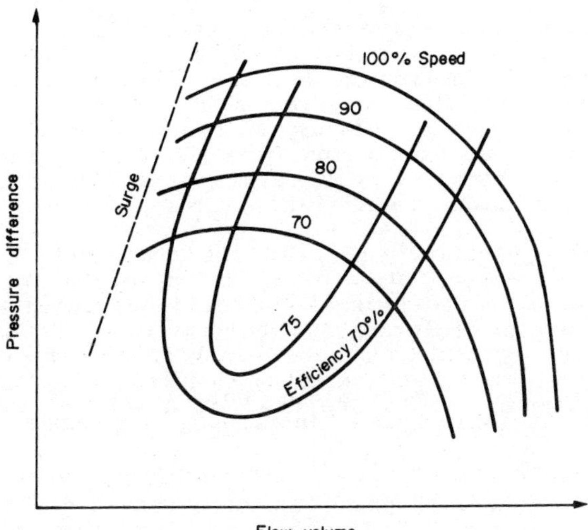

Fig.3.13 Efficiency of Typical Centrifugal Compressor

It can be seen that efficiency can be maintained over quite a wide operating range, which simplifies performance calculations. It should be borne in mind however that the mechanical efficiency does not also remain constant, but falls off with reduced flow.

3.3.5 Summary of heat pump compressors Essentially the compressors described here are well tried and proven units universally accepted in the field of domestic and industrial refrigeration. Certain differences remain between the refrigerator and the heat pump and these are listed below as a reminder to the reader:

(i) Heat pumps normally operate at higher condensing temperatures and greater pressure ratios than refrigerators. This puts more stress on both refrigerant and compressor.

(ii) Heat pumps generally operate for many more hours per annum at their maximum duty than refrigerators or air conditioning units (except in domestic applications) and this also tests the compressors' capabilities.

(iii) Because heat pumps must compete commercially with other heating systems such as direct gas or electric heating, or packaged boilers, there is great pressure to reduce the first cost of the system.

3.4 Prime Movers

The power required to drive a heat pump can come from a variety of different primary sources which are discussed in this Section. The most common drives are electric motors, but gas and Diesel engines have received a large amount of attention during the past decade. Steam turbines have been used in a number of industrial heat pumps, and the turbine has also featured in Rankine-Rankine cycle systems employing refrigerant-type working fluids in the turbine circuit.

3.4.1 Electric motors By far the commonest type of drive is the electric motor, 50 Hz or 60 Hz, single phase or 3-phase. The motor can be directly shaft mounted and refrigerant cooled as on a hermetic reciprocating compressor, it can drive via gears as on a hermetic screw or centrifugal compressor or the drive can be via belts or external couplings with an open compressor. Electric motors are used on all sizes of heat pumps from the smallest rotary vane compressors to the largest centrifugal.

The reasons for the widespread use of electric drive are simple: the capital cost is low and the maintenance required is also low. While greater capital outlay may be permitted if lower fuel bills are promised, it is not so easy to sell a system such as a gas engine-driven heat pump with higher maintenance demands and as long as this remains true, which may not be for much longer, the electric motor will maintain its supremacy.

The disadvantage of the electric motor is that it is less efficient in terms of primary energy. When the national interest (or world interest) is taken into account one should be studying systems whereby the fuels available to us are used in the most sparing way possible. The generation and distribution of mains electricity in the UK for example is achieved with an overall efficiency of the order of 30 per cent, and without recovering heat from power stations this is not likely to significantly improve in the near to medium term. In general this is reflected in the high cost of electrical power although this is not always true. Low cost hydro-electric power (e.g. in Sweden) and cheap nuclear energy (e.g. in France) have been used to promote widespread use of electric heat pumps having acceptable payback periods.

A secondary reason why electric heat pumps are less efficient is the necessary constant speed of operation. A domestic heat pump has to match a constantly varying demand against a variable supply temperature and this can best be achieved with a variable speed compressor. Although thyristor control is possible it is still expensive and not used in this application. For industrial process applications however the constant speed is not a severe restriction, and the previously mentioned low maintenance requirements are a positive boon.

3.4.2 Gas Engines Readers of the 1st Edition of this book may recall that this Section dealt in some detail with specific gas engine types, in terms of both manufacturer and model. With the exception of long-established manufacturers of industrial gas engines, the market is in such a volatile state that the validity of specific engine data may not be of relevance in a year or two. It is therefore believed to be of more value to the reader to present data on practical aspects of engine use as part of a 'case study'. The case study selected describes a heat pump/chiller employing a Komatsu gas engine, installed in Tokyo. (The study also provides useful data on other system components).

General Description. An air source gas engine driven heat pump/chiller unit was installed in the Technical Research Centre of Tokyo Gas Company in November 1981. The unit was commissioned by the 1st December 1981, and was operated over a period of about one year under full monitoring conditions. The building in which the unit was located was constructed in 1962, and had a total air conditioned area of 994m², split almost equally between three floors. The heat pump unit was installed on the roof of the building and its function was to supply hot and cold water to three air conditioning units located on each floor. The specification of the system is given in Table 3.5, (Ref.3.2).

The gas engine specified for the duty was a Komatsu 4 N 94 unit (unturbocharged) with four cylinders. The full specification of the engine is given in Table 3.6 and the data on the compressor, which is an open reciprocating type, is detailed in Table 3.7. The heat exchangers employed in the installation, including those to recover heat from the engine, are detailed in Table 3.8.

TABLE 3.5 Specification of Entire Unit

TYPE			AIR HEAT SOURCE HEAT PUMP CHILLER
Cooling	Capacity	Kcal/h	95,000
	Input	Kcal/h	79,100
Heating	Capacity	Kcal/h	133,000
	Input	Kcal/h	85,400
Outside dimensions		mm	2,200 length, 3,500 width and 3,200 height
Weight		kg	6,100
Power supply		-	200V, 3 phase, 50 Hz
Engine/Compressor power train system		-	Direct flexible, disc coupling
Capacity control system		-	Engine speed and compressor capacity combinedly controlled
Capacity control range		%	11 ~ 100
Protectors	Fuel gas system	-	Abnormal pressure rise and leak gas detectors
	Engine system	-	Engine overspeed, lubricant pressure drop, cooling-water temperature rise and improper startup
	Refrigerant circuitry	-	Refrigerant pressure rise, refrigerant pressure drop, lubricant pressure drop, water supply failure, water temperature drop, compressor safety valve, condenser safety valve and evaporator safety valve
piping size	Fuel gas piping	-	40A
	Water piping	-	65A
Auxiliary equipment power	Condenser fan	kW	0.55 x 6
	Radiator fan	kW	0.55 x 2 (in the cooling mode)
	Engine cooling pump	kW	0.75 x 1
	Sound-proof box fan	kW	0.4 x 1
	Crankcase heater	kW	0.125 x 1
	Control Power	kW	1.5

TABLE 3.6 Specification of Gas Engine

Type	-	Water-cooled vertical 4-cycle in line type engine without turbocharger
Model No.	-	4N94
Number of cylinders	-	4
Cylinder bore x stroke	mm	94 x 106
Continuous output rated	HP/rpm	42/1,750
Continuous torque rated	kg.m/rpm	17,2/1,750
Max. output	HP/rpm	55/2,400
Max. torque	kg.m/rpm	19.1/1,500
Compression ratio	-	12
Ignition timing	°BTDC/rpm	18/800, 22/1,750 (automatic ignition advancing type)
Igniter	-	Contactless semiconductor type
Ignition plug	-	Ni - Ni single elctrode made by NGK
Lubrication system	-	Trochoid pump forced lubrication system
Lubricant capacity		9.5/high level 8.0/low level
Lubricant subtank capacity		20
Applicable lubricant	-	Kyodo Sekiyu GE 30 Special
Starting system	-	DC starter (24V)
Speed governing system	-	All-speed electronic governor (Barber Colman Controller Model DYN1 10004 and Actuator Model DYNC 11004)
Range of working engine speds	rpm	800 ~ 1,750
Output shaft turning direction	-	Counterclockwise as viewed on the output shaft side
Fuel applied	-	LNG straight (8,540 Kcal/Nm^3 L.H.V.)
Dry weight	kg	350
Outside	mm	758 length, 558 width and 762 height

As shown in Fig.3.14 the system operates in two different modes. In the cooling mode, the water to refrigerant heat exchanger functions as an evaporator, and the cold water produced is distributed to the air handling units to control ambient conditions on the three floors of the building. Heat recovered from the engine and from the exhaust gas is dissipated via the radiator. When operating in the heating mode however, the water to refrigerant heat exchanger functions as a condenser and the heat recovered from the engine, including its exhaust gas, joins that from the condenser in headers before being distributed to the air handling units for space heating.

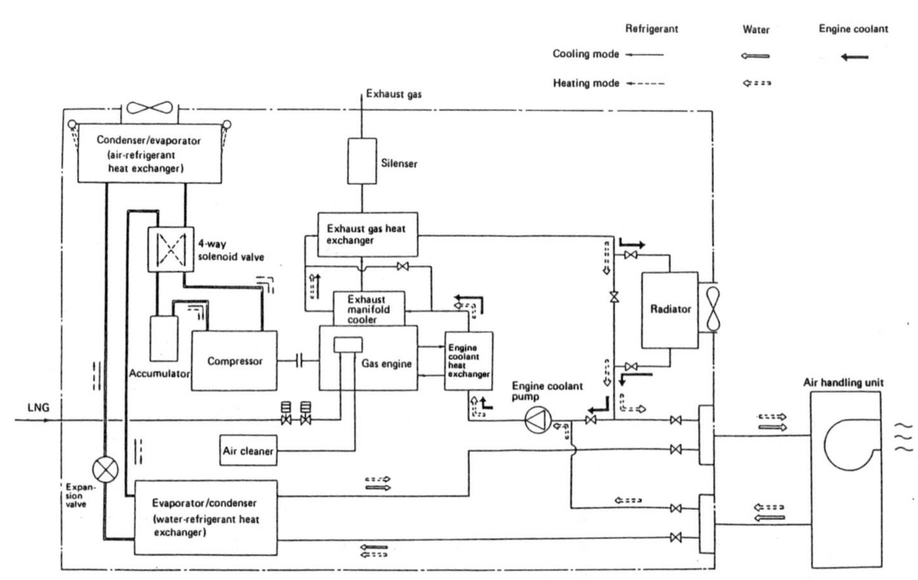

Fig.3.14 System Flow of Tokyo Heat Pump

The Control System. A number of manual and automatic control features are incorporated in this heat pump air conditioning system. In order to change from a heating to a cooling mode, a manually operated switch is used to change over the 4-way solenoid valve. For defrosting, the temperature of the outer surface of the refrigerant line immediately downstream of the evaporator is monitored and the gas engine heat pump system will operate in the reverse cycle should the surface temperature fall below 3°C. In this case, the engine waste heat is recovered as hot water and recycled as a heat source to reduce the defrosting time. Conversely, when the outside ambient exceeds 32°C in the cooling season, water is sprayed from nozzles adjacent to the condenser, thus called an 'evaporated condenser'

TABLE 3.7 Specifcation of Chiller

		Item	Unit	Specification
		Type	-	Reciprocating open type
		Model No.	-	5H40 (Toyo Carrier)
		Number of cylinders	pcs.	4
		Cylinder diameter	mm	82
		Stroke	mm	70
		Theoretical displacement	m^3/h	157
		Legal capacity	Legal RT	18.47
		Range of working speeds	rpm	400 ~ 1,750
		Capacity control range	%	100, 75, 50, 25
		Refrigerant	-	R22
Evaporator/condenser	Water-refrigerant heat exchanger	Type	-	Shell and tube type
		Flow rate	l/min.	150 x 2 changers
		Dimensions	mm	o22 and 1,380 length
	Evaporator/condensing coil	Type	-	Copper tube aluminium plate fin type with water jet
		Internal volume (refrigerant side)		22 x 4 coils
		Dimensions	mm	2,592 length, 90 width and 762 height
Expansion valve			-	Thermal type automatic expansion valve x 8 pcs.
Fan		Type	-	Axial flow type
		Capacity	$m^3/min.$	164 x 6 fans
Suction accumu-lator		Type	-	Vertical type
		Dimensions	mm	Ø355 and 600 height x 1 accumulator

TABLE 3.8 Specification of Heat Exchangers

Engine coolant Heat exchanger	Type	-	Shell/tube type
	Heat transfer	m^2	0.365
	Quantity of heat exchanged	kcal/h	23,100
	Dimensions	mm	396 length, 145 width and 221 height
Exhaust manifold Cooler	Type	-	Double wall type
	Quantity of heat exchanged	kcal/h	2,480
Exhaust gas heat exchanger	Type	-	Radial fin tube type
	Heating area	m^2	8.6
	Quantity of heat exchanged	kcal/h	16,600
	Dimensions	mm	1,500 length, 400 width and 400 height
Radiator	Type	-	Plate fin tube type
	Radiator front area	m^2	1.73
	Quantity of heat exchanged	kcal/h	39,700
	Dimensions	mm	1,500 length, 90 width and 1,150 height

by the manufacturer. This lowers the condensing temperature and keeps engine power requirements to a minimum for efficient operation.

The overall control of the capacity of the air conditioning system is implemented using a combination of engine speed and compressor capacity variations. The engine speed may be reduced from its maximum of 1,750 rpm to 800 rpm, without a noticeable effect on thermal efficiency. When the capacity rating falls below the 46 per cent of full load achievable using this technique, cylinders in the compressor are unloaded, up to a maximum of four, while maintaining a constant engine speed of 800 rpm. The speed of the engine is set according to a difference between the supply water temperature and a set value.

TABLE 3.9 Alarms and Protection Systems on Tokyo Unit

Engine:	Engine overspeed:	2000 rpm and above
	Lubricant pressure drop:	0.05 ± 0.1 kg/cm² and below
	Coolant temperature rise:	102.5+2.5°C and above
	Improper start:	Failing to start three times
	Fuel gas pressure abnormal rise:	500mm H_2O and above e
Refrigerant Circuit:	Refrigerant pressure rise:	24 kg/cm² and above
	Refrigerant pressure drop:	2.2 kg/cm² and above
	Lubricant pressure drop:	Operating refrigerant low pressure + 1.0kg/cm² and below
	Water supply failure:	Water supply suspended
	Water temperature drop (to prevent freezing):	5°C and below
	Compressor safety valve:	24 kg/cm² and above
	Condenser safety valve:	72°C and above at melting point
	Evaporator safety valve:	72°C and above at melting point
Gas Leakage Alarm:	Detection principle:	Heated catalyst and measured resistance
	Detectable range:	25% of LEL or more
	Alarm system:	Delay type automatic return
	Working temperature range:	-10 to 50°C
	Power consumption:	5W
	Power supply:	100W, single phase

Safety. A considerable number of features which should contribute towards the overall safety of the installation were incorporated. These catered for the engine, the refrigerant circuit, and gas supplies. Before start up, the engine room was ventilated for 20 seconds using a fan, to prevent the room filling with fuel gas prior to ignition. In addition a gas detector is provided in the engine room. Should the concentration reach 25% of the lower explosion limit, an emergency gas shut off valve would be automatically actuated. Data on the alarms and protection systems are given in Table 3.9.

Results of Trial. The overall performance data of the system operating in both the heating and cooling modes is given in Table 3.10. The unit operated as a heat pump for a total of 735 hours, commencing on December 11th 1981 and terminating on April 28th 1982. Operation in the cooling mode commenced on June 2nd 1982, and after a total of 526 hours utilisation, it ceased operation on September 30th 1982. The conditions under which the full load rated heated performance were obtained are given in Table 3.11.

The part load performance in terms of number of compressor cylinders operating is shown in Fig.3.15. These data are for an ambient temperature of 7°C and a supply water temperature of 45°C. They show how the heat pump COP can improve as the capacity is reduced below the maximum. At a capacity rating of 70 per cent, the COP for the heat pump is approaching 1.15 while the heating COP almost reaches 1.8. This improvement is attributed to a number of factors. The engine efficiency is maintained even though the speed decreases, as the system is operated with the throttle nearly fully open. Thus no increase in throttling loss is recorded at any set engine speed. As the compressor compression ratio decreases, (brought about by lower engine speed), the volumetric efficiency and mechanical efficiency increase. It is also suggested that by reducing engine speed, the volume of refrigerant circulating also decreases but the apparent heat transfer surface of the evaporator/condenser increases. This contributes towards an improvement in the system COP.

Measurements of the efficiency of the engine heat recovery system suggested that recovery of between 73 and 74 per cent of the total heat generated by the engine was achieved. A slightly higher recovery rate in terms of percentage was achieved at full rating compared to that at a partial rating. Interestingly, under partial load conditions the radiation loss from the engine accounted for 9 per cent of the total fuel input, while that under rated conditions amounted to only 3 per cent of the fuel input.

A considerable amount has been written about the start-up times of heat pumps, insofar as their ability to achieve operating supply temperatures of air and water is concerned. In many cases for a number of ambient air source heat pumps operating in the UK, start-up has been necessary at a very early hour in order to achieve satisfactory comfort conditions within a building. The start-up characteristics of the Tokyo heat pump are illustrated in Fig.3.16, measurements being made on March 26th 1982 at an ambient temperature of approximately 7°C. Data presented includes variations in ambient, engine speed and supply water temperature. The set delivery temperature of this was 45°C. It can be seen that the air handling units were put into operation within ten minutes of the heat pump start-up, and full supply water temperature was reached within 1 hour of start-up. During this period it was possible to reduce the speed of the engine from 1800 rpm to approximately

1200 rpm. It was also found that the control of supply water temperature was good, fluctuations within approximately +0.5°C around the set value being obtained for variations in heating load. Over a typical measuring period, engine speed would vary from maximum down to less than 1000 rpm, and heating capacity from 130,000 kcal/h to as little as 90,000 kcal/h.

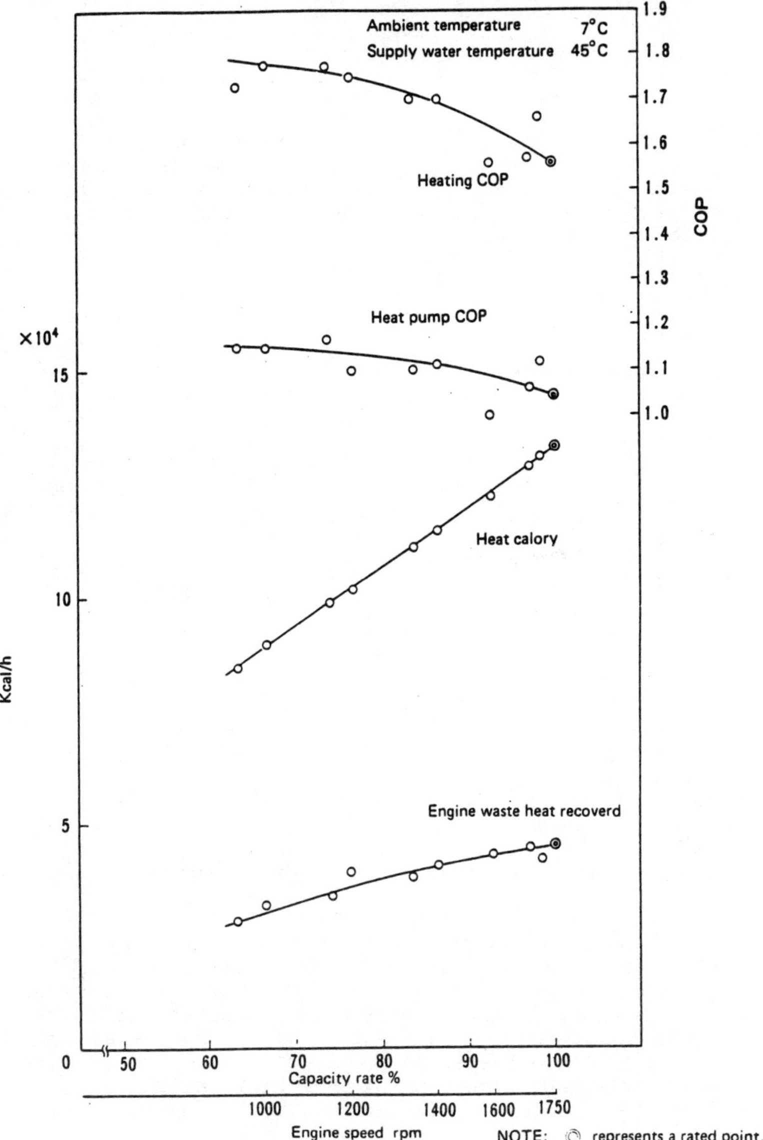

Fig.3.15 Capacity Rates vs. Heating Capability/ Efficiency

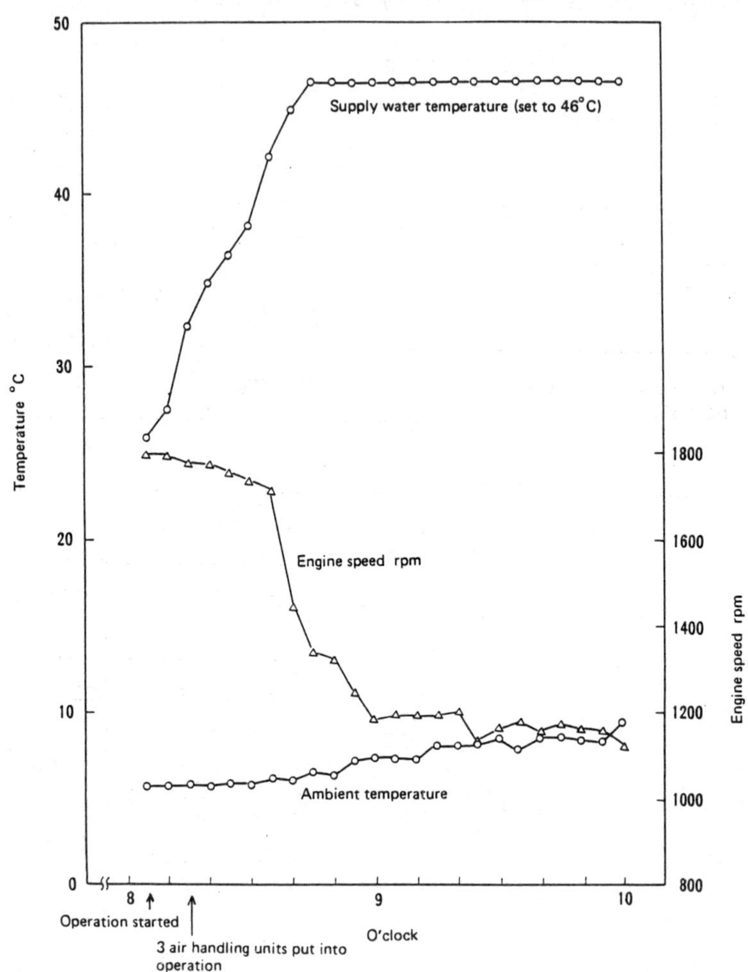

Fig.3.16 Heating Startup Performance

Operation of the defrosting system proved satisfactory. During a day when the ambient dry bulb temperature was 7.5°C and the wet bulb temperature 5.4°C, three defrost cycles were activated between 8 a.m. and 5 p.m. During the defrost cycle the supply water temperature dropped to about 33°C, but remained at less than 40°C for only about 10 minutes. Thus defrosting under these ambient conditions could be implemented quite rapidly without severe effects on the heat output of the system.

TABLE 3.10 Summary of field-proven Airconditioning Results

Capability (at a capacity rate of 100%)		Cooling	95,000 Kcal/h
		Heating	133,000 Kcal/h
COP	(at a capacity rate of 100%)	Cooling	1.20
		Heating	1.56
	(at a capacity rate of 65%)	Cooling	1.51
		Heating	1.78
Seasonal COP	Cooling	Operating time	526 hours
		Mean ambient temperature	$27.2^{\circ}C$
		Seasonal COP	1.21
	Heating	Operating time	735 hours
		Mean ambient temperature	$10.8^{\circ}C$
		Seasonal COP	1.65

TABLE 3.11 Full Load Rated Heating Performance of Tokyo Unit

Gas input	85,400 kcal/h
Gas Flow Rate	10.0 Nm³/h
Engine Output*	40.5 hp
Engine Speed	1750 rpm
Engine Thermal Efficiency"	30%
Heat Output	133,000 kcal/h
Heating COP	1.56
Ambient Temperature	7°C
Supply/Return Water Tmperature	45/40°C
Condensing Temperature	50°C
Evaporating Temperature	-4°C
Evaporating Duty	63,100 kcal/h

*Estimated values

TABLE 3.12 Full Load Rated Cooling Performance of Tokyo Unit

Gas input	79,100 kcal/h
Gas Flow Rate	9.3 Nm³/h
Engine Output*	37 hp
Engine Speed	1750 rpm
Engine Thermal Efficiency	29.6%
Cooling Capacity	95,000 kcal/h
Cooling COP	1.20
Ambient Temperature	30°C
Supply/Return Water Temperature	7/12°C
Condensing Temperature	38°C
Evaporating Temperature	-2°C

*Estimated values

With regard to performance of the system in the cooling mode, no significant difficulties or divergences from predicted data were noted. Although cooling performance is largely outside the scope of this report, the rated output of the system is given in Table 3.12.

Noise, Vibration and Exhaust. The acceptability of gas engine heat pumps depends as much on their environmental behaviour as on their heating and cooling capacity. Measurements were made of the noise levels achieved by this unit over its operating regime, including full and part load operation, but a number of points around the enclosure in which it was housed. It was found that at a distance of 1 metre from the enclosure, the overall noise level was 72 dB(A). This was claimed to be of the same order as that of an air source electric drive heat pump, and thus the acoustic properties of the enclosure were regarded as highly satisfactory. The vibration level, measured on the floor near to the heat pump during operation, was an average of 40 dB or less, the threshold value being

53 dB. This vibration level is defined as an overall acceleration level subjected to a sensual compensation. It was therefore claimed that the vibration level may be considered negligible.

TABLE 3.13 Maintenance Schedule and Costs - Tokyo Heat Pump (All prices in Yen)

Date	Hours Operated	Activity	Parts Cost	Labour Cost
2 May	2736	Change Engine Oil (18L)	7200	
		Replace oil filter	1880	
		Replace plugs	34200	
		Adjust valve clearance		
			43,280	38,000
5 August	4848	Change compressor oil (40L)	20,000	
		Check Refrig. Circuit safety valve pressure gauge		
		Change engine oil	7,200	
		Replace oil filter	1,880	
		Replace plugs	34,200	
		Adjust valve clearance		
			63,280	74,000

Analyses were done of the constituents of the exhaust gases from the engine both with and without catalytic treatment. Until a catalyst was applied, quite high NO_x levels were achieved, 3,200 ppm being recorded at an engine speed of 1,700 rpm and an exhaust gas temperature of 430°C. Carbon monoxide levels were only 1/6th of this, and the CO_2 percentage was 10.3. Following application of catalytic treatment, the NO_2 was reduced by about 80 per cent to 120 ppm. The catalyst would therefore prove effective as a means of reducing pollution by the exhaust gases, but no data was available on the life of the catalyst and its cost penalty.

Operating Routine, and Maintenance Data. Operation of the system is comparatively straightforward. At the beginning of each heating or cooling season the refrigerant valve has to be changed over, and this can be readily done in 30 minutes by one operator. Daily operations include turning on the secondary side of the air handling units, opening the gas supply master cock, and manually starting the heat pump. All subsequent operation is fully automatic, and it is claimed therefore that operation can be by inexperienced personnel. A microcomputer is used to control the output of the heat pump during its operating period.

The system has operated for a total of 1,261 hours during the monitoring period, and no failure was experienced. It was therefore claimed that the demonstration showed that the gas engine heat pump system was sufficiently safe and fully reliable.

TABLE 3.14 Equipment and Operating Costs for Heating and Cooling 1,200m² Office Building in Japan (Costs in Y1,000)

	Gas Engine Heat Pump	Chiller & Boiler	Absorption Heat Pump	Electric Heat Pump
Heat Pump/ Chiller	9300	2080	6250	4400
Cooling Tower	-	624	744	-
Boiler	-	1239	-	-
Water Pump	82	82	82	82
Cooling Pump	-	96	96	-
Foundations	100	120	100	100
Transport & Installation	300	200	200	250
Piping:	700	1500	900	600
of which Water	600	600	600	600
Coolant	-	700	200	-
Gas	100	100	100	-
Stack	-	100	-	-
Electric Power Supply	990	3510	1170	4500
Expenses	1721	1418	1431	1490
Total	13,193	10,869	10,973	11,422
As Percentage	100%	82.4%	83.2%	86.6%
Electricity Costs*	649	1392	599	2505
Gas Costs*	1185	1593	1874	-
Total Fuel Costs*	1834	2985	2473	2505
Total Running Costs**	3763	4926	4469	4357

*Y1000/year

**Includes water, maintenance and repair, and labour

Operating experience showed that the system could be operated on a maintenance free basis for a period of at least 1000 hours. Routine maintenance of plugs and oil used in the system resulted in the following comments: The plugs used in the gas engine had platinum clips welded onto the central and external electrodes, and these did not have to be replaced during the period of test. No great widening of the plug gap was experienced, and the plugs appeared to have a serviceable life of at least 2000 hours.

The mean oil consumption was as low as 0.32 cc/hph. An engine oil sub-tank having a capacity of 20 litres was used, and it was therefore claimed that the heat pump would be capable of operating for approximately 1,500 hours without the need to supply additional engine oil. The analysis carried out on the oil after a period of 1,260 hours suggested that no significant degradation had occurred, the moisture content remaining zero, and it was therefore recommended that the oil change frequency be set at the same as the oil supply (top-up) frequency, 1500 hours.

Data obtained by Komatsu on this and other engines operating over much longer periods indicate that an overhaul interval of 16,000 hours (based on full load working hours) in a heat pump duty is required. A cylinder head check is recommended after 8000 hours, including a check on valve clearances, gasket replacement etc. With the gas engine used by Tokyo Gas Company, the costs incurred on a mid-term overhaul would be Y200,000, and on a 16,000 hour overhaul Y700,000. On a similar gas engine driven heat pump delivered to the Komatsu Electronic Metals Company, the maintenance costs during a total of 6,984 operating hours are given in Table 3.13. It can be seen that the total maintenance cost over this period is approximately Y220,000. It is claimed that this is approximately the same as that for conventional systems such as electric heat pumps, boilers, and absorption chillers.

Capital cost data for the gas engine driven heat pump for an office of slightly larger floor area than that of the Tokyo Gas Company (1200 m^2) are given in Table 3.14. These costs are compared with three other systems, a chiller with a separate boiler for heating, gas fired absorption heat pump, and an electric drive reversable heat pump. It can be seen that the gas engine driven unit has somewhat higher capital costs, but the savings result from the fuel costs, and these are shown at the bottom of the table. (These costs take into account electricity as well as gas where appropriate).

3.4.3 Other drive systems The Diesel engine may of course be used to drive a compressor, and would normally exhibit a significantly higher efficiency, although resulting in less recoverable heat. The steam turbine has been employed in a number of industrial heat pump applications, one such example being in an edible oil processing factory in the UK. In this case the turbine replaced an expansion valve. Organic turbine have been proposed, but to date have remained unacceptably expensive except in the most specialized and site-specific instances.

As in many other applications, the Stirling engine has received attention as a heat pump prime mover. The work of Philips in the Netherlands is of historical interest in this context, and is detailed in Chapter 4.

3.5 Heat Exchangers

Design of heat exchangers for heat pumps is a major subject which would deserve a book dedicated only to this topic. This Section is restricted to a qualitative description of the main principles involved. The designer's time is divided equally between obtaining optimum heat exchange conditions and controlling the refrigerant/oil mixture. A quantitative approach to heat exchange is given in Ref.3.3, for general applications and in Ref.3.4 for refrigerants. Ref.3.5 contains an excellent detailed discussion on refrigerant/oil behaviour.

3.5.1 Heat transfer The main heat exchangers in a vapour compression cycle heat pump are shown in Fig.2.6 of Chapter 2. Heat transfer within these units is either single-phase or two-phase. Single phase means that all the fluid is liquid during the heat exchange, or it is all vapour. Two-phase means that the fluid is a mixture of liquid and vapour. Two-phase heat transfer is understandably much more complex than the single phase form. Boiling (evaporation) and condensing are two-phase processes while superheating, subcooling, water side and air side heat transfer are all single phase (except when the air humidity is significant).

Heat transfer in heat pumps normally is described as 'forced convection', which means that the fluid is being driven through the heat exchanger under an externally imposed pressure in order to promote good heat transfer coefficients. Under these conditions the most important parameter is invariably the Reynolds Number. $\frac{4M}{\pi\mu D}$

Where M is massflow
D is diameter
μ is viscosity

High flow rates of inviscid fluids down small diameter pipes give large heat transfer coefficients, for single phase and two-phase flows, unfortunately the Reynolds Number is also crucial in determining pressure loss, and so a compromise must always be reached.

Free convection heat transfer takes place where the driving force is not external, but is determined by the buoyancy of the fluid. Perhaps the commonest case is in pool boiling, in a heat exchanger of the 'kettle' type, or in a 'flooded' type evaporator.

A heat exchanger is configured to be counter-flow, cross-flow or parallel-flow depending on the temperature profiles required. Counterflow is the most common type, although perfect counterflow can only be achieved in concentric tube heat exchangers. Note that evaporation or condensation tends to level out the temperature distribution within an exchanger. Fig.3.17 gives examples of temperature distributions along heat exchangers.

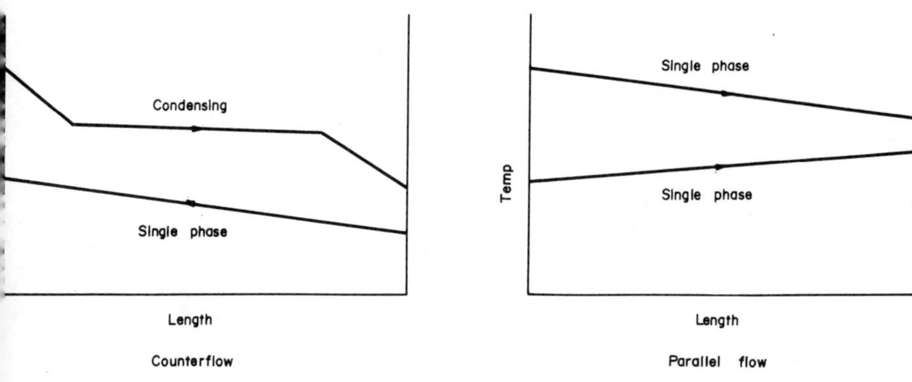

Fig.3.17 Examples of Heat Exchanger Temperature Distributions

Note that to fully design the condensers illustrated in Fig. 3.17 it is necessary, on the refrigerant side alone, to calculate heat transfer coefficients for desuperheating and subcooling together with coefficients for condensing at vapour qualities between 0 and 100 per cent. In comparison the single phase coefficient calculation is straightforward.

3.5.2 Refrigerant behaviour The heat exchangers must be designed for all conditions under which they operate. Their positions relative to each other and the compressor must be taken into account, together with the associated plumbing. There are two very important reasons for this: liquid refrigerant should not enter the compressor and any oil which leaves the compressor should be returned thereto.

Taking the second of these problems first, oil return is normally ensured by maintaining an adequate velocity of refrigerant in the pipelines. Less dense vapours, or evaporation at lower temperatures require higher velocities, and the minimum velocity must be maintained for all operating conditions, including part-load. Obviously a system might operate over such a wide range that the pressure drop on the suction line may become too large, and these systems often have twin suction lines.

Evaporators with the refrigerant evaporating within tubes ('Direct Expansion') must also be sized such that the tubes are purged by the leaving vapour. The oil enters the evaporator in solution in the oil, and is left behind after evaporation unless the velocity is sufficient to carry along the oil in a mist.

With larger, flooded, evaporators the oil remains in the liquid refrigerant until the concentration is such that it separates, and flooded evaporators have oil drains, or oil return systems. This will only work where the oil is significantly denser than the liquid refrigerant, so that it sinks to the bottom.

Work recently carried out at the New University of Ulster (Ref. 3.6) has included a quantification of the effect which varying concentrations of oil in the refrigerant can have an evaporator performance. In a project relating specifically to rotary sliding vane compressors, where the proportion of oil dissolved in the refrigerant can be substantially higher than in the case of small reciprocating compressors, measurements were made of the heat extraction rate (from source to refrigerant) for a variety of oil fractions of up to 20 per cent by weight. These data are shown in Fig.3.18.

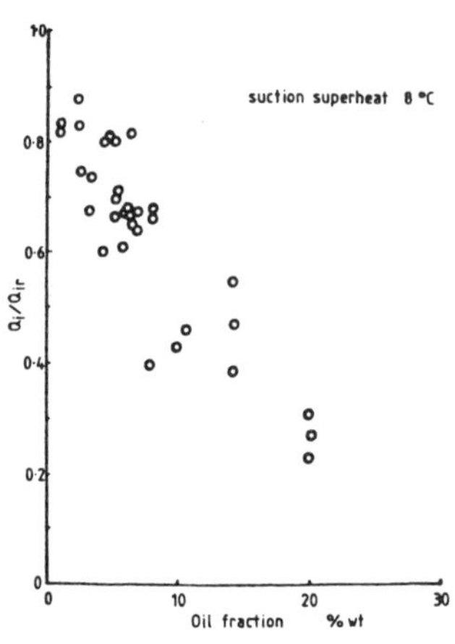

Fig.3.18 Extracted Heat as a Function of the Oil Concentration in the Refrigerant (Q is the Extracted Heat with a Pure Refrigerant)

It was concluded from the study that oil should be excluded from refrigeration systems wherever possible, and, additionally, significant improvements could be achieved if this could be done and also if superheat settings could be reduced from those currently considered the norm. Peak COP's for a given oil fraction were achieved at evaporator superheats of about 5°C, 2-3°C less than normally used.

Oil in the condenser is not such a problem, for unless the compressor is blowing out an excessive quantity, then it is dissolved by the condensing refrigerant and carried to the evaporator.

Liquid refrigerant behaviour can be more fickle than that of oil. Variable capacity systems, on/off operating and, worst of all, reversing heat pumps (in which the evaporator becomes the condenser and vice versa) can wreak havoc with the designers aims. Liquid should collect in the condenser, or in the liquid receiver below the condenser, and be fed to the evaporator via the expansion valve. In a direct expansion evaporator there is not a very large volume of liquid, so the greater part of the refrigerant charge is concentrated in the high pressure leg.

Liquid can reach the compressor by different means: it can condense in the sump during running, or it can condense while the compressor is stationary, which can easily happen if the compressor becomes cooler than the condenser. Liquid can drain or be forced into the compressor from the condenser if a pressure difference occurs during cooling. After reversal of a system, the component which takes on the role of evaporator was the condenser and can therefore be filled with liquid which would flood into the compressor if 'floodback protection' is not used. This protection is generally ensured by a liquid trap on the suction side of the compressor, from where the liquid is slowly drained or evaporated. This can take the form of an ingeniously designed intercooler (Fig.3.27).

The temperature distributions in the evaporator and condenser are somewhat modified if a mixture of working fluids is used. With a so-called non-azeotropic mixture, eg R21 and R114, instead of the essentially constant working fluid temperatures during evaporation and condensation, the temperature will decrease during condensation and increase during evaporation. This is illustrated in Fig.3.19, (Ref.3.7). The performance of heat pumps can be improved, particularly in cases where the heat source and heat sink exhibit high temperature gradients, ie where the heat pump operation leads to a substantial drop in temperature of the source and/or a substantial rise in sink temperature. The phenomenon may also be illustrated on a temperature-entropy diagram, as shown in Fig.3.20.

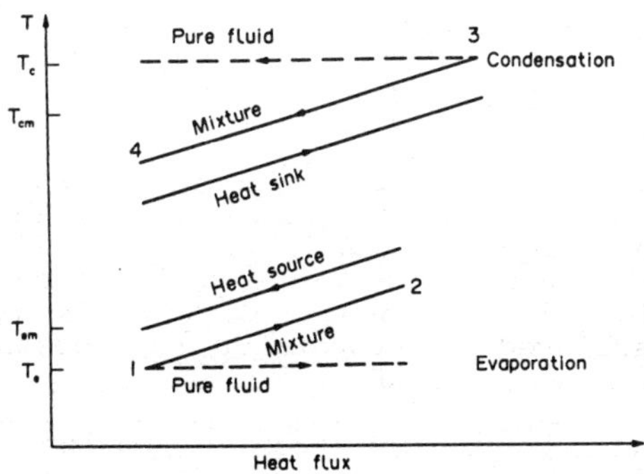

Fig.3.19 Temperature Variations of Working Fluids During Condensation and Evaporation (Pure Fluids and Non-Azeotropic Mixtures)

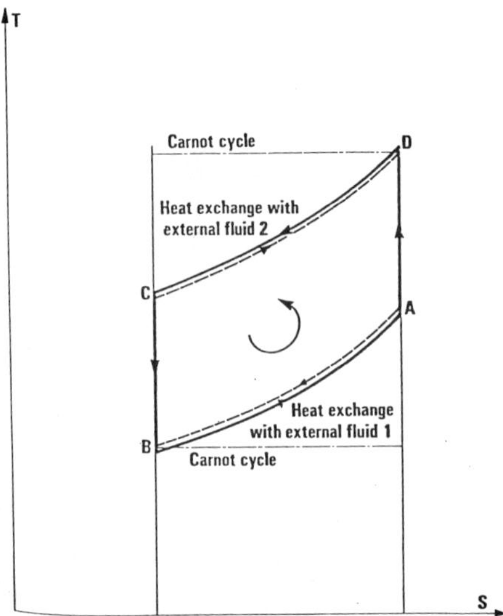

Fig.3.20 Temperature Variation During Vaporization and Condensation of a Fluid Mixture, Parallel to the Temperature Changes of the Fluid with which Heat Exchange Takes Place

3.5.3 Heat exchanger types This Section contains a selection of examples of the most common types of heat exchanger, showing how the different designs overcome the problems described. Air/refrigerant exchangers are described first, then water/refrigerant and finally refrigerant/refrigerant types.

One feature which most air-heating or air-cooling exchangers have in common is the extensive use of fins to extend the surface area on the air-side of the exchanger. This is because the heat transfer coefficients can be five or ten times worse on the air side and this must be compensated for by large areas. Typical air-cooled heat exchangers are shown in Fig.3.21.

Sometimes the finning runs across all the coils, the tubes being expanded into fin 'plates'; sometimes it is in the form of fins wrapped around each tube. Because a fan is required to force air through these coils fin design is critical and the final type will depend on many things: refrigerant flow and pressure, manufacturing methods, material compatibility, allowable temperature differences, allowable pressure drops, space available, cost limitations etc. Ref.3.8 gives some details on finned coil design.

Fig.3.22 shows the liquid refrigerant manifolding into the evaporator from the expansion valve. This type of exchanger is parallel flow and is used when the pressure drop through the refrigerant lines is sufficient to cause a significant temperature drop as evaporation proceeds. The last row of pipes before the suction header is, however, on the upstream side in order that the maximum superheating may be achieved.

This type of layout is more common for low temperature applications, otherwise a counterflow arrangement is commonly utilised. The orientation can be arranged to suit the application with changes only needed on the external plumbing.

Fig.3.21 Two Air Cooling Coils

Fig.3.23 shows a different type of heat exchanger, a refrigerant /water unit. The water flows outside the bundle of pipes and is constrained by the segmented baffles to flow in a zig-zag path crossing and re-crossing the tubes. This is designed to enhance the water side heat transfer coefficient, although it obviously does so at the price of higher pressure loss.

It should be noted that the outer shell shown in the lower part of Fig.3.23 incorporate an insulation jacket which explains the difference in diameter between the two components. This is provided because in a water chilling application the evaporator is well below room temperature. In heat pump applications this is not always so, and insulation is not always necessary.

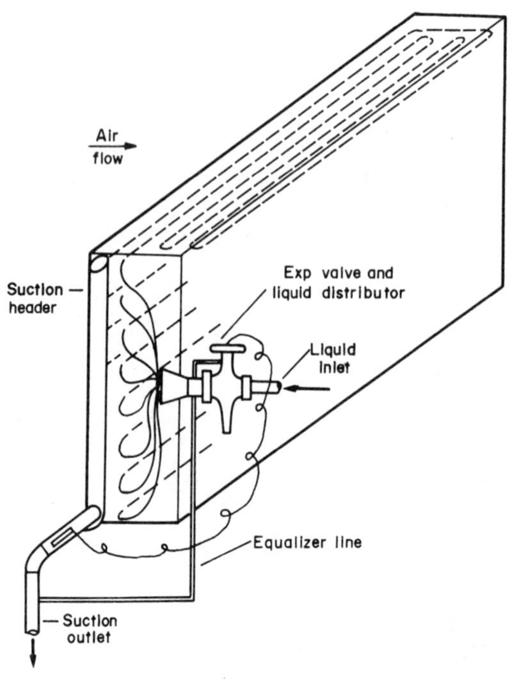

Fig.3.22 Layout of Direct Expansion Air-Heated Evaporator

The type of heat exchanger shown in Fig.3.23 is referred to as a shell and tube heat exchanger. In this configuration it is used as a direct expansion evaporator but this type can be used as a flooded evaporator - with refrigerant on the 'shell' side - or as a condenser, in which case the axis could be vertical or horizontal. The vertical axis would improve liquid refrigerant drainage from the condenser although it would not necessarily improve the condensing heat transfer coefficient. This is because the coefficient depends upon the film of liquid refrigerant on the wall of the tube. With the tube horizontal, this can drain around to one sector only, whereas with a vertical tube, the whole surface may be covered by the liquid, especially towards the outlet side.

Many variations are possible with shell and tube heat exchangers and two are illustrated in Fig.3.24 with finned tubes used to improve the refrigerant heat transfer coefficient.

The U-bend heat exchanger is also called a 'two-pass' exchanger and has the major mechanical advantage that the relative expansion of the tubes (which may be copper) and the shell which may be steel) does not need to be accommodated because there is no seal at the 'return' end. One-pass exchangers have either sliding seals, similar metals or expansion joints.

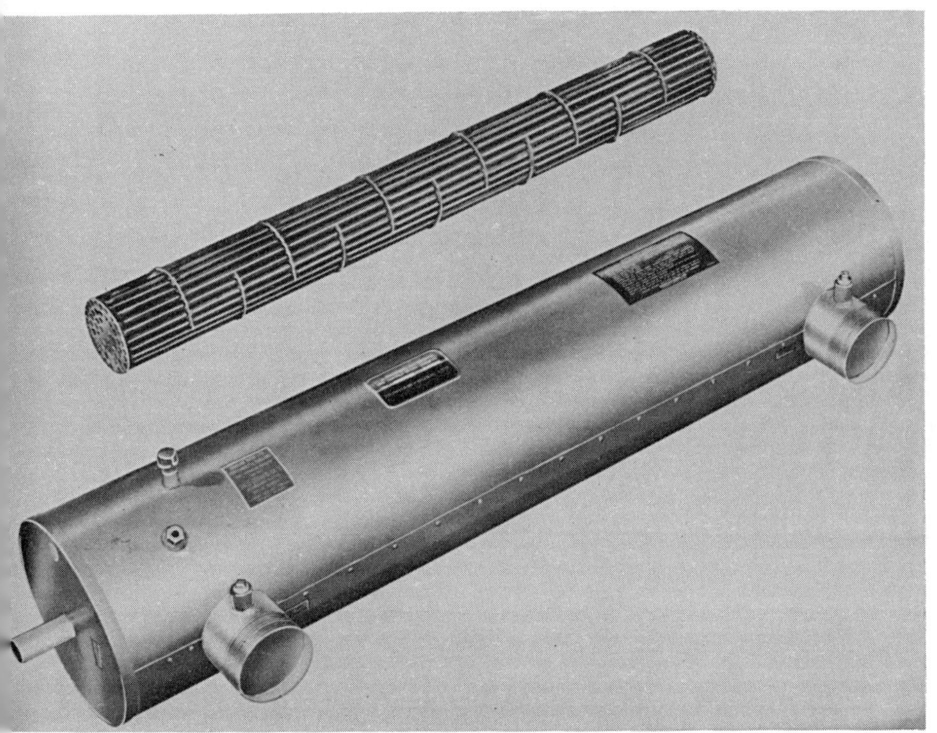

Fig.3.23 Direct Expansion Evaporator
(Courtesy of Dunham-Bush Ltd)

The shell and tube heat exchanger is by far the commonest used for refrigerant/water duty. Because it is so common, construction of small units is very cheap indeed with a variety of patented arrangements for simplifying construction or enhancing heat transfer. Two types of tube used sometimes in evaporators are illustrated in Fig.3.25.

The 'Innestar' tubing consists of an aluminium extrusion inside the copper pipes of a direct-expansion evaporator, increasing the surface area and the Reynold's number and thereby improving heat transfer, and, it is claimed, oil entrainment.

The Union Carbide tube works on a quite different principle. The tube is coated with a porous surface which provides a multiplicity of nucleation sites for boiling the refrigerant. This means that for a given heat flux the temperature difference can be dramatically reduced, something which is very important in heat pump duties. A 'wick' of fine metal mesh is sometimes used for a similar purpose.

A variation of the shell and tube unit is the tube in tube heat exchanger, consisting of two concentric straight tubes. This has the advantage that true counterflow can be achieved but this is negated by the relatively large size required.

Fig.3.24 Two Types of Shell and Finned Tube Condensers
(Courtesy of Dunham-Bush Ltd)

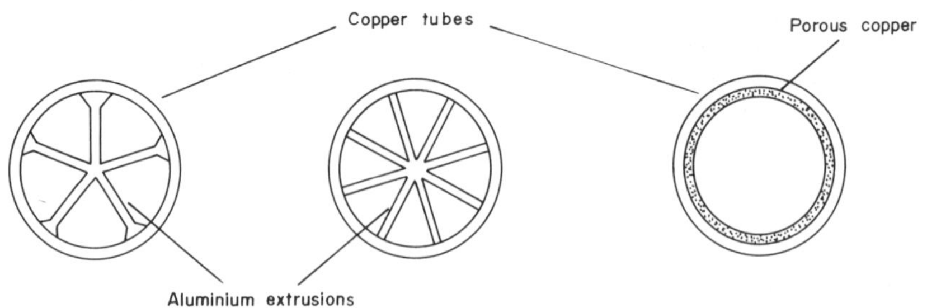

Fig.3.25 Two Types of Direct Expansion Evaporator Tubes

A more compact variety is the shell and coil heat exchanger, illustrated in Fig.3.26. This type can be extremely compact, and can also have good heat transfer coefficients both inside and outside the tubes as a result of turbulence created by the unusual shapes. Despite this they are not widely used, partly because their relative scarcity makes the costs marginally greater than for straight tubes.

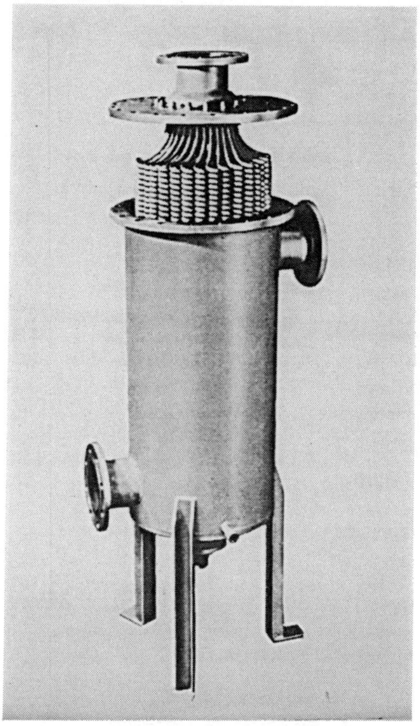

Fig.3.26 Shell and Coiled Tube Heat Exchanger

The intercooler, also called a precooler, interchanger, subcooler or superheater, is a refrigerant vapour/refrigerant liquid heat exchanger which is installed in the suction line between evaporator and compressor. For this reason the most crucial aspect of inter-cooler design is an unrestricted flow on the vapour side, and to this end relatively large passages are used, with fins extending the surface in the vapour side, as illustrated in Fig.3.27.

This heat exchanger can also be of the concentric tube type for small heat pumps, when size may not be as important as price, or assembled using soldered tube construction, which involves two parallel tubes soldered together.

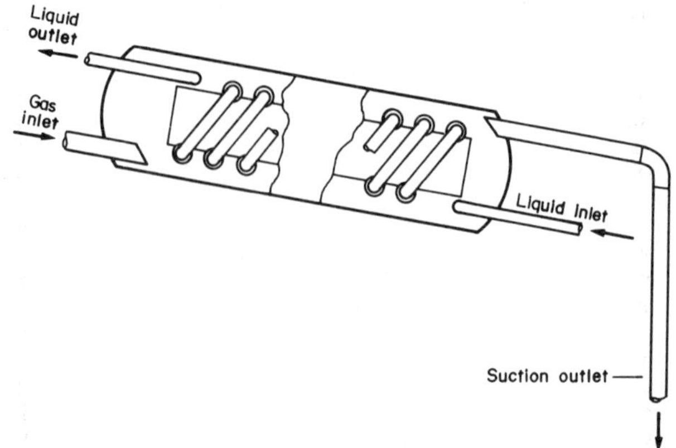

Fig.3.27 Shell and Finned Tube Intercooler (ASHRAE Handbook)

3.5.4 Sizing of heat exchangers The thermal design of heat exchangers is difficult enough without the need to incorporate extraneous factors such as cost-effectiveness, but this must be done for heat pumps. Because the whole raison d'être of a heat pump is to produce heat more economically than by conventional means it is important that the possible return on capital invested is painstakingly computed.

An increase in size of heat exchanger will cost more money, but it will also reduce the temperature difference across it, and will thereby reduce also the difference between evaporating and condensing temperatures. In effect, larger heat exchangers will increase the COP. The designer must achieve an optimum heat exchanger size to maximise rate of return on investment.

One interesting study which made this comparison was done at the Electricity Council Research Centre (Ref.3.9). The outcome of this is that for electrically driven heat-only heat pumps in the UK, the condenser should be substantially increased in size, perhaps doubled, in comparison with units which were designed primarily for the market in the USA, although they were not necessarily optimised for USA conditions.

It is outside the scope of this book to elaborate upon heat exchanger design procedures. However, a few organisations such as the Heat Transfer and Fluids Service (HTFS) and the Engineering Sciences Data Unit (ESDU) offer data and in some cases software packages for either mainframe or desktop computers which can help considerably in designing evaporators and condensers. Under development are even more sophisticated packages which use artificial intelligence techniques, (expert systems) linked to conventional software, to enable routine access on a computer by potential users to expertise in system selection and design - ie programming the thought processes of the expert for non-expert use.

3.6 Materials

All material selection for heat pump components is naturally of secondary importance to selection of the refrigerant itself, which was discussed in Section 3.2. Once the refrigerant is chosen attention must be switched to obtaining materials of construction which are compatible with it.

Copper is a material whose presence is almost compulsory. The high thermal conductivity makes it suitable for heat exchanger tubes, and it normally forms the electrical windings in hermetically sealed electrically driven compressors. The main problem with copper is the symptom known as 'copper plating' which occurs when copper goes into solution in the refrigerant and is subsequently deposited in another part of the system, where it becomes clearly visible on inspection. In extreme cases this can lead to compressor failure. It is now fairly widely understood that the products of oil-refrigerant reactions combined with high operating temperatures and air in the system can cause copper plating, although despite the use of advanced oils the problem still occurs.

The only major restriction on the use of copper in heat pumps is the fact that it is totally incompatible with ammonia. This means that for ammonia systems all heat exchangers must be aluminium or iron, and even copper-bearing brazes cannot be used. Hermetic driven ammonia systems are obviously not possible, therefore. This restriction creates problems for ammonia/water absorption systems for which an all-welded steel construction is normal practice, with a small quantity of inhibitor in the system to prevent rusting and generation of non-condensible gases, an accumulation of which can finally prevent a system from operating.

Although soldering or brazing a joint is cheaper and simpler than welding, problems can occur with flux which can be corrosive, and can remain inside the pipes until the system is charged with refrigerant. At this stage, solder drops, residual braze, weld spatter, swarf from pipes having been cut and any other debris can be carried by the refrigerant into the compressor, and it is in order to separate out this debris that a suction side screen is fitted. The possibility of this type of rubbish being present should not be forgotten when compatibility tests are envisaged, and this is why sealed tube tests generally include a mixture of assorted metals together with the refrigerant and the oils.

3.7 Controls

Most currently marketed heat pumps for domestic heating have on-off thermostatic control, basically very similar to the controls on domestic central heating units which the heat pumps are designed to replace (see also Chapter 4). There has, however, been a major growth in interest in "intelligent" controls which take into account the inside demand pattern and outside conditions both of which can be biased according to the day and month and even the location. These would, if available low cost, be a boon to users of central heating because of the potential improvements in comfort and economy. If the heat pump is to become economically competitive than the control systems must be equally versatile, and it is doubly important that this is so because of the extra parameters with which the heat pump controller must juggle.

The COP of the heat pump depends on a variety of factors, principally the temperatures of evaporation and condensation. The evaporation temperature ideally would follow the ambient, or heat source temperature closely in order that the heat pump might benefit from milder conditions. This can be arranged by various means, but it is always important that the evaporator is large enough to give only a small temperature difference between source and refrigerant.

Liquid feed to the evaporator is controlled by the expansion valve. A thermostatic expansion valve which is the most common type for medium sized systems will maintain constant temperature of super-heated vapour entering the compressor,but this only controls actual evaporating temperature indirectly.

Very small systems, or cheap constant temperature units (such as water chillers) simply use a fixed restriction, such as a capillary, to provide the pressure drop. While these are cheap and reliable (with an upstream filter) they are quite unsuitable for variable temperature, variable capacity operating conditions.

The expansion valve can be pressure-regulated, to give a constant evaporating pressure. Obviously this also does not give much scope for a changing source temperature.

With a flooded evaporator a low pressure side float valve can be used, analogous to the float valve on a domestic cistern. This simply maintains a constant fill of liquid refrigerant which will evaporate at a pressure and temperature appropriate to the source. The principal reason that this type of valve is not more widely used is that flooded evaporators are generally considered appropriate to large systems.

A high pressure float valve can be used, maintaining liquid level in a reservoir below the condenser. This will allow the condenser to always operate efficiently without filling with liquid and therefore the system capacity is governed by the rate of condensation, or heat demand.

Any one of the expansion valve types described above may be appropriate to a heat pump and the selection will depend on the relative emphasis placed on efficiency, cost, reliability, refrigerating power or heating power.

Research work has been undertaken (Ref.3.10) to develop a control system based on the use of a microcomputer-controlled expansion valve. The University of Dortmund in West Germany carried out experiments on an air-water heat pump where a thermoelectric expansion valve, linked to the control system, was used in an attempt to favourably control the superheat temperature, and optimise evaporator operation.

Another project examined the simulation of a whole house heating system and the interaction between the heat pump, radiators, and the external and internal environments. The aim was to develop a control system which would improve the match between the heat pump and the dwelling, based principally on controlling condenser temperature, but also using a two speed compressor.

Condensation is crucial because the delivery of heat is the raison d'être of the heat pump. In general the condensing temperature is determined by the cooling temperature (air or water), the flow of refrigerant and the size of condenser. Having sized the condenser, one can only vary the flows of refrigerant and coolant. In a warm water-heated house, for example the water may be delivered at 65°C and it may return at 55°C. On a mild day however a lower temperature might be acceptable, with a delivery of 55°C and a return of 45°C and this would be preferable to running intermittently at the higher temperatures. Note also that with a higher water flow rate the first case could be satisfied by reducing temperatures to deliver at 62°C and return at 58°C, which would also give a lower condensing temperature.

The flow rate of refrigerant is regulated by the compressor. Positive displacement compressors can be controlled by varying the speed, and this gives simple proportional control. Capacity control can also be achieved by the successive isolation of one or more cylinders of a reciprocating compressor, or for a screw compressor, by varying the face area of the inlet. A centrifugal compressor cannot be controlled by speed variation and variable inlet guide vanes are used for capacity control.

The ability to control capacity and 'throttle back' a heat pump to part-load operation helps the performance in several ways. Losses due to transient heating are eliminated, and the temperature differences across heat exchangers are held to a minimum because of the reduced flows. One American study (Ref.3.11) predicted that a 30 per cent saving could be made over conventional domestic heat pumps simply by the adoption of improved capacity control.

There is one further string to the bow of the efficiency controller; this is the option of using boost heating. Under certain combinations of unfavourable circumstances the PER of a heat pump may fall below that of a direct fired boiler. This can happen with air-source heat pumps during extremely cold weather with heavy precipitation. When this occurs, if the configuration of the heat pump will allow direct fired heating, the loss can be minimized. The absorption cycle heat pump can be particularly easily adapted to direct heating, because it incorporates a large burner or heating element.

Protection: Apart from the performance governing controls already described, it is necessary, for safety and to prevent mechanical damage, to protect the heat pump against unfavourable running conditions with a series of overload switches or governors.

Damage can result from any of the conditions listed below:

High discharge pressure	Structural damage
High discharge temperature	Corrosion of valves, breakdown of refrigerant
High suction pressure	Damage to thrust bearings
Low suction pressure	Ingress of air Failure of oil supply
High pressure difference	Mechanical damage to moving parts

It is not always necessary to protect against all these conditions, because with one refrigerant for example, the discharge pressure limit may preclude any risk of a high discharge temperature, or vice versa. With a hermetic compressor there is no net thrust on the crankshaft and therefore the suction pressure is less critical.

Other conditions likely to lead to damage can occur, for example if the heat pump is recovering heat from a water source which stops flowing then the evaporator may freeze solid. To avoid this a flow switch or a temperature switch can be used. The protection need not be by switching off, however, for an 'intelligent' control, sensing a high discharge pressure (high condensing temperature) condition would reduce the capacity until such time as the heat demand is increased.

In the case of the high temperature gas engine-driven heat pump developed at NEI International Research & Development Co Ltd., (see Chapter 6 for full details),considerable effort was made to ensure protection from damage if the system erred outside its operating envelope during unattended running. The automatic shut-down system was actuated by the following parameters:

(i) Low engine oil pressure.

(ii) High engine oil temperature.

(iii) Low engine coolant flow rate.

(iv) High engine coolant pressure and temperature.

(v) High compressor discharge pressure and temperature.

(vi) Low compressor suction temperature and pressure.

(vii) Low compressor differential oil pressure.

(viii) Low gas supply pressure.

Shut down was achieved by closing the gas supply valve and the engine throttle and earthing the low tension side of the engine ignition circuit. Excessive refrigerant condensation in the compressor after shut-down was prevented by a non-return valve in the discharge line and a solenoid-operated isolating valve in the suction line.

3.8 Absorption Cycle

Apart from a few brief references, all the comments so far in this Chapter are related to vapour compression cycle heat pumps. This Section gives information on the only other refrigeration (heat pump) system currently in everyday use, that based on the absorption cycle. The cycle has already been described in some detail in Chapter 2 so there is no further general description here, but rather a scrutiny of some key components.

3.8.1 Generator and rectifier For an absorption cycle as shown in Fig.3.28, using two volatile fluids (e.g. ammonia and water) the separation of the refrigerant is the key process. The generator is effectively a distillation column, with the rectifier giving the reflux and the feed coming from the absorber. A distillation column can achieve more difficult separations by taller constructions and by increased reflux. Increasing the reflux, however, leads to increased heat demand, so this would ideally be regulated depending upon the operating conditions, by regulating the cooling flow in the rectifier.

In some units the rectifier is 'solution cooled', e.g. cooled by the liquid feed from the absorber. This does detract from the amount of heat which the liquid heat exchanger will exchange so there is no overall improvement to efficiency. Various other ingenious configurations are possible, including 3-way heat exchangers, in which the liquid heat exchanger is formed inside the rectifier chamber, with the 'cold' flow from the absorber simultaneously providing reflux and cooling the hot liquid from the generator.

The separation which takes place in the generator column occurs by an intermixing of rising vapour of increasing concentration, and falling liquid of increasing weakness. A large surface is necessary, as well as carefully designed flow rates. A simpler but more bulky alternative is an 'analyzer' in which the vapour rises in the form of bubbles in a column which is flooded with liquid, and has baffles to encourage intimate mixing. This is illustrated in Fig.3.29.

Since ammonia-rich liquor is less dense than water, the concentration gradient in the liquid is stable, and some degree of separation can be achieved in the rising vapour. The surface area for mass transfer is however relatively small and this is the main handicap of the analyzer.

3.8.2 Absorber The absorber takes the reverse role to the generator. At low pressure vapour and liquid mix together, while heat is removed. Quite a variety of configurations are possible, as shown in Fig.3.30.

Note that the absorption can take place inside or outside the cooling pipes, and with counterflow or parallel flow. In each case the main design object is to provide sufficient surface area of liquid for the mass transfer to take place. To this end, some manufacturers use internal ribs, or even a metal mesh to ensure that the liquid is distributed around inside the tubes.

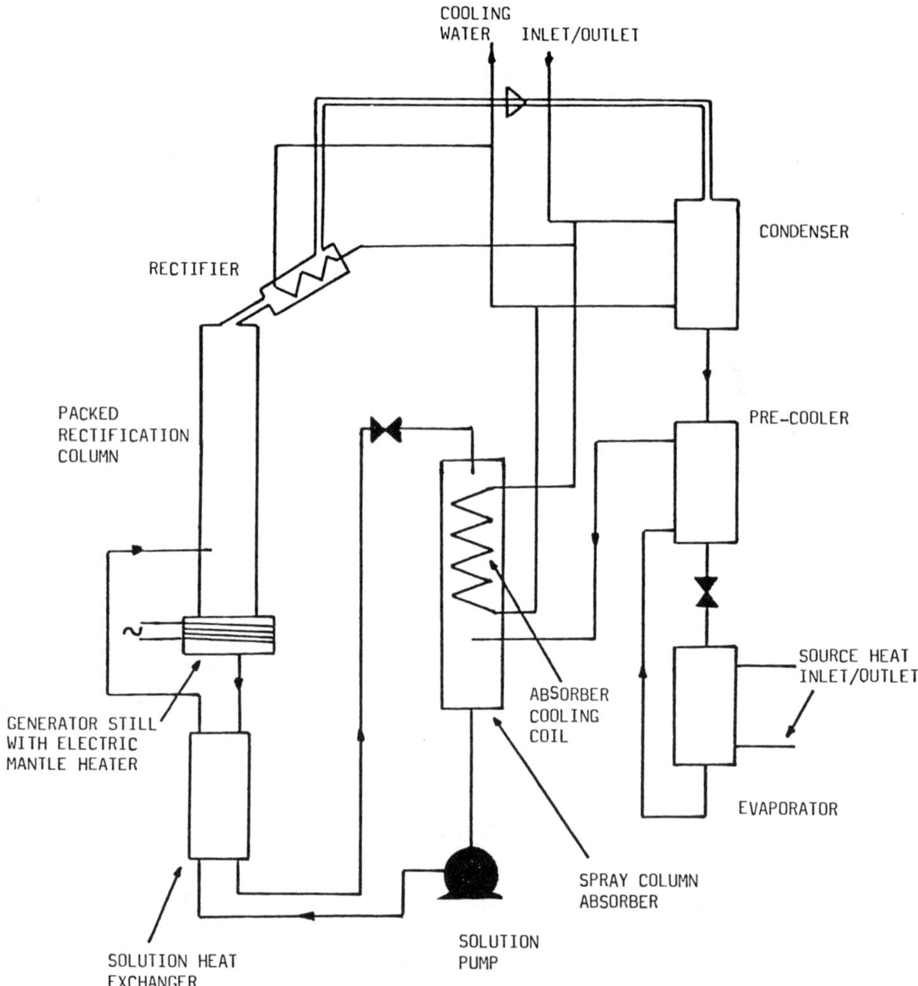

Fig.3.28 Water/Ammonia Absorption Cycle System

Absorption columns are common pieces of chemical engineering equipment but their normal mode of operation does not provide a basis for the heat pump design. Usually heat release is not an important feature whereas it is critical in the heat pump and where substantial quantities of heat are generated as, for example, in hydrochloric acid production specific engineering design procedures are used. A second and highly important consideration in relation to the heat pump design point requirements noted in the previous section is the extremely low pressure drop which can be accommodated. The vapour pressure in the evaporator of a system designed for low temperature industrial heat recovery which provides the energy for vapour transport may be only for an evaporating temperature of 30°C.

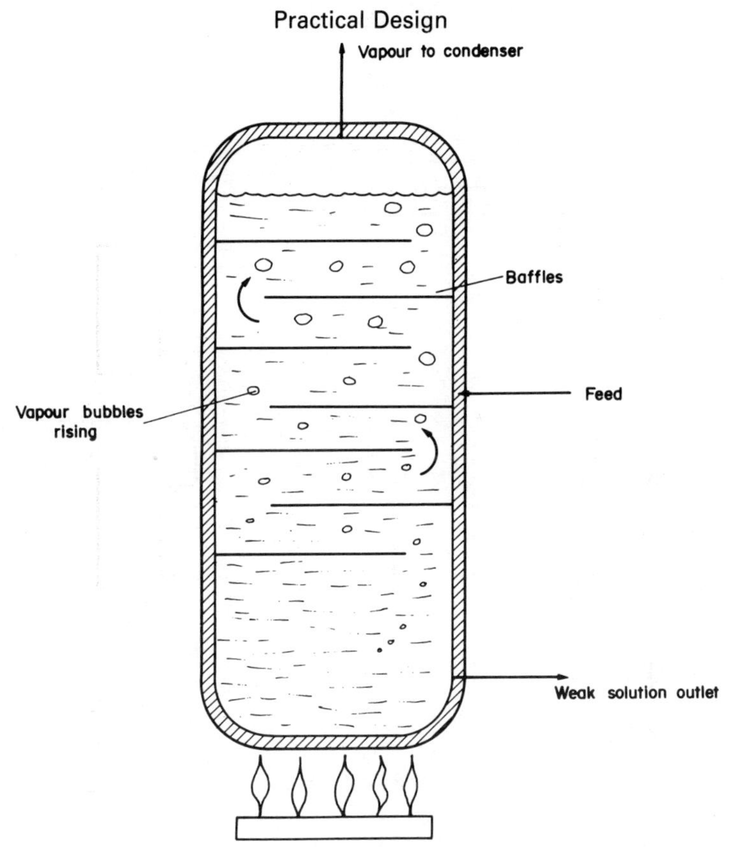

Fig.3.29 Schematic Analyser

Three basic forms of absorption column can be considered; (Ref. 3.12).

a) packed column
b) falling film
c) spray over coils

The packed column is essentially a vertical tube containing distributed elements such as raschig rings, pall rings, saddles etc. The liquid and vapour pass through in a counter-flow arrangement. The large surface area caused by the packing elements gives this column excellent mass transfer arrangements but this benefit is offset by a relatively high vapour impedance. Further, as the heat release occurs throughout the volume of the column it is difficult to dissipate in a fully controlled manner.

Commercial water/aqueous lithium bromide absorption chillers (e.g. the Sanyo unit in Fig.3.31) employs the spray over coil procedure. The lithium bromide solution is sprayed into a water vapour atmosphere. The droplets heat up in their passage

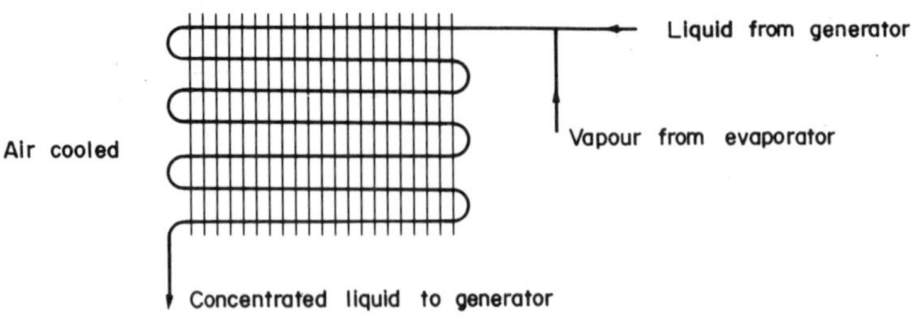

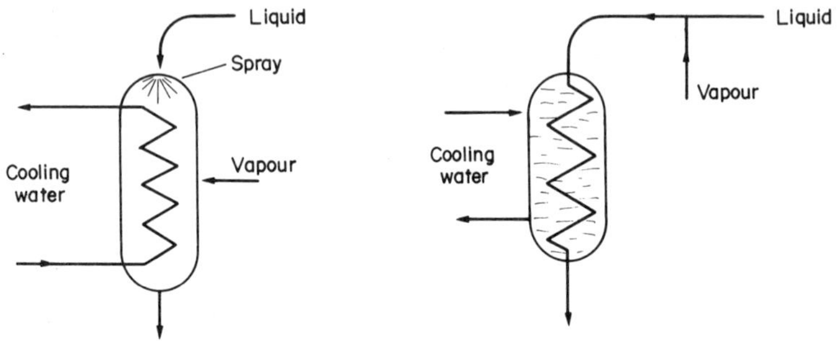

Fig.3.30 Three Absorber Configurations

through the vapour and they dissipate heat when they contact cooling coils. The vapour impedence of these devices can be very low and they demonstrably work in chiller applications. However a number of factors militate against the use of this design for the application being considered here

1) The lithium bromide solution enters the absorber close to its solubility limit and a minor disturbance in operation could cause crystallisation in the spray nozzle with serious consequences. This problem is far less likely to arise in chiller applications.

2) This type of absorber is used commercially in conjunction with a recirculation system. Control is achieved by recirculating a portion of the total flow. Although the recirculatiion pump is a relatively minor component it is clearly of benefit if it and its associated pipework, valves etc could be eliminated without introducing disproportionate disadvantages.

3) The mass and heat transfer processes within a spray column are extremely complex and it is unlikely that a suitable theoretical design model could be developed without an experimental basis.

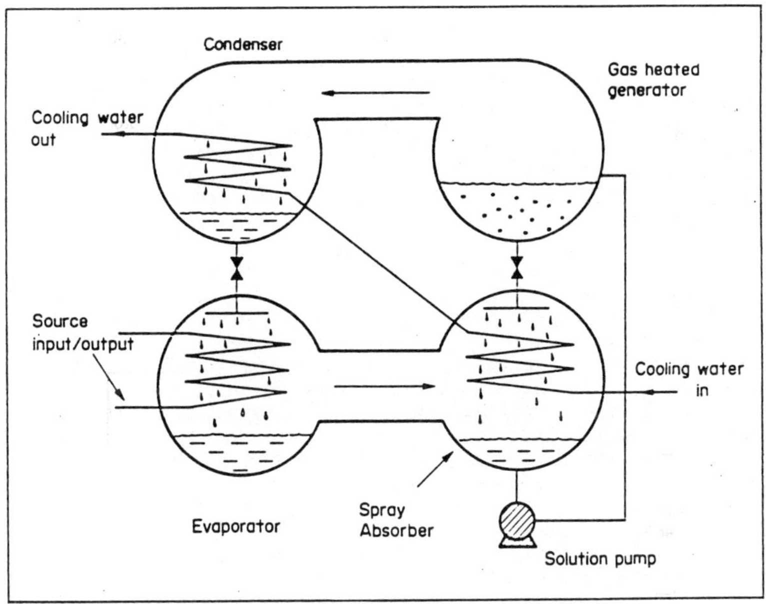

Fig.3.31 Basic Arrangement of Sanyo Industrial Lithium Bromide/Water System

The falling film absorber has the low vapour pressure impedence of a spray over coil column and it appears to offer the possibility of overcoming the latter's disadvantages. It consists of a vertical tube down which the absorbent flows, on its inner surface, as a thin film. The central space of the tube is occupied by the vapour. Heat dissipation is achieved by some cooling arrangement on the outer surface of the tube. Factors influencing its performance for a given combination of absorbent and vapour are the thermodynamic conditions which influence mass transfer e.g. concentration, temperature etc, the flow conditions in the tube, the tube geometry and the heat dissipation arrangement. It can be seen that these parameters are more amenable to analysis than those appertaining to a spray over coils arrangement.

A further factor in favour of the falling film absorber is the possibility of an exact match between the physical design and the thermo/chemical requirements.

The aqueous lithium bromide solution enters the absorber very close to its solubility limit and this restricts the rate at which heat can be extracted since it is important to avoid crystallisation. However, as the solution becomes more dilute its rate of cooling can be increased. The falling film absorber can be designed to have a variable rate of heat transfer along its length.

3.8.3 Evaporator The evaporator in the absorption cycle is strangely analogous to that used in the vapour compression cycle. In the latter case the question of oil carryover dominates, and the design must avoid trapping the oil on the heat exchange surfaces. In the absorption cycle the problem is the absorbent, which may enter the evaporator dissolved in the refrigerant, but which is considerably less volatile. One solution is simply to allow a proportion of the refrigerant to pass through the evaporator to the absorber as liquid.

If this is a large quantity it can affect efficiency, but a small steady leakage can prevent the accumulation of absorbent which would depress the evaporating pressure.

The consequence of a reduced evaporating pressure is that the absorber (which is at the same pressure) will produce a liquid output which is insufficiently concentrated. One group of workers (Battelle Institute, Frankfurt) have proposed the use of a small jet-pump to increase the pressure between evaporator and absorber.

In small scale absorption heat pumps, the heat exchangers are generally of the shell and coil or tube in tube type, a choice which reflects the relative cheapness of these arrangements compared with shell and tube units.

3.8.4 Solution pump The solution pump is a key component in an absorption cycle heat pump, because it can be expensive, and because it is the major moving part and is therefore the key to reliability.

One type which is suited to this high-pressure, low flow application is a hydraulically actuated diaphragm pump.

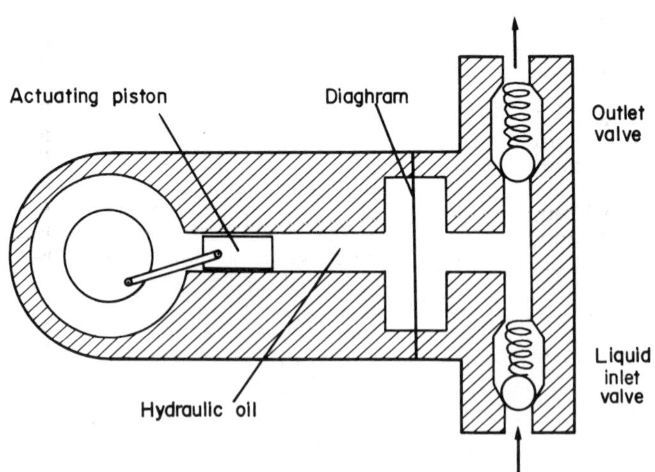

Fig.3.32 Solution Pump of a Type Suitable for the Absorption Cycle

This has the advantage that the refrigerant is positively sealed in by the diaghragm. The use of hydraulic actuation rather than mechanical gives higher reliability at the high operating pressures and minimizes the consequences of leakage. A solution pump is illustrated in Fig.3.32.

Development of cheap and efficient solution pumps for domestic absorption cycle heat pumps has been the subject of a number of projects. Stiebel Eltron in West Germany (Ref.3.13) has carried out research over a number of years in applying the hermetic compressor concept to the solution pump - the electric motor drive was completely submerged in the working fluid together with the pump. Unfortunately the conductivity of the liquid increased after some time to such an extent that hermetic operation was not possible. The eventual solution was to develop an electric motor with a casing between the stator and the rotor, the latter being submerged in the working fluid. The aim is to produce a solution pump selling for less than DM500.

REFERENCES

3.1 Downing, R.C. Characteristics of refrigerants, The Refrigerant Engineers Society, Service Manual Section 3, page RA 331, 1963.

3.2 Fujimura, Y. et al. Gas engine heat pumps - features and applications. Komatsu Technical Report, 1983, Komatsu, Japan.

3.3 Butterworth, D. Introduction to Heat Transfer - Oxford University Press, Oxford, 1977.

3.4 ASHRAE Handbook, Fundamentals, Chapter 2. Published by the American Society of Heating, Refrigerating & Air-Conditioning Engineers Inc., New York, 1977.

3.5 ASHRAE Handbook, Systems, Chapter 26. Published by the American Society of Heating, Refrigerating and Air-Conditioning Engineers Inc., New York 1976.

3.6 McMullan, J.T. Study of capacity control and the influence of lubrification oil on system and evaporator design, in heat pumps with rotary sliding vane compressors. Commission of the European Communities, Report EUR 9445 EN, CEC, Brussels, 1985.

3.7 Ofner, H. et al. The economy of non-azeotropic refrigerant mixtures in compression heat pumps. Heat Recovery Systems, Vol.6, No.4, pp 313-321, 1986.

3.8 Kays, W.M. & London, A.L. Compact Heat Exchangers, Third Edition, McGraw-Hill Book Company, New York, 1984.

3.9 Blundell, C.J. Optimizing heat exchangers for air to air space heating heat pumps in the UK. Energy Research, Vol.1, pp 69-94, 1977.

3.10 Pleininger, R. and Kuhn, M. Optimized operation of a frozen air heat exchanger by using a micro-computer controlled expansion valve. Paper 1.4, Proc. of Contractors Meeting on Heat Pumps, CEC Report EUR 8077 EN, Brussels, 1982.

3.11 C.C. Hiller, L.R. Glicksman. Improving heat pump performance via compressor capacity control. M.I.T. Energy Laboratory Report MIT-EL 76.001.

3.12 Heppenstall, T. A theoretical and experimental investigation of absorption cycle heat pumps for industrial processes. CEC Contract EEB-1-146-UK, EUR Report, Commission of the European Communities, Brussels, 1986.

3.13 Schracter, K.H. Development of a cheap and efficiently adjustable solvent pump for absorption cycle heat pumps. Report EUR 9916DE, Ibid.

CHAPTER 4
Heat Pump Applications — Domestic

It is in the role of domestic heating and air conditioning that the heat pump has been subjected to its most widespread use, largely dictated by the requirements of the market in the USA, where home owners need air conditioning throughout the year - cooling in the summer months and heating during winter. The heat pump in its reversible form, capable of fulfilling both these roles, has been developed over a period of about 40 years and has become a reliable and cost-effective piece of equipment.

In Europe, where climatic conditions are generally such that, at least for family dwellings, full air conditioning is either not necessary or impractical, the 'heating only' heat pump is seen as the most promising system, the aim being to optimise the heating capability so that capital and operating costs are at a reasonable level when compared with fossil fuel-burning devices such as wet or ducted air central heating systems. While in the United States the number of domestic heat pump installations numbers well in excess of one million, the rate of installation in Europe is still comparatively low, dictated both by economic factors and by a lack of really suitable equipment, in terms of both capital cost and efficiency.

The purpose of this Chapter is to describe the factors to be taken into account when applying heat pumps in the home, to present some examples of successful installations, and to comment on potential developments of significance in this area.

4.1 Role of the Domestic Heat Pump

Heat pumps installed in a family dwelling must be capable of fulfilling a number of roles, depending upon local climatic conditions. Because of their comparatively high current capital cost the more functions that heat pumps can perform, the better.

The primary energy uses in a house are for space heating, full air conditioning in climates which necessitate this, and the provision of domestic hot water for washing and other services.

A refrigeration duty associated with the storage of perishable foodstuffs is also a common requirement. Cooking, lighting and television are the other energy users, on a largely intermittent basis.

A packaged heat pump should be capable of providing the space heating and air conditioning requirements (possibly including the part-time use of supplementary heaters when weather and economic conditions justify it). It is also desirable to use the heat pump to provide at least a part of the domestic hot water requirements. Heat pumps can be linked to refrigeration duties - of course as mentioned in the Introduction, the systems are essentially identical - but the domestic refrigeration load of typically 0.5 kW, including a deep freeze, offers little scope for use of the heat rejected at the condenser. In commerical and industrial premises, however, as will be seen later, the linking of refrigeration and heating duties can be of considerable economic benefit.

4.1.1 Provision of space heating As discussed in previous Chapters, and illustrated later in this Chapter, a heat pump may use a wide variety of different low grade heat sources, rejecting this heat at the condenser to heat either a gas or a liquid (or in some cases a thermal storage medium which may be liquid or solid). For simple space heating, the domestic user is accustomed to either a 'wet' central heating system, in which hot water is circulated to radiators located in each room, or to a ducted air system, in which warm air is blown into each room. Individual room heaters, such as radiators, storage units, and convectors, are also widely used, often as a supplementary heat source to back up the central system.

The temperature of the heat distribution system varies quite widely between ducted air at only 40°C to very hot water, or even steam heating systems in the region of 100°C. A typical water temperature for a hydronic system might be 75°C. (This may be lowered, with a corresponding increase in 'radiator' size).

Because the heat pump efficiency is closely allied to condensing temperature, a lot of emphasis in heat pump applications has been placed on lowering the heat distribution temperatures. It is felt that by increasing surface areas (perhaps by using under floor heating) water at 50°C may be acceptable. Similarly by increasing air circulation rates the air delivery temperature can be lowered to 35°C. Developments of this nature in new housing stock can radically alter the acceptability of heat pumps, provided that the reduction in capital cost also needed, in addition to the COP increase, is realized.

Central heating systems operating directly on fossil fuel (the combustion system being installed in the home of the user) normally provide most, if not all, of the domestic hot water requirements. As discussed in Section 4.1.2, this is an important point to be considered when designing heat pumps, particularly for existing homes. The space heating requirement in the house, however, predominates in terms of total energy usage. In the United Kingdom, for example, space heating consumes between 60 and 65 per cent of the total domestic energy usage, whereas water heating is little in excess of 20 per cent of the total (ref.4.1).

Of the two currently popular central heating systems, the 'wet' system is the most popular in the United Kingdom, and in some European countries. Where a full air conditioning system is required, however, chilled or heated air distribution is the most practical method. It is ideal for incorporation in houses during construction, but is less attractive for 'retrofitting' than the wet central heating system, which utilises small bore pipes to transfer water from the boiler. Ducted air systems require ducts of significant cross-sectional area, making them difficult and expensive to install in existing property.

As a device for space heating, the heat pump need not function on the basis of a central system servicing a number of rooms. Individual room conditioners may be installed, each with its own compressor and condenser, with either an internal or external heat source for the evaporator. Systems of this type have been developed in the UK, including a 'through-the-wall' air-air heat pump room conditioner. Cost, however, has remained a problem, and the inability to offer satisfactory performance in cold weather always necessitates some consideration of supplementary heating.

A recent study by the Building Research Establishment in the UK, on behalf of the Department of Energy, (ref.4.2), surved commercially available electric drive air-air, air-water and ground-water heat pumps for space and domestic hot water heating The study was directed at highly insulated houses. At the time (1983-84), it was claimed that approximately 2000 domestic heat pumps were installed in the UK, the majority being retrofitted.

Interesting, if somewhat disconcerting, data were obtained on the costs of the heat pumps available. From the outset it was found that most systems could meet the low heat demand of a well-insulated house (3.6 kW with a mean inside temperature of 20°C and an ambient of -1°C). As a result some interpolation had to be made in order to obtain prices of 3.6 kW output heat pumps.

At the time of the study, average retail prices of heat pumps were £1800 for a 5 kW output machine and £2900 for a 10 kW machine. Retail prices of heat pumps having a 3.6 kW output were £1470 for an air-water unit and £1310 for an air-air system The additional cost of providing domestic hot water was £220.

The study concluded that much had to be done before heat pumps could become "common domestic heating technology". It was believed, however, that the cost-effectiveness of heat pumps could become more acceptable in the medium term. (Note that the study was conducted in the context of the UK only).

4.1.2 Domestic hot water heating While the provision of space heating or air conditioning is the most important role of the heat pump, or indeed any energy user, in the home, the ability to meet the domestic hot water requirements may be the most important factor governing its widespread acceptability

A number of reports examining the future role of heat pumps in houses see their primary role as providing space heating, at least as far as the United Kingdom is concerned. Reference 4.3, for example, contains arguments for this, but it is also pointed out that water heating and heat recovery functions will become more important as the trend towards 'low energy' housing increases, with the opportunity this gives for a fully integrated system, based possibly on heat pumps. (An e xample of such a house, constructed by Philips, is given later in this Chapter).

This however tends to neglect one of the major problems in adapting existing housing to be compatible with heat pumps - the desirability of replacing a single unit which is currently capable of providing both space heating and domestic hot water, namely the fossil-fuelled central heating boiler, with a heat pump also able to perform both of these functions. As will become evident, the difficulty is associated with the economic viability of using a low temperature external heat source to provide high grade heat in hot water.

The cost of electricity prohibits its use for significant amounts of boost heating, while capital costs would generally preclude a 'tandem' heat pump/fossil fuel boiler system. This means that the heat pump would ideally discharge water heated to the required temperature.

Of course, there are several ways around this problem, but it does highlight one difficulty in a society where electricity costs, with the exception of some off-peak tariffs, to the consumer are high, the majority of heat pumps are currently based on electric drives, and supplementary heating, both for space heating and hot water generation, would be electric. Until the advent on a large scale of 'low energy' housing, this may remain a European dilemma, at least.

4.1.3 Refrigeration and heat recovery As mentioned earlier in this Chapter, the use of heat pumps as domestic refrigerators in conjunction with recovery of the useful heat from the condenser, appears unlikely in view of the low duties involved, and the low evaporator temperatures. It is interesting to note here however that a system for doing just this was marketed in 1954 by the Ferranti Company in the United Kingdom. Although evaporator temperatures were somewhat higher than in the deep freezers used in homes today, the 'fridge-heater' was able to cool a larder by up to 11°C, using the rejected heat to heat water in a 136 litre storage tank. With an overall coefficient of performance of 3, the power consumed by the compressor was approximately 0.4 kW. Although the system apparently performed adequately, it was classed as 'luxury goods' and as such was subjected to a 60 per cent tax levy; (it cost £141). The system is described more fully in reference 4.4.

Another role which the heat pump can play, and one which may become increasingly popular, is in the recovery of waste heat. One normally associates waste heat with industrial processes and large buildings, but the domestic consumer also discharges large amounts of heat from his home. Although largely intermittent, each discharge of hot water, after washing dishes, bathing or showering, is a potential source of medium grade heat, and in fully air conditioned houses, the number of air changes may be such that recovery of heat in exhaust air would be feasible. Such features are common to many of the new houses being designed around heat pump systems, but the cost-effectiveness of such comprehensive systems is currently unattractive.

4.2 Heat Sources

In domestic applications one of the main claimed attractions of the heat pump is that it can make use of what are effectively 'free' heat sources to provide comfort conditions in the home. Air is the most obvious source of heat for general use; it is universally available and has attractions from the mass marketing point of view. Water, when available, has several advantages over ambient air as a heat source, and these are discussed below. The heat contained within the ground has been used as a heat source for heat pumps, and is still the subject of extensive study. A number of projects in this area have been funded in Europe, and ground source domestic heat pumps are marketed by an enterprising company in the USA. Energy may also be collected from various waste heat source, originating either within the dwelling or from outside (for example sewage plant and power station discharged water), and the use of solar collectors in conjunction with heat pumps in houses is the subject of interest in Europe and North America

4.2.1 Air source heat pumps Most commercially available domestic heat pump systems use outside air as the heat source, and this has been the case since the heat pump was introduced into the domestic scene. Most air source heat pumps working on the vapour compression cycle also use air as the heat sink, and the layout is typified by the Lucas Climasol system, a domestic air-air heat pump marketed in the United Kingdom in the 1960's. Illustrated in Fig.4.1, the outside air is induced over a finned evaporator coil through which the heat pump working fluid is circulated. A similar heat exchanger is used in this installation to transfer heat from the condenser to the air circulating throughout the house.

As with other heat pump systems, the components may be located in one 'package', or the system may be 'split', with the indoor air coil some distance away from the evaporator. These are illustrated later in the Chapter.

While ambient air is the most convenient heat source, for obvious reasons, it also suffers from a number of disadvantages which necessitate careful optimisation of the design, depending upon location. These are largely concerned with the fact that ambient air can vary considerably in its temperature.

As described in earlier Chapters, the performance of a heat pump, or more particularly the COP, varies as the temperature difference between the evaporator and condenser, decreasing as this temperature difference grows larger. This has especially unfortunate implications for domestic heating using air source heat pumps - as the ambient air temperature reduces, the demand for internal heating increases but the ability of

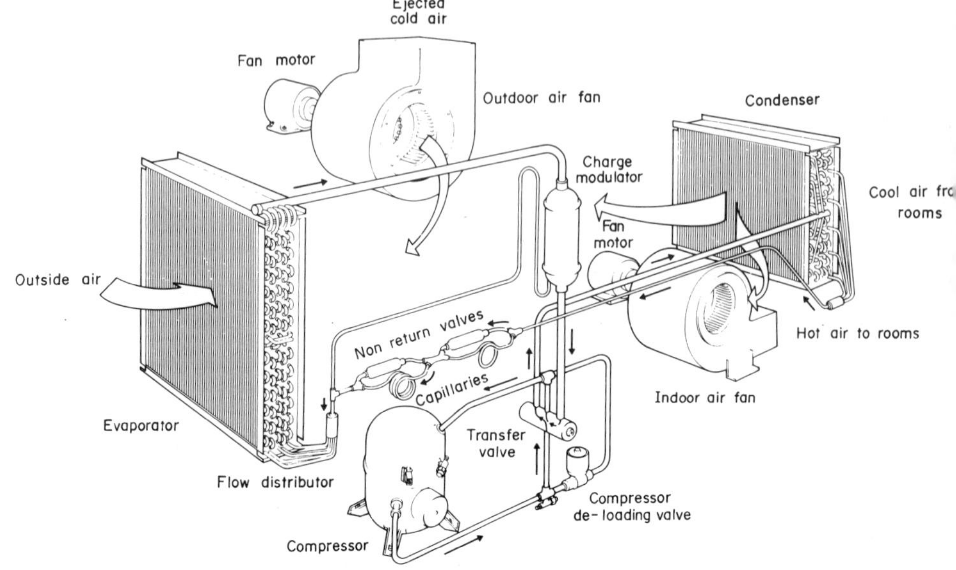

Fig.4.1 Layout of an Early Domestic Air-Air Heat Pump - The Lucas Climasol

the heat pump to maintain even a constant heat output by itself is substantially impaired. Fig.4.2 shows how the COP reduces for a typical air source heat pump, as a function of the outside ambient air temperature, (ref.4.5). Much development work on heat pumps and ancilliary systems is directed at minimising the effect of this on overall performance, and techniques for doing this are described below.

Supplementary heating: The capital cost of a heat pump, whatever the heat source, is considerably greater than that of the conventional central heating boiler (discussing here the requirement for a 'heating only' system as appropriate to the United Kingdom and much of Europe), assuming equivalent powers, and therefore the capital cost differential would be even further extended if a heat pump was sized to provide full heat load, even if practicable for the house. Therefore air source (and other) heat pumps are sized according to the overall economic assessment of the duty required, normally meeting a proportion of the annual heating demand. The balance is provided by supplementary heating, which is often electrical in the USA but may be based on fossil fuel combustion in the UK and Europe, depending on the balance between capital and running costs. If the heat pump has to fulfill a refrigeration air conditioning duty during the summer months, the heat pump sizing should also take the capacity required here into account -

oversizing for cooling duties is also wasteful in terms of energy and operating costs.

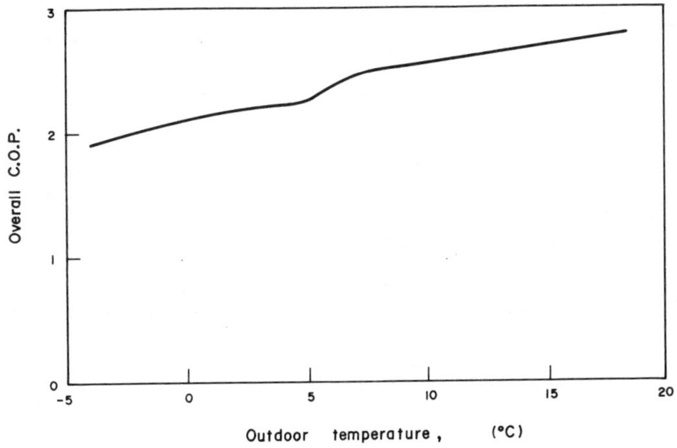

Fig.4.2 The Variation of COP of a Typical Air Source Heat Pump with External Ambient Air Temperature

Typical of the supplementary heating requirements for domestic use is the characteristic shown in Fig.4.3.

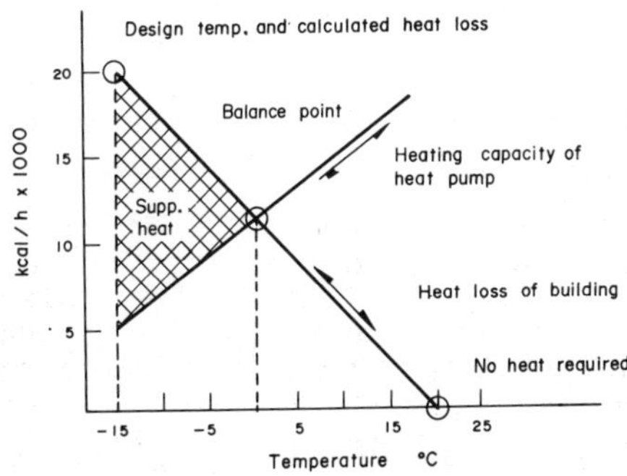

Fig.4.3 Balance Point and Supplementary Heating Requirements of the 'Weathertron' Air Source Heat Pump

This graph, applicable to the Weathertron air-air heat pump manufactured by the General Electric Company in the USA, shows that supplementary heating would be required if the ambient outside air falls below freezing point, i.e. when the heat loss from the building exceeds the heat output from the heat pump necessary to maintain comfort conditions. In order to obtain optimum economic benefit from the system, the manufacturers stress the importance of energizing the supplementary heaters (in this case electric) only when the heat pump output is not sufficient to meet the total demand.

The most effective technique for controlling the supplementary heaters in this case is to use indoor and outdoor thermostats, the indoor unit having two stages of activation, the first controlling only the heat pump. Primary control of this second stage indoor thermostat is governed by the activation of the external thermostat in bringing in the supplementary heaters as the outside temperature falls below the balance point (see Fig.4.3).

Defrosting: Low external ambient temperatures lead of course to one of the more critical areas of air source heat pump design - the incorporation of evaporator defrosting mechanisms. As the ambient air temperature falls, frost can accumulate on the evaporator coil exterior due to freezing of the moisture which condenses out of the air as it is cooled. If this frost is permitted to accumulate, the effectiveness of the evaporator in removing heat becomes severely reduced, and the pressure drop across the heat exchanger increases. It is therefore necessary either to prevent frost formation, or to remove the ice rapidly after it begins to form. In order to do this most air source heat pumps have the capacity to operate on a 'defrost' cycle.

The mechanism of frosting and defrosting has been the subject of extensive studies, both of an experimental and theoretical nature. Of particular interest is the work carried out by Barrow at the University of Liverpool (ref.4.6). Barrow established that the insulating effect of the frost layer on heat pump evaporator surfaces was negligible, the only important reason for a reduction in the overall heat transfer being attributable to an increase in the air-side convective resistance. The fan characteristic was found to be an important feature in the overall operation of the system, changing as it does with the changes in surface geometry due to frost formation. The thickness of the frost layer was the principal parameter as far as the heat transfer is concerned.

The results of two interesting earlier studies are reported in references 4.7 and 4.8.

While the use of solar energy as a heat source (normally indirectly applied via solar collectors) is discussed later in Section 4.2.4, one interesting concept for an air source heat pump which has been applied to houses in Germany is illustrated in Fig.4.4, (ref.4.9).

The air is drawn in under the eaves of the roof of the house into the roof space, which is warmed by the sun. The evaporator coils, not shown on the drawing, are mounted in the apex of the roof. While of course maximum benefit will occur only when solar heating is available, the residual heat of the roof structure will help to minimise the regularity of the defrosting cycle.

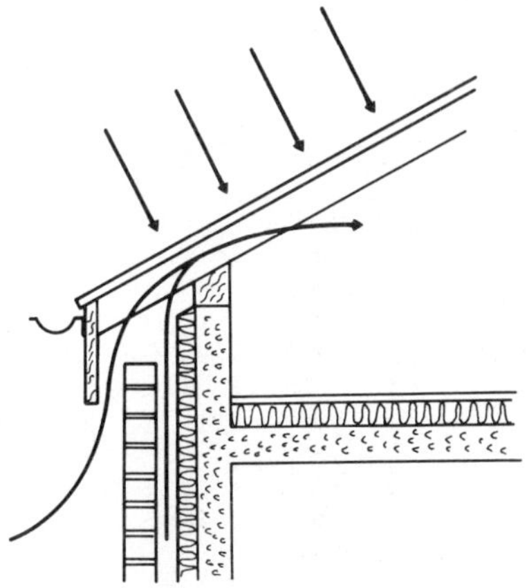

Fig.4.4 Use of Solar Gain in the Roof Space to Increase Ambient Air Temperature Approaching the Evaporator

Table 4.1 summarises the relative merits of the possible heat sources for domestic heat pumps, based largely on conditions applicable in the United States (ref.4.10). As emphasised above, the major disadvantage of air is shown to be the fact that less heat is available as the demand increases. However, convenience and universal availability will outweigh this as far as commercial exploitation is concerned.

4.2.2 Water source heat pumps There are a considerable number of water source heat pumps on the world market, including models for domestic use. However depending upon the particular source from which the water (and its heat content) is obtained, it can be at best inconvenient and at worst prohibitively expensive, as indicated in Table 4.1.

Taking the various sources of water in turn, city, or mains, water would be an ideal heat source for a heat pump as it is essentially continuously available and is at a temperature of typically 4-12°C, i.e. it never falls below freezing point. However, water is becoming an increasingly expensive commodity in industrialised societies, and the cost of providing such water for heat pump use would be prohibitive and impractical. Well water is attractive also because of its near-constant temperature, varying between about 10°C in northern climates and rising to about 15°C in areas nearer the equator, however good quality well water is becoming rare, because of other demands on it. Poor quality well water can lead to scaling or corrosion in the heat exchanger. Another drawback to its regular use on a commercial basis is associated with the cost of the installation. Drilling may be needed to open up a suitable source, and substantial amounts of piping may be needed. Pump capacities may be high, and problems could arise when it is recognised that the water must also be disposed of when it has passed through the heat exchanger.

TABLE 4.1 Relative Merits of Heat Pump Heat Sources and Sinks

Heat Source	Air	Mains or City	Well Water	Surface Water	Waste Water	Earth	Solar
Source classification	Primary	Primary or auxiliary	Primary	Primary	Primary or auxiliary	Primary or auxiliary	Auxilliary
Suitability as heat sink	Good	Good	Good	Good	Variable with source	Usually poor	May be used to dissipate heat to air
Availability (location)	Universal	Cities	Uncertain	Rare	Limited	Extensive	Universal
Availability (time)	Continuous	Continuous except local shortages	Continuous check water table	Continuous	Variable	Continuous, temperature drops as heat is removed, slowly rises when pump stops	Intermittent, unpredictable, except over extended time.
Expense (original)	Low, less than earth and water sources except city	Usually lowest	Variable, depending on cost of drilling wel	Low	Variable	High	High
Expense (operating)	Relatively low	High, usually prohibitive	Low to moderate	Relatively low	Low	Relatively moderate	Unexplored. Promising as auxiliary for reducing operating cost.
Temperature (level)	Favourable 75–96% of time in most of United States (eg)	Usually satisfactory	Satisfactory	Satisfactory	Usually good	Initially good–drops with time and rate of withdrawal	Excellent

Temperature (variation)	Extreme	Variable with location (4°C to 1Ø°C)	Small	Moderate	Usually moderate	Large - less than for air, however	Extreme
Design Information	Usually adequate	Usually adequate	Usually adequate	Usually adequate	Adequate if source is constant in supply and temperature	Inadequate	Practically available
Size of equipment	Moderate	Small	Small (except for well)	Small	Variable (usually mod-erate)	Small (except ground coils)	Available in some areas
Adaptability to standard product	Excellent, can be factory assembled and tested	Excellent	Excellent (except for well)	Excellent	Poor	Poor	Poor
Sources it may Augment		Air, earth					
Special problems	Least heat available when demand greatest. Coil frost-ing requir-es extra capacity, alternative source, or standby heat. May require ductwork.	Scale on coils. Loc-al use restrictions during shortages. Disposal. Water temp-erature may become too too low to permit further heat removal.	Corrosion, scale may form on he-at transfer surface. Disposal may require second well. Water loca-tion, temp-erature, composition usualy un-known until well drilled. Well may run dry.	Water may cause scale, corrosion and algae fouling.	Usually scale form-ing or corosive. Often in-suficient supply. Very limit-ed aplica-tion, hence required individual design. Freeze-up hazards.	Limited by local geo-logy and climate. Installa-tion costs difficult to estimate Requires considerable ground area, may damage lawns. Leaks difficult to repair.	Probably will re-quire heat storage equipment at either evaporator or conden-ser side.

Sometimes also known as groundwater when it can be obtained from shallow bore holes usually less than 50 meters deep, studies recently carried out in the UK have suggested that such a source could, if effectively used, reduce the payback period of space heating heat pumps by at least 10 per cent, for the same heating duty, (ref.4.11). Initial estimates suggest that up to 50 per cent of the land area in England and Wales could provide suitable sites for the location of groundwater heat source heat pumps.

Natural surface water such as lakes, seas, and rivers is also a source of heat for heat pumps, and has been used in many installations, both in domestic and commercial buildings. However, difficulties similar to those with air source heat pumps can arise in the winter. Although the water source may be at 6 or 7°C entering the evaporator heat exchanger, care must be taken not to cool it to 0°C, because freezing can occur. By increasing the flow rate of the water, the temperature drop through the evaporator can be kept as low as 1-2°C. The use of water as a heat source was tried in an early domestic heat pump constructed by Sumner (Ref.4.12) with some success.

Slightly warmed water can make a more attractive heat source than that at ambient temperature. The use of waste heat contained in domestic water effluent is discussed later in this Chapter. For example water from washing machines and baths has been investigated as a heat source. The use of solar energy a separate heat source in the context of this discussion, can also be conveniently linked to a water source heat pump system, as described in Section 4.2.4.

The use of waste heat from power stations, generally in the form of warm liquid effluent, has attracted considerable attention as a heat source for many uses, including heat pumps (ref. 4.13). One of the most interesting studies related to this heat source, although now a little dated, was carried out by Kolbusz at the Electricity Council Research Centre, Capenhurst, (ref.4.14).

As shown in Fig.4.5, Kolbusz claimed that by using a back pressure turbine (1) to drive the heat pump compressor (2), waste heat from condensing power stations could be usefully recovered. The waste heat extracted in the condenser (3), which also serves as the heat pump evaporator, is used by the steam-driven heat pump as the first stage (4) in heating water. The second stage of heating is carried out by the condenser of the back pressure turbine. Kolbusz showed that a COP of 6.6 could be achieved, and for each tonne of coal burned, 2 tonnes of coal equivalent could be made available to a district heating scheme. (Items 6 and 7 are the boiler and feedpump respectively). A heat source at 28°C in this system could, via the heat pump, supply water at 70°C.

For domestic use, it is either in the role of individual small solar-assisted water source heat pumps, or schemes for district heating of housing estates as outlined above, that the main uses of water exist. There will, however, always be a role

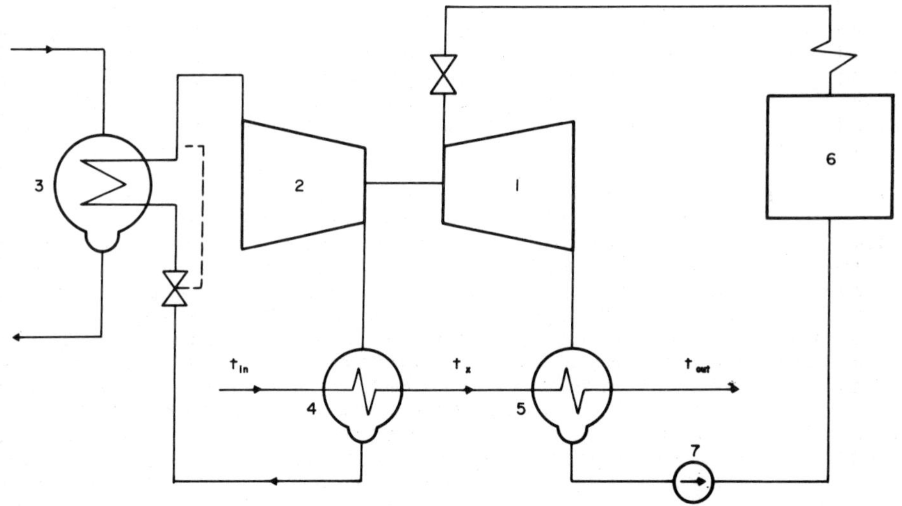

Fig.4.5 Flow Diagram of Combined Heat Pump and Back Pressure Turbine District Heating Scheme

for the simple water source heat pump, based on river or other surface water areas, where they can be conveniently sited. It is worth pointing out here, however, that in commercial and industrial premises, discussed in Chapters 5 and 6, the water source heat pump really comes into its own, displacing, particularly in the latter case, air source systems.

4.2.3 The ground In the 1950's a considerable number of ground (or earth) source heat pumps were installed in the United States, and the technical press was rife with photographs of major excavations in large gardens! It is a popular misconception that the variation in temperature in the ground is minimal throughout the year, but as shown in Fig.4.6, the temperature variation can be significant and in winter the temperature approaches zero even at a depth of 1 metre (ref.4.15).

(This data is for Brussels - in Europe considerable interest has been shown in ground-source heat pumps). One may present these results in a slightly different way, as illustrated in Fig.4.7.

Here the variation (expressed as a mean amplitude) in temperature of the soil throughout the year is expressed as a function of depth at which measurements were made. The advantages of digging to a depth of only 1 metre to lay one's evaporator coils are not therefore particularly great as far as achieving a near-constant temperature is concerned.

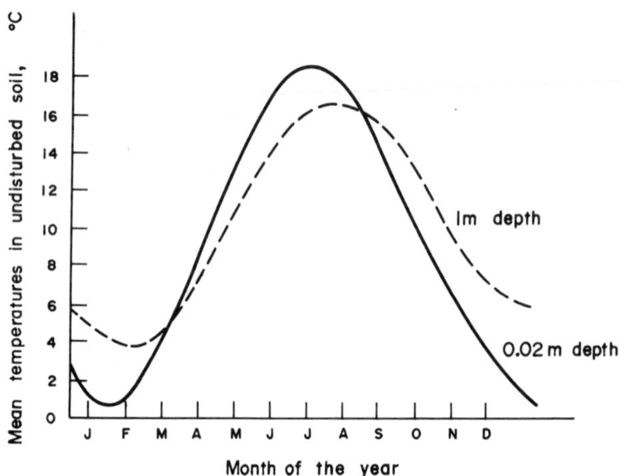

Fig.4.6 Seasonal Variation of Soil Temperature at Depths of 0.02m and 1.0m, Measured in Belgium

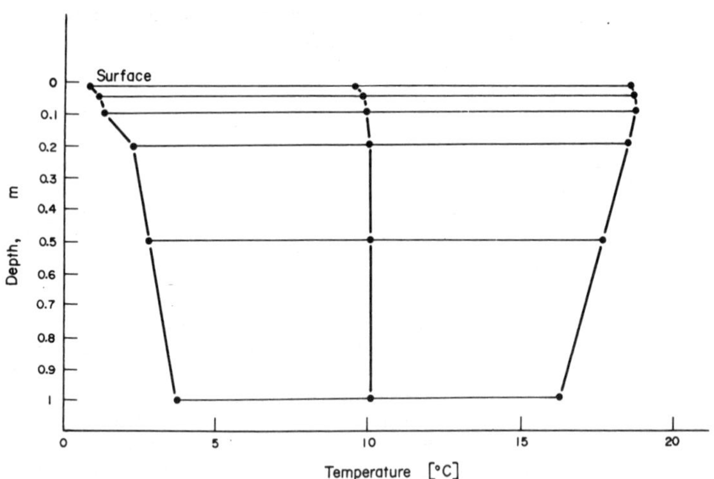

Fig.4.7 Mean Amplitudes of Temperature Variations as a Function of Depth Below the Ground in Undisturbed Soil

A number of proposals have been put forward as to the optimum depth of ground coils. (In a ground source heat pump, the evaporator normally takes the form of a serpentine pipe network, using pipes of typically 25 mm diameter, extending at a constant depth over an area of several hundred square metres). On economic grounds, at least from a first cost point of view, location of the coil as close as possible to the surface is desirable. Detailed studies of the ground as a potential heat source for heat pumps in Denmark has revealed the following (ref.4.16).

(i) The normal heat removal rate for an evaporator coil located in the ground is between 20 and 25 W/m. (It is interesting to note in this context the wide variation in ground coil performance obtained by workers. Values as low as 10 W/m have been obtained (ref.4.17), the maximum in Europe being 50-60 W/m, reported by von Cube in Germany (ref.4.18).

(ii) The optimum depth and pitch of ground coils are 1.5m and 2.0m respectively, although interference effects do extend beyond the 2.0m limit. (The Danish report does state that coils can be buried at a smaller depth, but emphasises that the output of the heat pump can be reduced by 5 per cent for each degree reduction in coil temperature).

(iii) With a brine solution circulating in the coil (as an alternative to evaporating in the ground coil, a water/refrigerant heat exchanger may be used), the average working temperature in winter is of the order of -3°C.

(iv) Changes in soil conditions do not have a significant effect on soil performance. Two notable exceptions to this exist, however. If the water content is high, the performance improves because of the increased thermal conductivity and good contact with the coil. Should the soil be of high gravel content, the effect is detrimental to coil performance.

(v) The Danish study looked at the possibility of using vertical pipes as opposed to horizontal ones. It was concluded that vertical units were unsuitable for heating only use, as the temperature conditions in the ground could not be fully restored during the summer. It was pointed out, nevertheless, that some form of regeneration could be expected if in summer the heat pump was reversed and used for a cooling duty.

A further interesting point revealed by this study was the fact that the minimum temperature of the ground, while always higher than that of the air, occurs about two months later than the minimum air temperature, i.e. in a situation when the heating load requirement is falling.

The comments on the use of vertical tubes are of interest. While horizontal tubes perform well for heating only, and the ground 'recovers' during the summer, some form of regeneration in the warmer months could make vertical tubes, together with their obvious economic and space advantages, acceptable. Research work at Queen's University, Belfast on vertical 'U' tube ground coils, while carried out during the heating season only, suggested a considerable heat up-take (ref.4.19). Having found that an area of 150-200 m^2 was required, in terms of surface under which horizontal evaporator pipes were buried, to achieve a useful output of 12 kW, the researchers tried a U tube coil penetrating to a depth of 8 metres, contained within a 127 mm diameter hole. From two holes, an output of 1.2 kW was achieved, resulting in a ground coil surface area reduction by a factor of between ten and twenty, when compared to a horizontal coil, for the same heat uptake.

Apart from the problems of coil design, and the vertical coil does have limitations of the type found in the Danish study, the other disadvantages listed in Table 4.1, particularly the adaptability to the heat pump as a standard package, weigh against its widespread adoption. It can be used in conjunction with solar energy, and this is briefly discussed below. For further data on ground as a heat source, the reader should consult refs.4.20 and 4.21.

4.2.4 Solar heat sources All heat pump sources are influenced by solar energy to a greater or lesser extent, but solar energy may be harnessed for heat pump use by solar collectors employing a circulating fluid, by the use of solar energy to raise the temperature of the air approaching the evaporator, or by solar concentrators. Systems are commercially available in Europe and North America using solar collectors or solar air heating, and these are described later. The use of solar concentrators may be more appropriate for application on absorption cycle heat pumps (see Chapter 2), which are not yet cost-effective for domestic use, but are the subject of much development work. High grade heat, at a temperature not normally available from flat plate collectors, is needed to raise the temperature in the generator of an absorption cycle heat pump, although flat plate collectors can be used for solar powered air-conditioning, partly because the required temperature is lower than for heat pumping and partly because air conditioning is a summer duty and the available solar collector temperature is therefore higher.

As mentioned earlier, the use of solar energy, normally collected using roof-mounted flat collectors, to supplement other heat pump heat sources is commonly practiced. Solar collectors have been studied in conjunction with ground coils (see also Section 4.2.3) and are also extensively used, both without and in conjunction with heat pumps, in association with heat storage media. These act as sources for heat pumps when solar energy is not immediately available, for example on days of cloud cover or overnight.

The main advantage of solar energy, whether it be collected via flat plate collectors, concentrators, or, to a less extent, supplementing existing air sources for heat pumps, is its temperature. By being able to provide heat for the evaporator at a higher temperature than that normally available in 'conventional' air, ground or water sources, the heat pump COP may be raised.

Most of the heat pump systems currently available which make use of solar energy collected via flat plate collectors employ an indirect heating system, in which a secondary fluid, normally water, is circulated through the collector. However, as illustrated in Fig.4.8, a flat plate solar collector may also function directly as the evaporator of a heat pump, in this case the refrigerant being circulated through the tubes within the collector.

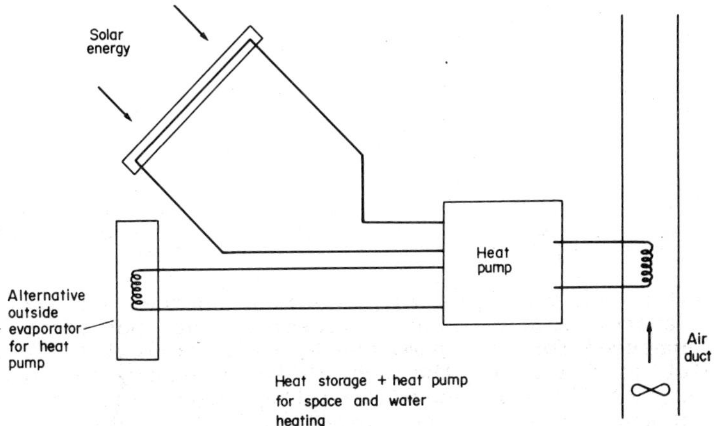

Fig.4.8 Use of a Solar Collector as the Evaporator of a Domestic Heat Pump Installation

For a 'heating only' system, a conventional collector may be employed, with minor modifications. Should it be desired to operate on a cooling cycle as well, for full year-round air conditioning, the glass sheet which normally forms the front of the collector should be removed. This then allows the heat which it is required to remove from what has become the condenser to be effectively dissipated to the atmosphere. Such designs were considered in the United States as early as 1955 (Ref.4.22).

The more common use of flat plate solar collectors, as mentioned in the last paragraph, in conjunction with heat pumps, is to use them in a separate circuit which provides heat to either the evaporator directly or, a more popular concept, to a heat store into which an evaporator coil is immersed. (See also Section 4.3).

A more comprehensive layout of a solar-assisted heat pump is shown in Fig.4.9. This, and other systems, described in more detail in a report published by the Building Research Establishment in the United Kingdom (ref.4.23), was one of a number of proposed low energy houses designed at the Establishment,

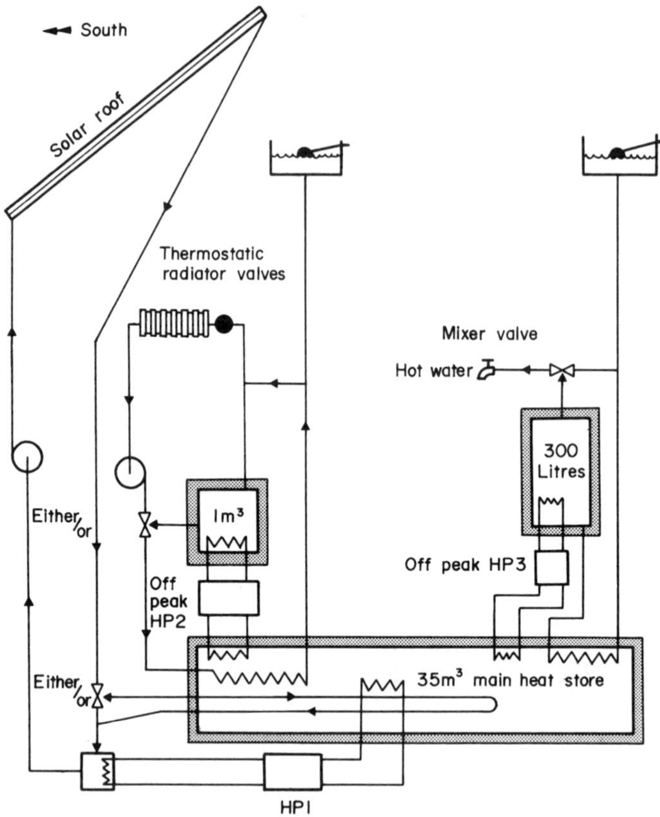

Fig.4.9 An Experimental Low Energy House using a Solar-Assisted Heat Pump Array

who operate a number of houses on their site for test purposes. The house, a flow diagram for which is shown in the figure, used three heat pumps, one to upgrade heat from the flat plate solar collector into storage when radiation is low; one to upgrade the heat from the store for space heating; and a third heat pump to upgrade heat from the store for provision of domestic hot water.

One of the drawbacks of conventional flat plate solar collectors, particularly when used in regions where solar radiation levels are low, is the size and cost. Examples of these installations are given later, but attempts have been made to reduce the size of collectors for liquid heating by using concentrators. The operation of such a collector is shown in Fig.4.10.

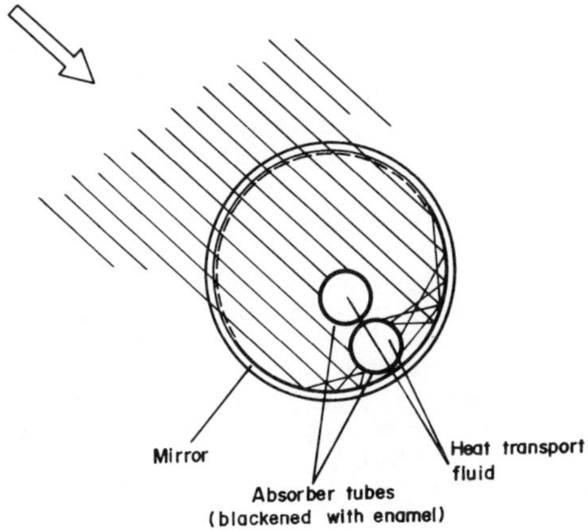

Fig.4.10 A Module of the Philips Solar Concentrator Used on Their Aachen House

Incorporated in the Philips house at Aachen, (discussed later in the context of heat pump installations), the figure shows one module of the collector, which consists of an evacuated glass tube, silvered over half its radial surface, containing two blackened water carrying absorber tubes positioned symmetrically within the silvered portion as shown, each collector tube being a quarter of the diameter of the glass tube. The collector is thus a flat plate collector, with conventional water filled collector tubes, but the transport of heat from the absorber plate is not by thermal conduction but by radiation. The vacuum prevents convection losses, and re-radiation to the glass is prevented by coatings of tin or indium oxide. Using such collectors, much higher liquid temperatures can be achieved - in many cases obviating the need for a heat pump. In the Aachen house, for example, 20 m^2 of solar collector of this type if expected to collect 36 to 44 GJ per year at an average efficiency of 50 per cent to be stored in a 40 m^3 tank at temperatures of up to 95°C.

Discussed in more detail in a subsequent section, the heat store is an important part of any solar-assisted heat pump system, and the advantages of high collector and storage temperatures can be seen in Fig.4.11, which shows the influence of storage temperature on the heat pump coefficient of performance, (ref.4.24).

The reader is also advised to consult McVeigh (ref.4.25) for a comprehensive treatment of solar collector design and applications.

The use of solar collectors in conjunction with ground source heat pumps has also been studied (ref.4.16). One layout adopted is shown in Fig.4.12, in which a solar collector is used to supplement the heat input to the evaporator, linked to a ground source coil. The particular study under review was solely theoretical in nature (although ground coil performance data had been previously obtained experimentally). The results of the study are shown in Fig.4.13, which illustrates the relationship between work input and ground coil/solar collector area, assuming a total heat pump duty of 12,260 kWh per annnum.

The research group concluded that it was impractical to install a solar collector if the size was less than 3 m^3 per kW loss of heat from the dwelling. The results suggested that expenditure weighted in favour of the solar collector favoured the performance of the system. For a collector area of 30 m^2 and a ground coil area of only 100 m^2, a COP of approximately 3.4 - high for a domestic heat pump - could be achieved. On the basis of their work investigating the isolated use of ground coils, the group found that a coil area of 300 m^2 would be necessary to attain a COP of only 2.7.

In spite of the increased COP obtained with the aid of flat plate solar collectors, the energy savings achieved were insufficient to justify the extra cost of the solar collectors, however.

With financial assistance from the European Commission, Laboralec, a Belgian organisation, has carried out research into the combination of horizontal ground coils and unshielded solar absorbers as heat sources. The horizontal ground coils comprised a 2 layer system at depths of 60 cm and 120 cm, and the combination of coil and solar input prevented soil freezing around the former. However, the overall COP was not improved.

An interesting feature of this project, which was directed at the heating of a building with a floor area of 500 m^2, was the form taken by the absorbers. These in one trial consisted of large conventional domestic 'radiators' connected in series around the outside wall of the building. Absorption levels of 20 W/m^2K were measured. An alternative configuration, the use of a horizontal stack of such radiators with a gap of a few cm between each radiator, gave higher coefficients, up to 80 W/m^2K. This was a strong function of wind speed, however, (ref.4.25). The layout of the ground coils is shown in Fig. 4.14, while data on the temperatures throughout the various systems during the winter of 1981/82 are shown in Fig.4.15.

Other work on air solar collectors (ref.4.27) has also indicated that collectors of some considerable size are needed to be technically effective. Approximately 20 m^2 of surface for a 6 kW output domestic heat pump was projected in this particular study. The influence of heat storage, both on overall COP and on system cost, is also a factor of prime importance, and while demonstrations are under way in a number of installations world-wide, data is still lacking on commercial system economics.

4.2.5 Waste heat as a heat pump heat source The heat pump has two principal roles in the domestic situation as far as heat recovery is concerned. It may be used for whole-house heating, as generally discussed in the previous sections of this Chapter, or it may be applied on a particular energy user within the home, such as a washer or tumbler dryer. In this section the former situation will be briefly considered, and later in the Chapter an example of the heat pump employed on a domestic item of plant will be given. Proposed 'total energy' houses, in terms of heating and/or cooling capabilities, are also illustrated later, and in the section on commercial systems, one waste heat recovery system will be detailed.

Reference has already been made in Section 4.2.2 to the proposals of Kolbusz for the use of waste heat from power stations as a heat source for domestic heat pumps, but there are a large number of waste heat streams available which originate from within a house or block of apartments.

These may be simply illustrated with reference to Fig. 4.16 (ref.4.23), in which both conventional air-air heat recovery systems (ref.4.28) and heat pumps are used in a house to minimize energy consumption for space and water heating. (Although conventional heat recovery systems are proposed for the heat contained in extract air, there is no reason why this heat

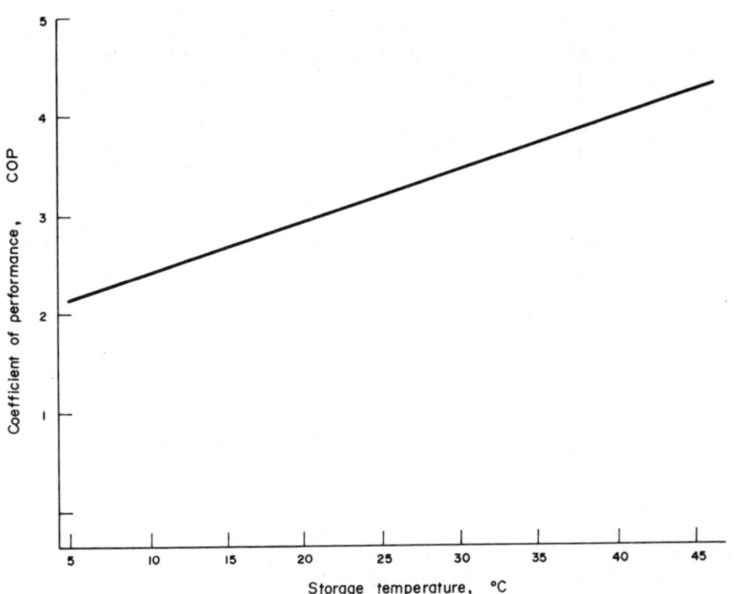

Fig.4.11 Overall System COP as Affected by Heat Storage Temperature for a Typical Installation

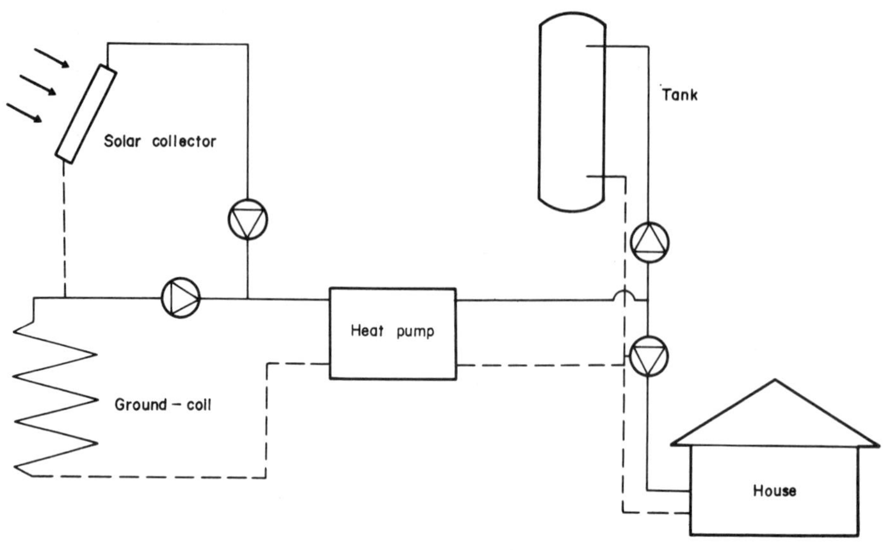

Fig.4.12 Layout of a Solar-Assisted Ground Source Heat Pump for Domestic Heating

should not be used to supplement the input to a heat pump, as will be shown later). A number of different sources of warm air at various temperatures and frequency of availability can be considered. While heat available from, for example, flue gases or cooker extractor hoods, or even general kitchen exhausts, may be of a sufficiently high grade to justify the use of conventional waste heat recovery techniques as illustrated in Fig.4.16, such extracts may be mixed with lower grade exhaust heat from room ventilation systems to provide a heat source for a heat pump evaporator. This may be done either by directly passing the air over the evaporator, or, as will be illustrated later, using a heat exchanger to preheat fresh make-up air as it approaches the evaporator coil.

The most common proposals for waste heat recovery in the home associated with heat pumps normally use waste water as a heat source. As illustrated in Fig.4.16, the intermittent nature of this effluent necessitates the use of a heat store. Heat sources feeding this store include a bath or shower, dishwasher, and washing machine, and the heat is upgraded to feed the domestic hot water tank. Other suggestions (ref.4.29), have included the recovery of heat from domestic sewage to supplement these sources. Many proposals do of course provide for further heating of the storage medium by solar collectors.

Recent work at Ontario Hydro in Canada (ref. 4.30) has borne out much of the earlier studies and experiments carried out in the UK and elsewhere. In a study of residential heat pumps

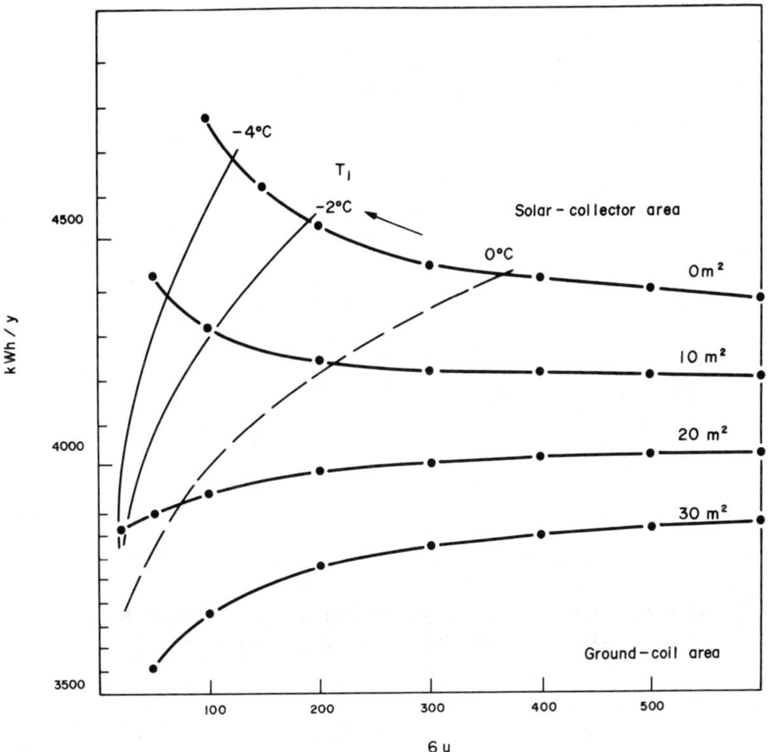

Fig.4.13 Compressor and Circulating Pump Work as a Function of Ground Coil and Solar Collector Sizes

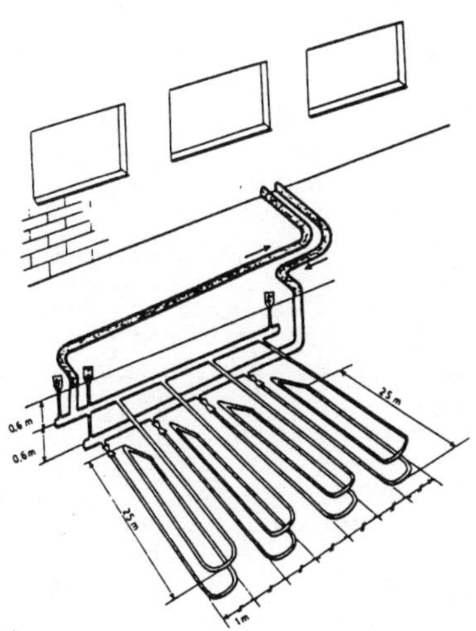

Fig.4.14 Lay-out of the Horizontal Soil Heat Exchanger

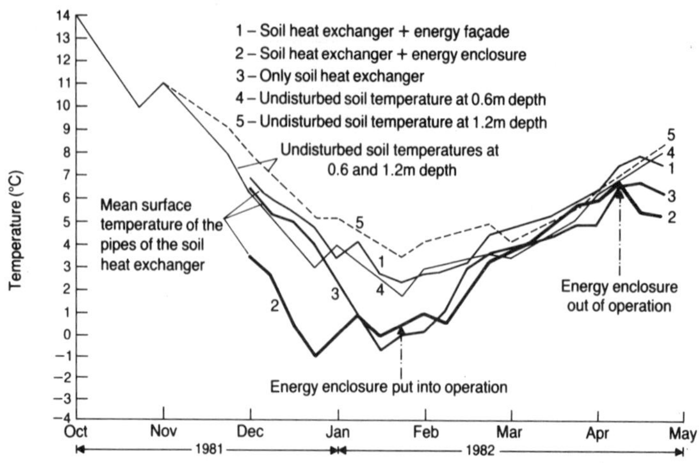

Fig.4.15 Temperatures at Various Locations in the System

and energy efficiency in general, it was concluded that heat recovery from exhaust air to heat domestic hot water could be an effective alternative to conventional electric resistance water heating (the immersion heater). Using ventilation extract systems similar to that shown in Fig.4.16, an air-water electric heat pump of 1.4 kW has been evaluated where the heat sink is a water storage tank. The tank was also provided with electric heating elements, giving a total water heating capacity of 2.9 kW.

A water temperature for domestic use of 60°C was assumed, and a COP of 2.98 was achieved when the average tank 'cold' temperature was 10°C. In the mode of operation involving the heat pump alone, the recovery rate of the water temperature was significantly slower than with conventional immersion heaters.

Energy savings varied from 3.15 kWh per day to 7.77 kWh per day, depending on whether the electric heating elenents were used to accelerate tank warm-up rates. The peak energy saving represented a cut in consumption of 42 per cent, and corresponded to a daily water usage of 227 litres.

The researchers also studied air-air heat pumps for space heating, again based on heat recovery from ventilation extract air. Interestingly, a desuperheater in the heat pump circuit was able to provide 1.7 kW towards domestic hot water heating. A similar concept is described in Section 4.4.2.

As with many heat pump systems in the domestic environment, a major reliance is put on 'top-up' direct electric heating. Where electricity is relatively cheap, this may be acceptable, but in the UK and much of Europe the commercial success of such systems must await a further 'energy crisis'.

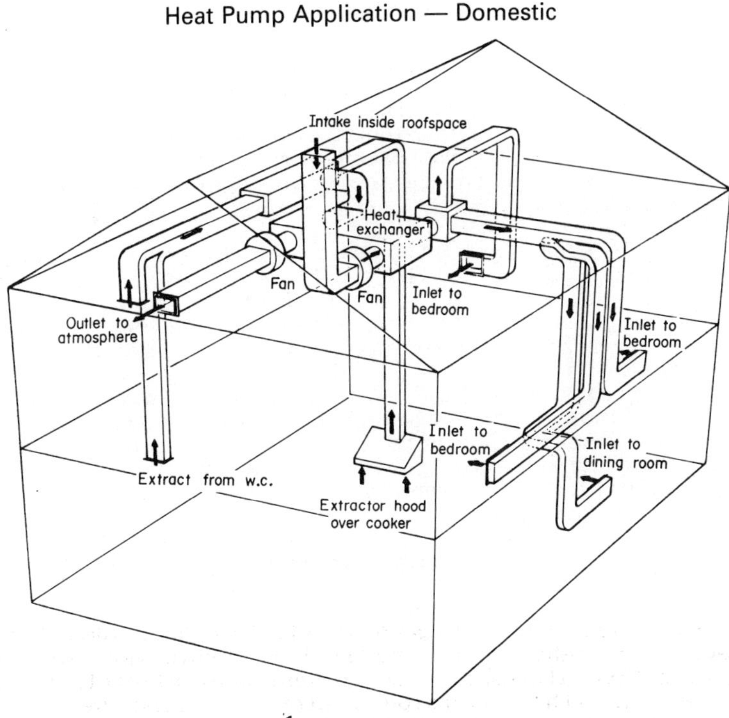

Mechanical ventilation system

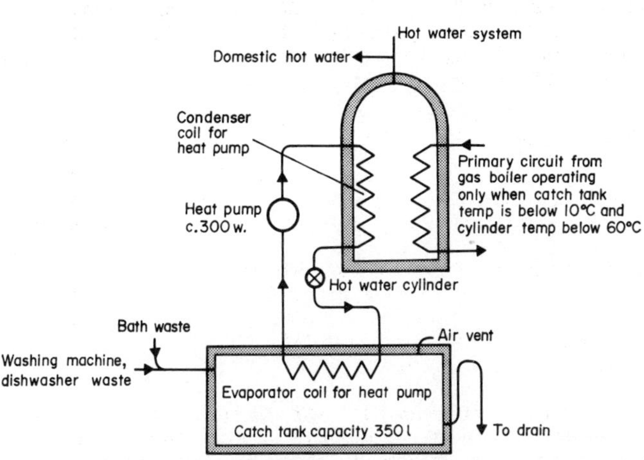

Heat pump operates whenever the cylinder requires heating (i.e. cylinder temp < 60°C) provided that the catch tank temp is not below 10°C.

Hot water system

Fig.4.16 The Building Research Establishment Low Energy House Proposal, Utilizing Internal Waste Heat Sources for Conventional Heat Recovery and Heat Pump Systems

4.3 Heat Storage

The storage of heat, and its economic implications for the domestic heat pump, are subjects of some considerable debate, and no clear answer can be given yet to the question: Should heat storage be used in the home to supply heat to the heat pump evaporator?

As has already been shown, the heat sources available for domestic heat pump systems, if they are to be commercially attractive to manufacturers wishing to produce in quantity, are susceptible to seasonal and daily temperature variations which unfortunately lead to lower COP's at times when the building heating demand is greatest (in cold weather with low solar gain and during the night). The expense of direct supplement electric heating, as advocated by many heat pump manufacturers, has led to the search for alternative solutions to this problem, and the storage of heat collected during the summer, or on winter days when solar energy can be effectively utilised, is a promising technique.

It is not proposed to go into great detail on the subject of heat storage in this book, and for a more detailed discussion the reader is referred to the Bibliography. However, the two basic types of heat storage will be briefly described, and, using work carried out in the USA, an illustration of the use of heat stores in conjunction with a heat pump is given.

The basic forms of heat store, and their relative properties, are given in Table 4.2 (ref.4.25), with a 20°C temperature rise.

TABLE 4.2 Storage Systems - a Comparison of Stores for 1 GJ Capacity

	Rock	Water	Heat of Fusion Material
Specific Heat (kJ/kg°C)	0.837	4.187	2.09
Heat of Fusion (kJ/kg)	-	-	232.6
Density (kg/m³)	2242	1000	1602
Weight (kg)	59737	11941	3644
Relative Weight	16.4	3.27	1
Volume (m³)	26.6	11.94	2.274
Relative Volume	11.69	5.25	1

4.3.1 Sensible heat storage

Sensible heat storage media are normally either liquid or solid, and do not undergo a phase change when heat is introduced to or removed from them. Instead they undergo a temperature change.

The most common sensible heat storage medium is water. Apart from being readily available, water as a heat storage medium has one unique advantage in that it can be removed from the store for use without the necessity for a heat exchanger (unless it is used to heat a ducted air system). In the context of heat pump use, the store may therefore be heated by heat removed from the condenser for later direct use for radiators or domestic water services. Alternatively a heat exchanger served by a water-cooled solar collector in the store could boost store temperature, which would then serve as a heat source for the evaporator of a heat pump. It is possible to ring the changes with heat storage media, and many system concepts may be considered.

Water as a heat storage medium can be stored in metallic, concrete or plastic storage vessels. Where temperature rises are small, the volume of water required to store sufficient heat for practical domestic requirements can be large, and volume and cost requirements have inhibited adoption on any large scale. It is interesting to note in this context that many of the 'low energy' house proposals incorporate the storage tank, whatever the medium used, below the floor of the dwelling. This would be very difficult to implement in most existing buildings.

Solid sensible heat storage media have been routinely used for many years in the home. Heat storage radiators, normally employing a brick material as the storage medium, interspersed with electric heating elements, have in the past been very popular in the United Kingdom. The main argument for the adoption of these heating systems centred around the fact that the electricity generating authorities were able to offer significant reductions in the cost to the consumer of electricity used during 'off-peak' demand times, when the stores could be charged up. As heat storage media in such system, they were charged to a higher temperature than possible with water, but there is no reason why solid storage media should not be used for storage of low grade heat in conjunction with heat pumps. The ground of course is such a storage medium, operating on a much longer time cycle than the electrical system described above.

As shown in Table 4.2, however, a rock as such is less attractive than water because of the larger volume and mass needed to store 1 GJ at low temperatures.

4.3.2 Heat of fusion storage media The use of the heat of fusion of materials for heat storage, as with sensible heat storage techniques, is determined largely by the volume capacity of the material. In the case of latent heat of fusion materials, however, the volume needed is a function of the heat of fusion of the material used. Unlike a sensible heat storage medium, the heat of fusion system gives up or absorbs heat at a constant, or near constant, temperature. In order to do this, however, it must change its phase and this may also be associated with a change in volume. When receiving heat the material melts, and it rejects this heat when it resolidifies. Latent heat of fusion materials in general offer a greater heat storage capacity per unit volume than sensible heat stores, and typical performances of a number of the more common media of this type are given in Table 4.3, (refs. 4.32, 4.33).

It can be seen from the table that heat of fusion media can be selected to cover a reasonable range of operating temperatures appropriate to domestic use. However, with the notable exception of Glauber's Salt, sodium sulphate decahydrate, they are comparatively expensive and are unlikely to find widespread application for some time. Development of new heat of fusion media is a subject being followed by many laboratories, however.

4.3.3 Heat pumps employing heat storage Many studies have been carried out on the affect that heat storage might have on the performance of a heat pump for domestic heating and air conditioning, and most of these have also been associated with solar collectors. One of the most comprehensive studies reported, incorporating a full economic analysis, is that by Gordon concerned with a proposed installation in Buffalo, New York (ref.4.34). Applied to the house shown in Fig.4.17, the system is designed to make use of the most readily available and broadly applicable solar and HVAC components, in conjunction with electricity energy.

Fig.4.17 Artist's Impression of a House Employing Solar Collectors Serving a Heat Pump/Heat Storage System

The primary system components include:

(i) Approximately 65 m² of single-glazed flat black-coated liquid circulating copper solar collectors

(ii) A non-pressurized water storage tank having a capacity of about 9000 litres, the tank being a lined concrete vessel.

(iii) An air heating coil using circulating water for direct solar heating.

(iv) An air-air split system heat pump for solar-assisted heat pump operation and supplementary heating.

The Buffalo area of New York was selected for this study because it is representative of northern climates where solar heating might be applied. The primary heating load is of the order of 4000 degree days and the availability of solar energy is relatively low, averaging about 800 kJ per square metre per day on a horizontal heating surface during the primary heating season (October-April). The emphasis on electrical energy for system operation and supplementary heating is created by local conditions restricting the distribution of natural gas.

The heat pump selected for the house was the Westinghouse Hi-Re-Li split system type HL036C and AG010, being able to operate with evaporator temperatures varying between -27°C and 32°C. Fig.4.18 shows the operation of the system when using the heat store to supply heat to the heat pump.

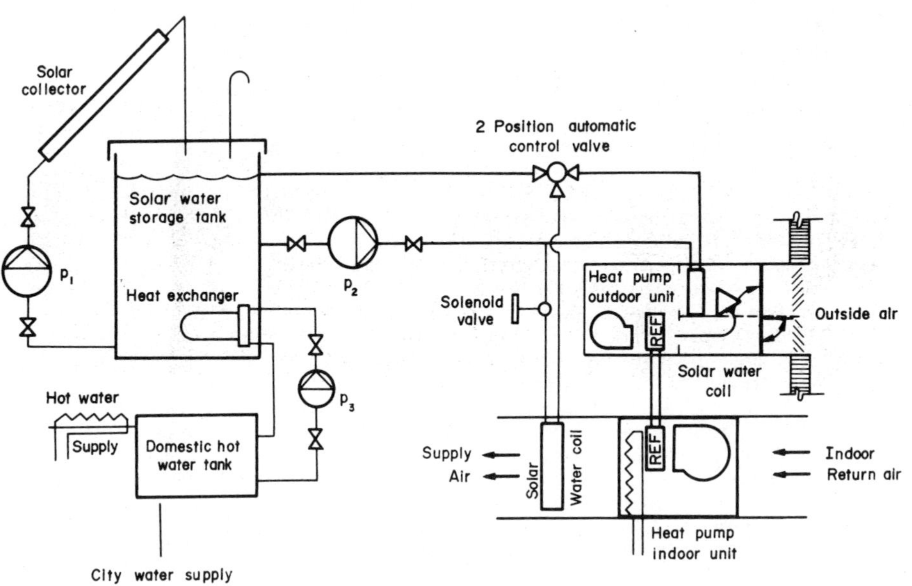

Fig.4.18 Use of a Water Storage Tank, Heated by Solar Energy, to Assist Air-Air heat Pump Operation

(The heat pump may operate conventionally, using the air source, or the solar collector may, as a third alternative, feed the air heating coil directly). In the mode under discussion, pump P_2 circulates water from the storage tank through the water-air coil located in the ductwork of the evaporator side of the heat pump. The adjustable damper is positioned to recirculate a stream of air from the centrifugal fan of the evaporator section of the heat pump over the evaporator coil. The heat thus rejected at the condenser is circulated in the form of ducted warm air in the dwelling. In this mode the automatic control valve is set to return water from the coil to the water storage tank.

TABLE 4.3 Heat of Fusion Storage Media

Material	Melting Temp °C	Density kg/m³	Heat of Fusion kJ/kg	Heat of Fusion J/cm³
$NaC_2H_3O_2,3H_2O$	58	1297	265	340
$Na_2S_2O_3,5H_2O$	48	1650	209	344
$Ca(NO_3)_2$, $4H_2O$	47	1858	154	283
P116 (wax)	47	785	209	163
$FeCl_3$, $6H_2O$	36	1617	223	359
Na_2CO_3, $12\ H_2O$	36	1522	265	400
$LiNO_3$, $3H_2O$	30	-	307	440

The energy requirement for operation of the solar assisted heat pump was estimated to be 11342 kWh per year, compared to a consumption of 36832 kWh per year for electric heating alone. With a unit electricity cost of 2.29 US Cents, the solar system with heat storage saves about $653 per year giving a payback period of the order of 14 years. The solar system, acting alone or in conjunction with the air-air heat pump, provides up to 70 per cent of the annual space heating requirements.

4.4 Experimental Domestic Heat Pump Systems

As described in the next Section, (4.5) a large number of heat pumps are currently available for use in the home. Most have originated in the United States, where reversibility provides year-round air conditioning, but some are for heating only. However, while these systems are relatively efficient, they are designed largely for retrofitting to existing structures, and it is impractical to use as many heat sources as possible in what may become a 'total energy' house as far as heating and air conditioning are concerned. Recently the emphasis has been on the design of low energy houses, where full use is made of double glazing, ventilation control, thermal insulation etc to minimise heat losses. It is here that new heat pump systems have a major role to play, and some of the experimental and demonstration-type installations investigated during the past ten years are described here. Further examples can be found with reference to the Bibliography.

4.4.1 The Pennsylvania Power & Light Co house The Pennsylvania Power & Light Company propose an experimental house, the aim of which is to achieve energy savings of 50 per cent over conventional homes. The heat pumps used in this experiment are water-air and air-water units, and the overall layout is illustrated in Fig.4.19, (ref.4.29).

The heat pump, of course, as discussed above, forms only part of a total system which is designed to achieve energy economics. This has necessitated design and siting of the house to take advantage of sunlight, shadows and natural breeze ventilation. Efficient lighting (fluorescent and mercury vapour) and collection of waste heat. The home will be located in Pennsylvania, the indoor living space being 147 m².

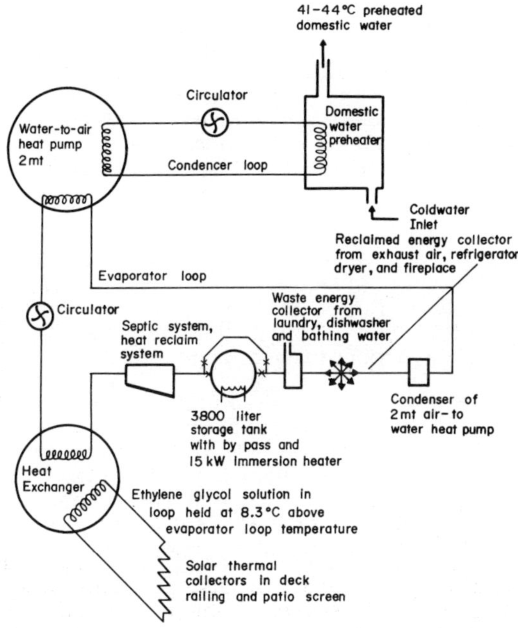

Fig.4.19 The Pennsylvania Power and Light Company Experimental House, Incorporating Heat Pumps

The overall system is designed around the two heat pumps, supplemented by solar collectors. The utility believes that one third of the energy requirements will be furnished by the heat pumps, one third by the heat recovery systems, and the balance by the solar collectors. The predicted energy savings are based on the assumption that the conventional home would also be 'all electric'.

Appliances used in the house include clothes washer and dryer, dishwasher, and electric cooker. Insulation, in the form of 2.5 cm thick Styrofoam wall fillings with urethane around the framing areas, will be provided. Special features incorporated in the energy recovery system include:

(i) Fence-line solar collector to provide supplementary heating

(ii) A heat recovery system to use waste heat from the refrigerator and clothes dryer (in the form of warm air)

(iii) A heat pipe (ref.4.31) to assist recovery of heat from bath and shower waste water

(iv) Heat recovery from the sewage system.

The water source heat pump is the basis of the system, providing heat for domestic water and the HVAC system. This primary heat pump transfers energy from a water loop which draws heat from a number of sources, including the solar collectors, sewage system, storage tank (having a capacity of 3785 litres, supplemented using a 15 kW immersion heater), waste water from bath and shower, a heat pump condenser, and a heat exchanger recovering heat from air exhausted by the refrigerator and clothes dryer.

4.4.2 The Lennox solar-assisted heat pump   Reference has already been made in this Chapter to the use of solar energy to provide heat for the evaporator of an air-source heat pump. An experimental installation utilizing a heat pump manufactured by Lennox Industries Ltd should demonstrate the benefits to be gained with such a preheating system.

The argument for a solar collector to aid the air-source heat pump is solely based on the fact that as the ambient air temperature falls, the performance of the conventional heat pump becomes less attractive, and solar energy is one heat source which may be used to counteract this deficiency. In an air-source heat pump, used in conjunction with a solar collector, the solar collector is an integral component within the house construction. As illustrated in Figs. 4.20 and 4.21, in its simplest form it is an air duct under the roof cladding.

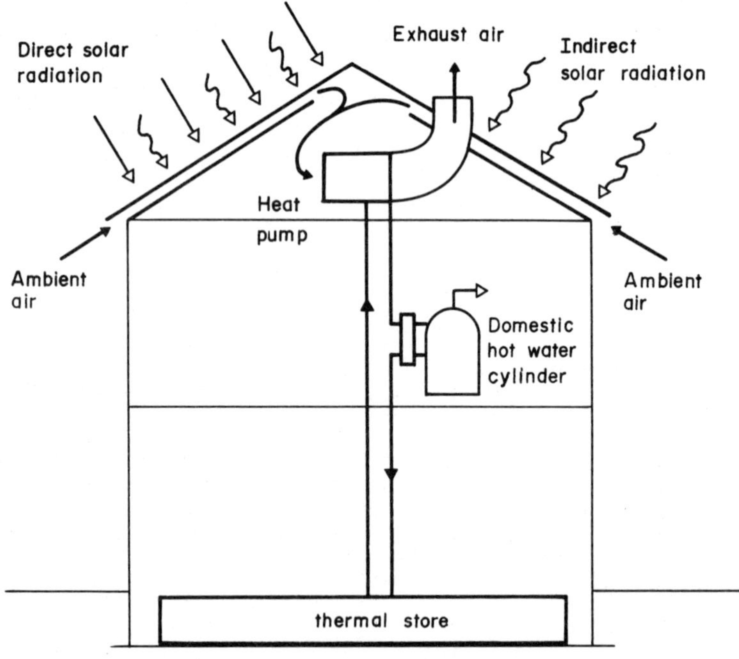

Fig.4.20 Schematic of the Lennox Solar-Assisted Heat Pump (Air Source) Applied in a House

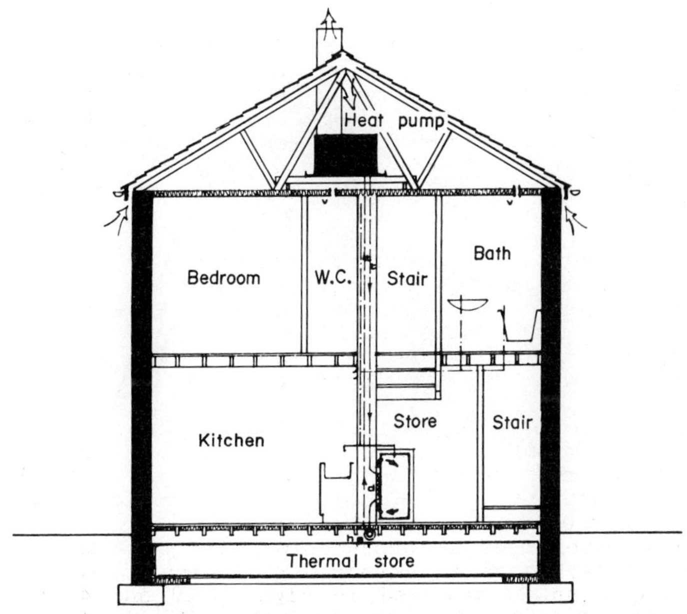

Fig.4.21 Architect's Drawing of a Lennox Installation

It is argued that since collection takes place at or near ambient temperature, re-radiation losses are considerably reduced, compared to other collection systems, and relatively high efficiencies can be achieved, at a low cost, particularly if designed into the building at an early stage. Lennox suggest that single glazing could be added on one roof face to provide considerably increased efficiency, at a higher capital cost.

The heat pump in this system extracts heat from the air and uses this energy to raise the temperature of a thermal store located below the house (see also Section 4.3). As illustrated in Fig.4.22, the thermal store permits the heat pump to operate only when the ambient temperature is sufficiently high to justify economic operation.

As this occurs, even during the heating season, when the solar gain is also at a relatively high level for a substantial part of the time, a useful rise in the air temperature fed to the evaporator can be achieved. An average rise of 2.5°C above outside ambient using this solar collector is feasible, leading to an increase in heat pump efficiency of between 8 and 10 per cent. This represents a payback period of less than 5 years.

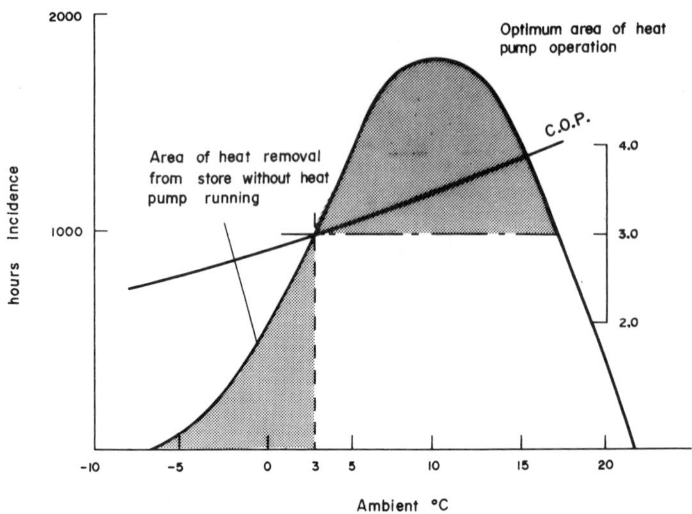

Fig.4.22 The Effect of a Thermal Store on the Economic Operation of an Air-Source Heat Pump

The function of the heat store is to provide space heating during periods of low ambient temperature and this would of course be done without running the heat pump (e.g. during the night). It is claimed that such a store could maintain comfort conditions within a house of the type illustrated for a nine day period in an ambient temperature of 0°C. (Even in the UK, however, an installation of this sort would still include some form of supplementary heating). Other advantages claimed for the system are that it will ease peak demand on electricity generating stations, and could, at low cost, protect unoccupied properties from damage due to frost and damp over extended periods.

Heat removal from the heat store, could be via a liquid or air, to heat low temperature ceiling panels or radiators, or a ducted warm air system. The architectural company, Triad, was associated with the overall system design of this installation. The system is in many ways similar to that described in Section 4.2, used in Germany, although the Lennox unit is somewhat smaller. Described later, solar assisted heat pumps using conventional flat plate collectors are now commercially available from a number of manufacturers.

4.4.3 The Philips experimental house The Philips experimental house in Aachen, West Germany, is a major part of a programme on solar energy set up by the Philips Research Laboratory and RheinischeWestfalische Elektrizitatswerke, of Essen, specifically to cover systems for the utilisation of solar energy and for the efficient use of energy in building. One of the major motivations behind the project was that an analysis of the consumption of energy in West Germany showed that 50 per cent of all primary energy is used in the form of low grade (less than 100°C) heat. The greater part of this heat is used for space heating in buildings, the balance being for the supply of hot water. About half of the electrical energy generated is also used to provide low grade heat, the major part being for domestic use.

It was argued that the development of energy saving techniques for space heating in houses could therefore significantly benefit the overall energy situation in Continental Europe, and four main measures, all used in the experimental house, were proposed. These were:

(i) Reduction of building heat losses

(ii) Heat recovery from waste water and air

(iii) Use of alternative sources of energy which are not harmful to the environment (e.g. solar and ground)

(iv) Development of integrated energy systems for buildings.

The layout of the Philips Experimental House is shown in Fig. 4.23. Much in the way of energy savings was achieved by thermal insulation, controlling ventilation, and double glazing. This, combined with heat pump and other systems led to an energy demand for the house only one sixth that of a normal dwelling, and one third that of a modern well-insulated home.

The heating energy requirement was of the order of 8300 kWh, and was provided by solar energy and heat stored in the ground. Solar collectors feed a water storage tank having a capacity of 42 m^3. Between 10,000 and 12,000 kWh can be stored in this tank annually. By recovering heat from waste water, the electrical energy requirement for domestic water heating can be reduced by about 50 per cent.

In this house the heat pump used a number of heat sources, visible in the figure. A ground collector consisting of 120 m of plastic pipe buried beneath the floor of the house provided heat for one evaporator at 7°C, and this was upgraded to supply the domestic hot water tank at 50°C. A 1.2 kW heat pump, operating with a COP of about 3.5, was able to supply something over 3.5 kW. (Cooling of the house was carried out in summer, but not by means of the heat pump - in this instance the warm air is cooled by circulating it along the celler walls). Waste water was also used as a heat source for the heat pump. By recovering heat using both the heat pump and conventional liquid-liquid heat recovery equipment, the annual energy usage for water heating (involving service water, washing water, and the dishwasher), can be reduced from 3980 kWh to 980 kWh. (It is interesting to note that the total energy consumed by electrical equipment which may be regarded as sources of warm air was 2915 kWh per year in this house - this equipment includes a tumble dryer, deep freeze, refrigerator, cooker, lighting and a television set). Warm air heating of the rooms was aided by heat recovery from the exhaust air using a rotating regenerator.

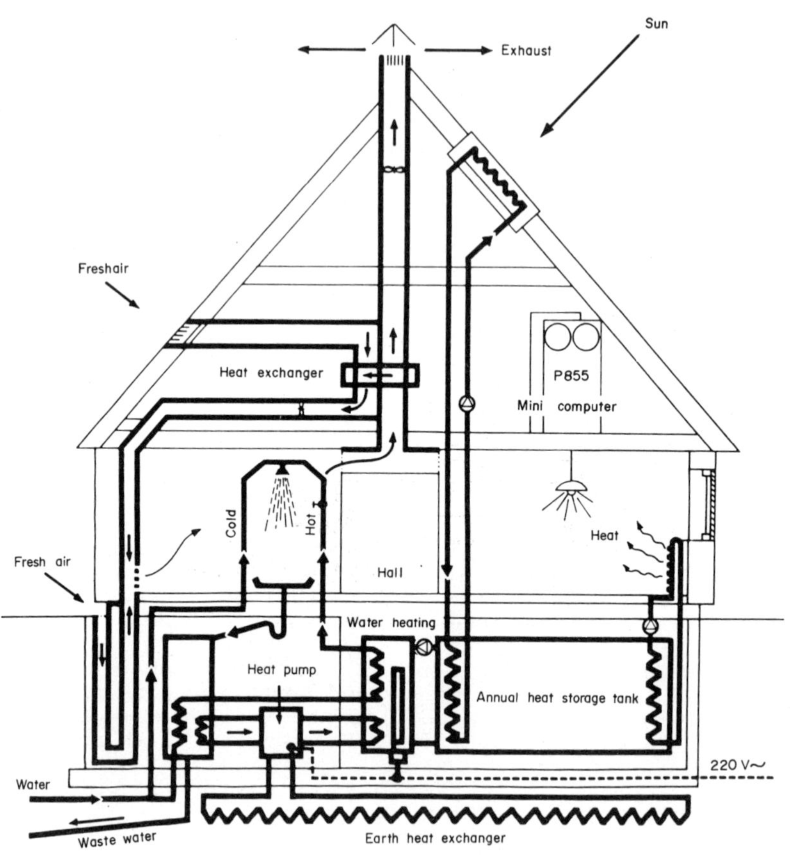

Fig.4.23 The Philips Low Energy House in Aachen - a Layout of the Air and Water Distribution Systems, the Latter using Heat Pumps

4.4.4 Heat pump development for domestic appliances As discussed earlier in this Chapter, the heat pump may be used to recover waste heat from individual domestic appliances. Indeed some of the earliest domestic heat pumps introduced onto the market made use of heat recovery from refrigerators, or the more basic cooled larder, for space heating.

The uses of upgraded waste heat in the home are threefold - use for space heating, water heating, and, more rarely, reuse within the appliance itself. Also discussed in Chapter 6, in the context of industrial drying processes, work published in the Netherlands on the application of heat pumps in domestic tumble dryers is of particular interest as an example of the last category above, where the heat pump is used to recover heat from a high humidity exhaust for preheating dry air entering the tumble dryer. By means of a closed cycle system, significant energy savings may be achieved, (ref.4.35).

Current designs of tumble dryers use hot air, either in a closed cycle in which the humid exhaust is cooled via water or cold air prior to reheating and recirculation, or in an open cycle, where the hot exhaust is rejected directly to atmosphere. The operation of a close cycle tumble dryer is shown diagramatically in Fig.4.24 and Table 4.4 details the most important operating characteristics of such a dryer, (1st column). Note that the additional 10 minutes operating time over and above the 62 minutes used for drying is allocated to cooling of the garments, and some energy is consumed during this period in driving the cooling air fans. The energy to provide heat for drying is in the form of electricity.

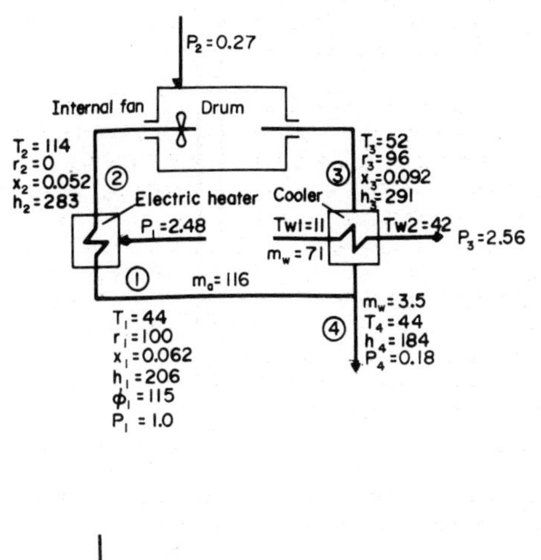

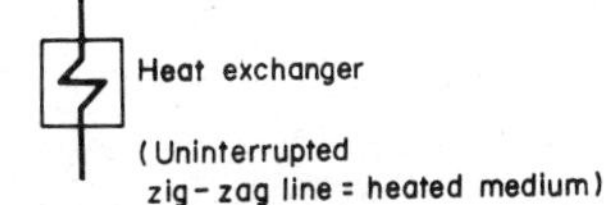

Fig.4.24 Flow Diagram of a Conventional Closed Cycle Domestic Tumble Dryer

By applying a heat pump, as shown in Fig.4.25, the energy consumption can be considerably reduced.

TABLE 4.4 Operating Characteristics of a Tumble Dryer

		Standard Tumble Dryer Close Cycle	Tumble dryer with heat pump	
			Design (Closed Cycle)	Experiment (Open Cycle)
Laundry Load	kg	3.7	3.7	3.7
Starting Humidity of Laundry	%	71	71	71
End Humidity of Laundry*	%	1.0	0.0	1.3
Amount of Water to be Evaporated	kg	2.6	2.6	2.6
Drying Time	min	62	62	55
Total Operating Time	min	72	72	65
Power Demand	kW	2.75	1.36	1.72
Energy Consumption	kWh	2.88	1.49	1.56
Specific energy consumption	kWh/kg	0.78	0.40	0.42
Maximum air Temperature	°C	114	114	72

* A humidity of 0% means that the cotton is in equilibrium with an atmosphere of 20°C and a relative humidity of 65%

The humid hot air leaving the clothes compartment of the tumble dryer is passed over the evaporator coil of the heat pump, where it is cooled. This cooling process, together with supplementary cooling, as applied on the standard dryer, removes moisture from the exhaust air, making it acceptable for recirculation. Thus the evaporator coil is able to pick up latent as well as sensible heat from the exhaust air. The recirculated air is passed over the condenser coil where it absorbs the heat recovered by the evaporator and the heat of compression. The second column in Table 4.4 details the theoretical performance capability of such a closed cycle tumble dryer incorporating a heat pump, and it can be seen that an energy saving of about 48 per cent is achievable.

Tests have been carried out on an open cycle dryer incorporating a heat pump, although some of the benefits are lost if recirculation is not employed. As may be seen from the table, the maximum air temperature in a standard tumble dryer is of the order of 114°C (measured at the dryer inlet). In order to achieve such temperatures using the heat pump, a condensing temperature on the refrigerant side of the coil of about 125-

130°C would be needed. In the experiments it was originally proposed to use either R11 or R113, both of which have high critical temperatures, but no hermetic compressor unit was available and the specific heat output would be low. R12 was eventually selected, although its low critical temperature (112°C) meant that the conventional operating temperature would be unachievable. The air flow through the heat pump dryer was therefore increased to compensate for the reduced operating temperature, a maximum air temperature of 72°C being reached.

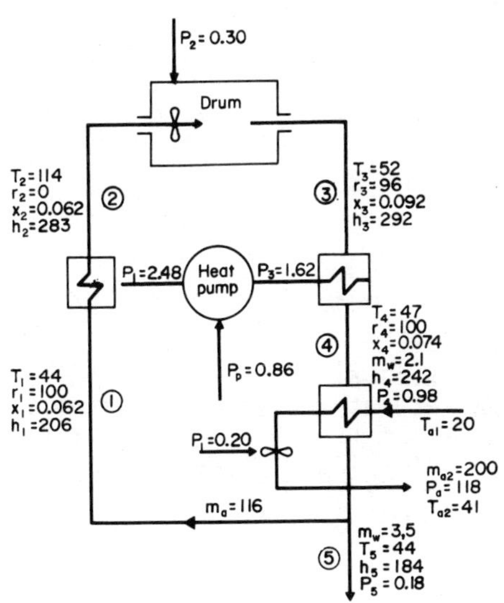

Fig.4.25 Flow Diagram of a Tumble Dryer Operating on a Closed Cycle and Incorporating a Heat Pump to Preheat Air using the Exhaust Humid Air

4.4.5 The absorption cycle domestic heat pump When the First Edition of this book was written, the forecats for the advent onto the domestic market of absorption cycle heat pumps (see Chapter 2) were suggesting massive penetration in the 1980's. This has not happened.

A number of reasons exist as to why the absorption heat pump has not proved successful in this context to date. Firstly, energy prices have not increased at the rate predicted in the 1970's. Secondly, high efficiencies have been achieved using condensing boilers, which are available at a comparatively modest capital cost premium. This means that the energy savings of a high cost absorption cycle heat pump having a PER of 1.3 has to be compared with boilers having efficiencies of 90 per cent or more, rather than non-condensing boilers of 60-70 per cent efficiency. The cost and, to a lesser extent, the complexity of early domestic absorption cycle heat pump models have also been instrumental in negating the optimism expressed some years ago.

Several major research and development programmes are now under way in Europe, Japan and North America directed at the ultimate production of lower cost, efficient and reliable absorption cycle heat pumps, normally fired using natural gas. The reader is recommended to consult references 4.36 and 4.37 for data on some of these developments.

4.5 Commercially Available Systems

As listed in Appendix 3, the number of manufacturers of heat pump systems, particularly in the domestic and commercial building sectors, remains high. The purpose of this Section is to present data on a number of commercially available heat pumps used in houses and apartments, with a broad geographical spread as far as applications are concerned. The emphasis on innovation in some of the European products is not to detract from the mass-produced units available from the United States. It serves more to highlight the techniques necessary to ensure that domestic heat pumps used in European and UK conditions can function economically and have a reasonable return on the investment, where utilization, restricted in many cases to the heating season, is lower than reverse cycle systems designed for year-round air conditioning.

4.5.1 The simple air-air reversible heat pump There is no doubt that the largest market for air source domestic heat pumps of the reversible type exists in the United States, and a large number of manufacturers are able to meet the market requirements. Two facets of this type of heat pump are illustrated by the Carrier Corporation units.

As discussed earlier, the heat pump may be in the form of a single package or may be 'split'. The Carrier product range offers both configurations. The single package unit is illustrated in Fig.4.26, showing the compressor, evaporating and condensing coils in a single ground-mounted case adjacent to the outside wall of the dwelling.

Space conditioning is carried out in this case via underfloor ducted air heating, with return and supply air ducts leading to the outside condenser coil. The heat pump package may alternatively be wall or roof-mounted, for convenient connection to the interior ductwork terminations. A defrost system is incorporated, responding to ambient temperature and being time-controlled. Also included as standard in the factory package are the evaporator and condenser fans, crankcase heater (to prevent dilution of the oil by the refrigerant during shut-down), and suction line accumulator to inhbit flood back of refrigerant affecting the compressor.

Optional extras on such a heat pump are largely associated with the control side. However, electric resistance heaters, for use as supplementary heating, are available from the manufacturer if no other internal heat supply exists. In the case of the Carrier system, these electric heaters would be mounted, depending on size, either inside the heat pump cabinet or on the outside, attached to it. Several thermostat options include a unit for bringing the supplementary heaters on line when the heat pump may be affected by breakdown. An outdoor thermostat package also energises the electric heaters when the outdoor temperature drops to a certain level. A further thermostat can be incorporated indoors to permit selection of the heating or cooling cycle, or continuous or automatic fan operation.

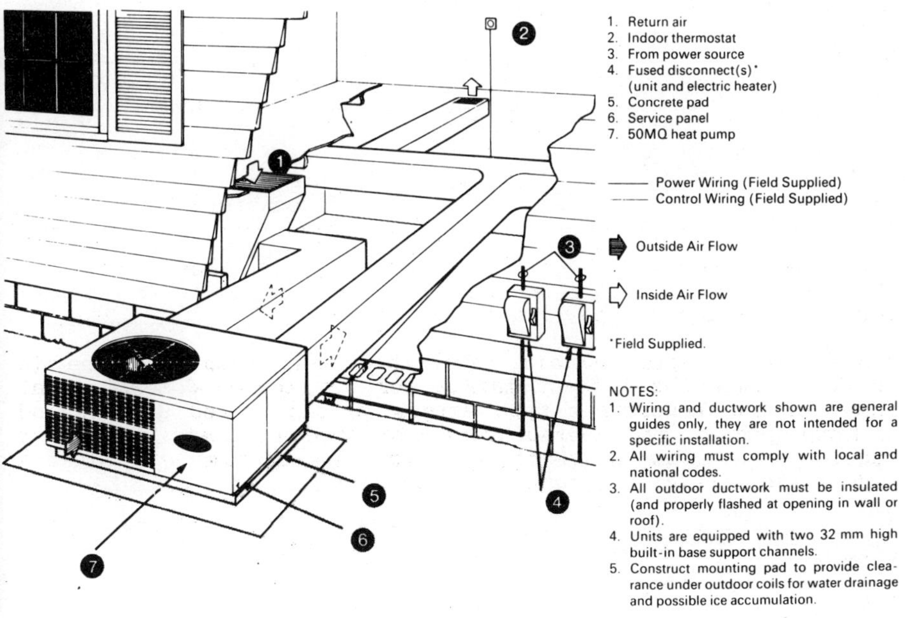

Fig.4.26 Single Package Carrier Domestic Reversible Heat Pump Installation

Performance and physical data on this range of heat pumps is given in Table 4.6. (Note that 1 kcal/h = 1.163×10^{-3}kW). The performance is based on the following conditions.

Cooling - Dry bulb temp. indoors = 26.7°C
Wet bulb temp. indoors = 19.5°C
Dry bulb temp. outdoors = 35°C

Heating - Dry bulb temp. entering indoors = 21.1°C
Dry bulb temp. outdoors = 7.2°C
Wet bulb temp. outdoors = 6.1°C

TABLE 4.6 Performance and Physical Data on Carrier Single Package Units

Unit		50MQ027	50MQ037	50MQ047
Normal cooling capacity	kcal/h	6680	9200	12600
Nominal heating capacity	kcal/h	7450	9200	12350
COP (heating)		2.5	2.5	2.4
Operating weight	kg	152.5	169.6	174.6
Refrigerant			R-22	
Operating charge	kg	3.0	2.9	4.0
Flow control			Bi-flow AccuRater	
Compressor			Hermetic, 2 cylinders, 2900 rpm	
Outdoor fan			Propeller, direct drive, 1200 rpm	
Discharge			Vertical	
Motor power input	kW	0.19	0.19	0.19
Outdoor coil			Plate fin	
Rows...fin spacing	mm	2...1.5	2...1.5	2...1.5
Face area				
inner coil	m²	0.73	0.85	0.66
middle coil	m²	-	-	0.85
outer coil	m²	0.77	0.89	0.89
Indoor fan			Centrifugal, direct drive	
Discharge			Horizontal	
Nominal air quantity	m³/h	1690	2340	3190
Air quantity range	m³/h	1360-2170	1870-2720	2470-3400
Motor - power input	kW	0.19	0.37	0.37
- speed	rpm	1100/975/900/825	900/850/800	900/850/800
Indoor coil			Plate fin	
Rows...fin spacing	mm	3...2.0	3...2.0	3...2.0
Face area	m²	0.31	0.43	0.43

In addition to this basic data, manufacturers generally present tabulations showing how performance varies with airflow and outdoor wet bulb coil temperatures. Also, correction factors are presented to take into account changes in outdoor dry bulb temperature.

Identical data is given for split heat pump systems, the layout of which is given in Fig.4.27.

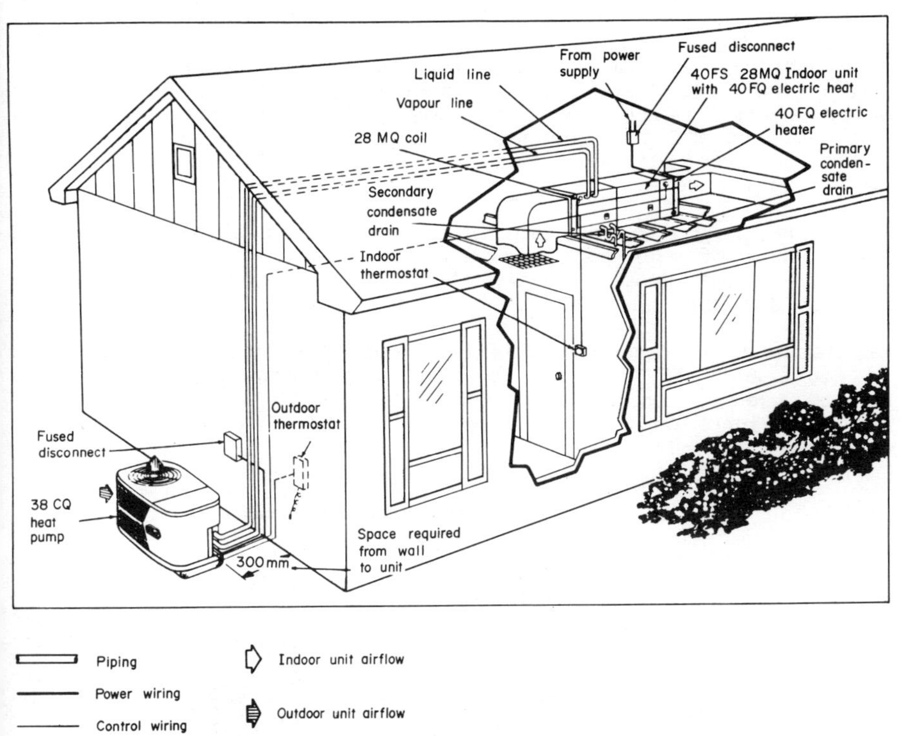

Fig.4.27 Split Heat Pump Layout - Showing a Carrier Installation

4.5.2 Air-water and water-water systems A number of heat pump systems are available for domestic use, based on the use of water as the heat source and/or water sink. Friedrich in the USA manufacture such a unit which can use water in wells, streams, lakes or even city mains as the heat source or, for summer cooling, the heat sink. As in systems described in Section 4.5.1, distribution of the heat within the dwelling is via a coil served by ducted air.

The availability of water as a heat source is not universal, as discussed in earlier sections of this Chapter, and general application is therefore somewhat limited. One area in which the Friedrich heat pump does offer innovation is the provision of a hot water generator for providing domestic hot water. The unit does this by using a refrigerant to water heat exchanger located in the refrigerant line between the compressor and the condenser. Claimed savings on domestic heating costs, when using electricity for both air conditioning and water heating in a conventional system, amount to up to 30 per cent. The heat exchanger is only used when the domestic water temperature falls below a preset level. The division between space heating and water heating is typically 8:1. For the heat pump operating in the space heating mode only, with electric immersion heating for the water, the COP varies between 2.2 and 2.5 (taking into account the energy expended in heating the water also). With the heat pump fulfilling both functions, the combined COP can increase to 2.9. More substantial energy reductions are achieved when the heat pump is utilised in the space cooling role.

By using an air source which is maintained at a temperature above that of the outside air, for example air in the loft or extract air from a house, the Metro heat pump is also able to heat domestic hot water.

Manufactured in Denmark, this heat pump can be mounted as shown in Fig.4.28. In addition to the example shown, the heat pump can be used in connection with a tank with a heat exchanger, or in conjunction with an existing electric water heating system.

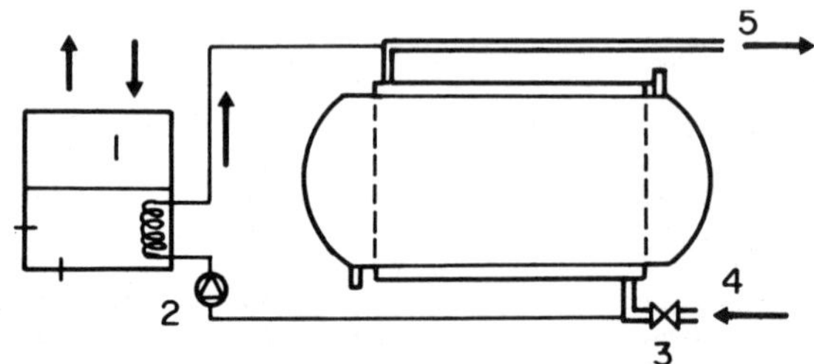

1. Heat pump
2. Circulating pump
3. Stopcock
4. Central heating, in
5. Central heating, out

Fig.4.28 The Metro Air-Water Heat Pump for Supplying a 'Wet' Central Heating System

The quantity of air used is about 300 m^3/h and the water is heated typically to a temperature of about 52°C. Boost heating in one form or another is needed in cold weather. The performance of the system for various ambients is given in Table 4.7.

TABLE 4.7 Performance of Metro Air-Water Heat Pump (Type 3000 MW)

Ambient Temp. (°C)	Heat Pump Energy Consumption (W)	Heat Output (W)	COP
25	580	1500	2.6
10	430	980	2.3
-5	325	560	1.75

Other heat pumps in this series are available as water-water units using either solar collectors or ground coils to supply heat to the evaporator via circulating water. Similar performances are achieved.

4.5.3 Tests on multiple heat pump installations Two case studies involving the systematic examination of a number of heat pumps under operation, are of interest. The first gives data on electric drive units used in Hawaii for water heating (involving a variety of heat sources). The second study involves the testing of 40 internal combustion engine-driven heat pumps, as part of a demonstration project in West Germany funded with assistance from the European Commission.

Heat Pumps for Water Heating in Hawaii: The escalation in oil prices was of particular significance to power generation authorities who relied principally on this fuel as their source of energy. In the case of the Hawaiian Electricity Company, 90 per cent of the energy consumed in the area covered by this operation derives from crude oil or petroleum distillate fuels imported via tankers. As a result of the increase in energy costs, a significant amount of work was done by the Consumer Services Department of the local electric company to introduce and evaluate waste heat recovery and the use of heat pumps for water heating purposes, (ref.4.38).

The typical air conditioned apartment block in Hawaii has the following energy consumption pattern. Individual apartments use 30 per cent of the total energy consumption, common areas use 10 per cent, and 20 per cent is allocated for central water heating systems. The greatest proportion of the energy is used by a central air conditioning unit, this representing the 40 per cent balance. The waste heat rejected by the central air conditioning systems is typically several times greater than the energy required by the building's water heating system. It was therefore felt logical to use condenser cooling water as a heat source for water-to-water heat pumps, and the tempera ture of this water would enable them to operate with comparatively high Coefficients of Performance. If the energy consumption of these systems is compared with the energy input to existing oil or gas water heaters with overall operating efficiencies of about 67 per cent, something approaching a 90 per cent reduction in energy consumption could be achieved.

The other heat pump system adopted by the utility was the air-to-water unit. These were anticipated as having coefficients of performance between 2.5 (residential units) up to 4 (large commercial units). The residential and commercial air-to-water heat pump water heaters also rely on stored solar energy as a heat source. The climate in Hawaii is such that in general no annual heating degree days are recorded, but between 200 cooling degree days in February and 500 cooling degree days in August and September are common place.

The metered performance of 16 commercial heat pumps used for water heating installed in Hawaii are listed in Table 4.8. With one exception, that installed in a laundromat, all of the heat pumps were retrofitted to apartment buildings and hotels. Full audits were carried out on the system, including typically 12 months pre-installation auditing and ten months post-installation audits. The data collected included the number of kilowatt hours of electricity consumed by the heat pumps, and the gas useage before and after the date of commissioning of each heat pump.

The efficiency ratio was considered to be the most important parameter in comparing the performance of heat pumps with other types of water heating systems. The efficiency ratio, defined in the notes appended to Table 4.9, was said to be the factor that the consumer was most interested in and is claimed to be an accurate indicator of the energy saving that occurred in practice.

The water-to-water heat pumps which are recovering heat normally dissipated by cooling tower loops in central air conditioning systems are detailed in Table 4.9. Average efficiency ratios of 6.7 were found to exist, although the first unit on the list, with an efficiency ratio of 8.1, had a comparatively lower hot water delivery temperature than the other systems, giving a COP approaching 5.

While a number of higher efficiency ratios were recorded, some installations exhibited low values. One installation with an efficiency ratio of 3.5 included the simultaneous addition of circulating water pumps for the building's hot water loop. As a result there were additional hot water piping losses which had to be made up for by the new heat pump. Thus the actual operating efficiency ratio of this heat pump was probably between 4 and 5.

A more detailed examination of the reasons for the low efficiency ratios in some instances led to the conclusion that lack of maintenance was a contributory factor in at least one case. It was recommended that periodic preventative maintenance was an absolutely essential item in a comprehensive energy conservation programme. A gradual deterioration in the efficiency ratio of several installations was noticed and this is attributable to the lack of maintenance. This was also revealed in a rising monthly capacity factor as the heat pump operates for longer periods each day in order to overcome resistances due to fouled or iced up heat exchangers.

TABLE 4.8 Air-Water Heat Pump Performance in Hawaii

Installed Capacity (kW)	Hot Water Temp. (°C)	Efficiency Ratio	Monthly Capacity Factor kWh/kW	Gas Cost (Cents/ Therm)	Electricity Cost (Cents/ kW.h)	First Year's Savings (% of cost)
15.2	49	6.5[1]	388	145	10.7	38
60.3	57	6.1	592	140	11.2	69
60.3	57	5.9	446	141	11.5	36
60.3	60	5.7	493	138	11.6	28
146	57	5.1	352	136	11.3	38
83.5	54	5.1	357	140	10.9	25
38.6	57	5.0	408	144	11.0	23
122	54	4.5	393	138	11.0	25
30.5	60	3.9	493	147	10.6	44
37.5	54	3.5[2]	561	144	10.5	19
30.5	60	3.2	468	148	10.6	20

[1]Laundromat with 5.5°C temperature reduction
Estimated efficiency ratio of 5 at original temperature

[2]Includes additional hot water 100p circulation losses
Actual efficiency ratio estimated to be between 4 and 5

TABLE 4.9 Measured Performance of Water-Water Heat Pumps in Hawaii

Installed Capacity (kW)	Hot Water Temp. (°C)	Efficiency Ratio	Monthly Capacity Factor kWh/kW	Gas Cost (Cents/ Therm)	Electricity Cost (Cents/ kW.h)	First Year's Savings (% of cost)
100	49	8.1	240	137	10.0	41
296	57	6.5	390	136	10.8	72
296	51.5	5.8	360	136	10.2	49
296	51.5	6.4	416	117	9.6	51
573[1]	60	6.2	299	133[2]	10.1	51

[1]Comprises 1 water-water heat pump rated at 351 kW plus 4 air-water heat pumps rated at 55.5 kW each

[2]Gas/electricity cost ratio 13.16, value of same ratio in UK currently 6.4, thus electric heat pump less attractive in UK

Efficiency Ratio - Ratio between the reduction in purchase of utility gas or LPG and the electric energy consumed by the heat pump. (Also defined as COP of heat pump divided by efficiency of gas appliance).

The converse of this could be seen however, in the case of the oldest of the water-water heat pumps in Hawaii. This had operated with a constant efficiency ratio of 6.5, saving its owner in excess of $5,000 per month over its full life. The maintenance contract which catered for periodic cleaning cost less than $30 per month, while a complete preventative maintenance contract for a unit of this size (300 kW) would amount to less than $250 per month, or less than 5 per cent of its monthly savings.

In terms of payback period, it can be seen that the return on investment in water-to-water pumps can be little in excess of 1 year to typically 2 years. In the case of air-to-water heat pumps paybacks are somewhat longer, the most successful being less than 2 years while 3 to 4 years seems more typical. In the exceptional cases a payback of 5 years has been recorded. The difference between the water-to-water and air-to-water systems is principally due to the fact that the heat source for the air-to-water heat pumps is at not such a high temperature as that for the water-to-water units. In addition the air systems require fans for the evaporator coils, requiring an electricity input, although in the case of water-to-water units it appears that in this analysis the cost of any pumping required for evaporator or condenser water supplies has been neglected.

Typical installation data are given in Table 4.10. The Hawaiian Electric Company conducted a number of studies in which the energy consumption of several hundred customers who had installed solar or heat-pump-type water heaters in their houses was recorded. All of the houses were single family homes with all services supplied by electricity, and measurements were made over a 12 month period prior to installation of the new systems, and a similar period of monitoring following commissioning. The Utility actually inserted in the electricity bills to the consumers the estimated heat savings resulting from the installation of heat pump (or solar) energy saving systems. In a house which used typically slightly in excess of 1000 kW/h per month prior to heat pump installation, an average monthly saving of 247 kW/h was achieved, resulting in an annual savings worth $325. (This was based on a cost of electricity of $0.11 per kWh).

The energy and cost savings possible using heat pumps in Hawaii is to some extent unique to these islands. There are not many locations where the installation of an electric heat pump could reduce a $8000 gas bill to a $3000 electricity bill, but the degree to which the installations have been monitored, and the large number of case histories which are available for study, indicate that in many instances heat pumps are viable for space and water heating and their viability rests solely on the relative price of the different fuels available.

TABLE 4.10 Typical Installation Data

1) 1980 retrofit on a 428 unit apartment block. Water-water reciprocating compressor heat pump.
Heat source - centrifugal chillers' cooling tower circulating water loop.
Heat pump rating - 328 kW, COP of 4.3 at 29°C
Pre-installation gas usage - 6617 therms/month
Gas cost $1.36/therm or $9000/month
Energy input to water - 434 x 10^6 Btu/month
Recorded heat pump energy consumption - 26,727 kW.h/month
Recorded gas consumption after heat pump installed - 625 therms/month
Electricity cost - $2885/month @ $0.108/kW.h
Net savings - $63,000/year
Installed cost - $88,000 ($268/kW)

2) 1982 retrofit of 800 room hotel. Centrifugal compressor - type water-water heat pump. Heat source as in (1) above
Heat pump rating - 1.15 MW, COP 6.2 @ 26.6°C source leaving temperature and 54.4°C delivery temperature
Energy input to water - 984 x 10^6Btu/month
Heat pump energy consumption - 46,500 kW.h/month
Estimated gas savings - 15,000 therms/month
Net savings - $196,800/year
Additional saving from reduced cooling tower fan operation - $6700/year
Total installed cost - $260,000 or $226/kW

3) Retrofit of existing residential electric water heater with an air-water heat pump water heater extracting stored solar energy from atmosphere.
Heat pump rating - 3.8 kW, COP 2.5
Water heating requirement - 50000 kW.h/year
Heat pump consumption - 2000 kW.h/year
Electric power cost saving - $330/year
Installed cost - $1000

A Demonstration of 40 Engine-Driven Heat Pumps: A considerable number of domestic heat pumps were the subject of a demonstration project in West Germany, partially funded by the European Commission. The heat pumps were based on units developed by Fichtel and Sachs, these being internal combustion engine driven heat pumps with heating powers of typically 20 kW, fuelled using oil or natural gas. Each heat pump is controlled by a microprocessor based system, and they were installed in houses in the Schweinfurt region. The theoretical heat demand of the houses varied between 20 and 60 kW. The first heat pump was commissioned in December 1981, and commissioning of the remainder was completed by September 1982. During the demonstration project the average utilization of the heat pumps was 2,500 hours per year, the highest utilization being 6,000 hours overall. The target life of the internal combustion engine for these units is approximately 40,000 hours, representing 10 to 15 years of operation, (ref.4.39).

The breakdown of heat pump type and building type is given in Table 4.11. The majority of the heat pumps were fired using fuel oil, and were located in older types of houses. Again, the small number of houses with swimming pools were fitted predominantly with fuel oil fired heat pumps. For larger houses although these are not identified in the table, bivalent systems were installed. However, the basic monovalent heat pump with the capability of meeting heat demands of up to 20 kW was de-designed to produce this output at an ambient temperature of -15°C.

The commissioning of the heat pumps operating on fuel oil was highly satisfactory, but the commissioning of the 13 natural gas heat pumps was delayed by up to three months due to special safety regulations which had to be considered. Initially it was also requested that the gas heat pumps be submitted to a special investigation by the gas authority every two months. However permission was granted to extend this period to six months during the demonstration. The location of the evaporators on all heat pumps is outside the house, for example in the garden, the yard or on the roof. All heat pumps are thus split units. Where possible all internal equipment has been integrated into existing heat systems.

As can be seen from Table 4.12, the number of operating hours accumulated by the 40 heat pumps up to the end of June 1983 was in excess of 100,000. This can therefore be regarded as a very significant test in terms of experience of this type of heat pump system.

A number of aspects of operation were investigated, and of particular significance in the domestic context is the noise level of internal combustion engine driven heat pumps. A target was set at the outset of 55 dB (A). In general this target was achieved in practice, but in some houses noise problems occurred due to the influence of adverse building materials, in particular where wood and concrete were used in the construction. It was discovered that vibration frequencies of less than 20 Hertz could be transmitted into the building, necessitating changes to the mounting of the engine, the foundations, and the exhaust pipes. It was found that these changes aleviated the noise problems.

The exhaust gases from the engines created some problems, particularly those emanating from the oil fired units. Carbon monoxide and sulphur dioxide emission values met current standards, but it was found necessary to recommend further work in minimising the soot output and the level of NO_x in the exhaust. It was pointed out in the report that the overall acid rain level would be dramatically reduced, a topic of particular interest in West Germany, because of the overall reduction in primary energy consumption.

The primary energy ratios were monitored for the boilers in the houses prior to installation of the heat pumps, and also following heat pump installation. As stated earlier, all the houses selected had heat demands of 20 kW or higher. However, the effective heat demand of some houses was smaller because of the behaviour of the householders, and the trend towards significant reduction in heating level for periods of longer than 8 hours overnight. These energy-saving activities led to significant numbers of boiler on-off cycles, and the measured efficiencies of the boilers in these houses were all less than 70 per cent. Also, the types of heating networks employed in the houses, and the operating temperatures, varied. In the case of radiators, the delivery and return temperatures were 90°C to 70°C down to 60°C and 40°C. Where underfloor heating was used, delivery temperatures could be 50 or 40°C, returning at 40 or 30°C respectively. Where private swimming pools were used, the water temperature varied between 24°C and 31°C. These different conditions all contributed towards different primary energy ratios.

Table 4.13 shows the measured primary energy ratios of the 40 heat pumps. It can be seen that significant variations occur, the majority of heat pumps having primary energy ratios between 1.11 and 1.4. It was found that the controller which switches on and off the power of the internal combustion engine did not function in such a manner as to ensure optimum energy use, particularly when ambient temperatures fell below -5°C. In such cases the full power of the compressors was not used, and this led to a reduction in PER of some heat pumps. Unspecified modifications were also made to the compressor as a result of the demonstration, leading to a 10 per cent improvement in PER.

While the technical performance of the heat pumps was in general satisfactory, and the demonstration led to recommendations concerning improvements which could have some influence on the overall economics, the attractiveness of the heat pump as an investment remained open to question. The average boiler efficiency prior to heat pump installation in the houses was 73 per cent. Following installation of the heat pumps, a reduction in fuel oil or natural gas useage of typically 44 per cent was achieved. The average consumption of the engine driven heat pump amounted to 0.8-1.2 litres/hour of fuel oil or 0.9-1.5 m^2 per hour of natural gas, depending of course on the speed, number of on/off cycles etc. In general it could be stated that a fuel oil heat pump saved one litre of fuel oil per hour when compared to a boiler system.

It was concluded that in houses where the effective heat demand was less than 20 kW, the heat pump had too long a payback period to be attractive. However, where bivalent operation could be justified, and where oil savings of approximately 4,000 litres/year could be made, a payback period of less than 5 years would be possible. It was found that the application of greatest interest would be where the internal heat demand of the house would be up to 60 kW, and operating hours of up to 6,000 per annum. It was appreciated that this was unlikely to occur in many domestic dwellings, but was more likely in commercial buildings and sheltered housing etc. Here it was believed payback periods of less than 4 years could be achieved.

TABLE 4.11 Heat Pump and Building Types at Schweinfurt

Building	Fuel Oil	Natural Gas
Old	25	11
New	2	2
Total	27	13
Of above with swimming pools	8	1

TABLE 4.12 Total Operating Experience at Schweinfurt

Total Operating Hours (to 31 June 1983)	118,766
'Heat Pump Years'	47.5
Average Hours Operation per Heat Pump	2,503/annum

TABLE 4.13 Performance of Schweinfurt Heat Pumps

PER	Number of Heat Pumps		Total
	Fuel Oil	Natural Gas	
1.0-1.1	0	1	1
1.11-1.2	5	3	8
1.21-1.3	12	7	19
1.31-1.4	8	1	9
1.41-1.5	2	1	3
Total	27	13	40
Average PER	1.32	1.25	

REFERENCES

4.1 Anon. Energy conservation: a study of energy consumption in buildings and possible means of saving energy in housing. A BRE Working Party Report. Building Research Establishment Current Paper CP 56/75, Dept. of the Environment, June 1975.

4.2 Butler, D.J.G. The cost effectiveness of heat pumps in highly insulated dwellings - an assessment. BRE Report Under Contract to D.En. Contract Number EC/2732 (Study 3), July 1984.

4.3 Anon. UK Workshop on Heat Pumps. ETSU Report R1, HL 77/683, Department of Energy, UK, March 1977.

4.4 Butler, C. Ferranti 'fridge-heater'. Arch. J. Information Sheet 28.J.1, May 31, 1956.

4.5 Heap, R.D. Heat pumps and housing. Proc. Conference on Housing and Energy, University of Newcastle upon Tyne, 15-17 April, 1975.

4.6 Barrow, H. A note on frosting of heat pump evaporator surfaces. J. Heat Recovery Systems, Vol.5, No.3, pp 195-201, 1985.

4.7 Stoecker, W.F. How frost formation on coils affects refrigeration systems. Refrigerating Engineer, p42, Feb. 1957.

4.8 Anon. Heat pump electricity usages and other operating characteristics experienced under various weather conditions. Southern Research Institute Report No. 2371-288-SVIII, July 22, 1955.

4.9 Klaus, H. Energieeinsparung bei der Raumheizung und Brauchwarmwasserbereitung im Mehrfamilien hausbereich unter Einsatz von elektrisschen Warmepumpen. Proc. EEC Contractors Meeting on Heat Pumps, Brussels, 28-29 Sept., 1978.

4.10 Anon. ASHRAE Handbook and Product Directory 1976 Systems Volume. American Society of Heating, Refrigerating & Air-Conditioning Engineers Inc., New York, 1976.

4.11 Martin, D. Heat pumps for heating in buildings. Energy Technology Series, No.5, Energy Efficiency Office London, Oct. 1985.

4.12 Sumner, J.A. Domestic Heat Pumps. Prism Press, UK, 1976.

4.13 Reay, D.A. Industrial Energy Conservation. A Handbook for Engineers and Managers. 2nd Edition, Pergamon Press, Oxford, 1979.

4.14 Kolbusz, P. The use of heat pumping in district heating schemes. Electricity Council Research Centre Report ECRC/M700, Feb. 1974.

4.15 Godard, O. and Poppe, H. Temperatures in the soil in Belgium and in Luxembourg. Bulletin of the Belgian Society of Astronomy, Meteorology and Earth Sciences, Vol.76, No.9-10, Sept/Oct. 1963. (In French).

4.16 Fordsmand, M. Analysis of the factors which determine the COP of a heat pump, and a feasibility study on ways and means of increasing same. Proc. EEC Contractors Meeting on Heat Pumps, Brussels, 28-29 Sept, 1978.

4.17 Schar, O. Warmepumperheizung. Elektrizitatsverwertung, Jg 53, No. 1/2, S15-18, 1978.

4.18 von Cube, H.L. Warmeguellen fur Warmepumpen. Warmepumpen - Vulkan Verlag, Essen, 1978.

4.19 Goulbum, J.R. and Fearon, J. Heat pumps using ground coils as an evaporator. Proc. Heat Pump Workshop, Rutherford Laboratory Report RL-77; 145/C, Oxfordshire, Dec. 1977.

4.20 Ball, D.A. et al. State-of-the-art survey of existing knowledge for the design of ground source heat pumps. Battelle Columbus Laboratories Report, Battelle, Ohio, USA, 1983.

4.21 Fordsmand, M. The use of soil as a heat source and heat storage medium for heat pumps. Final Report No. EUR9852EN, CEC, Brussels. European Heat Pump Consultors Denmark, 1985.

4.22 Sporn, P. and Ambrose, E.R. The heat pump and solar energy. Association for Applied Solar Energy, Proc. World Symposium on Applied Solar Energy, Nov. 1955.

4.23 Freund, P. Leach, S.J. and Seymour-Walker, K. Heat pumps for use in buildings. Building Research Establishment Current Paper CP 19/76, Feb 1978.

4.24 Jardine, D.M. and Kuharich, R.F. Operational report on an inegrated solar-assisted optimized heat pump systems. Trans. ASHRAE. Vol.82, No.2, pp 426-432, 1976.

4.25 McVeigh, J.C. Sun Power. An Introduction to the Applications of Solar Energy. Pergamon Press, Oxford, UK, 1977.

4.26 Geeraert, B. Space heating with heat pumps using soil and an energy roof as a heat source. Report No. EUR 9117 EN, CEC, Brussels. Laboralec, Belgium, 1985.

4.27 Hanby, V.I. Sizing of heat exchangers for heat pumps for domestic applications. Proc. of Heat Pump Workshop, Rutherford Laboratory Report RL-77-145/C, Oxfordshire, Dec. 1977.

4.28 Reay, D.A. (Editor). International Directory of Heat Recovery Equipment. J. Heat Recovery Systems, Vol.7, No.1 (Special Issue), Pergamon Press, Oxford, 1987.

4.29 Smith, C.B. Efficient Use of Electricity, pp 283-285. Pergamon Press, Oxford, 1976.

4.30 Perlman, M. and Mills, B.E. Residential heat recovery, ASHRAE, New York, June 1986.

4.31 Dunn, P.D. and Reay, D.A. Heat Pipes. 3rd Edition. Pergamon Press, Oxford, 1983.

4.32 Altman, M. Conservation and better utilisation of electric power by means of thermal energy s torage and solar heating. National Science Foundation, Univ. Pennsylvania Report UPTES-71-1, Oct. 1971.

4.33 Lorsch, H.G. Thermal energy storage devices suitable for solar heating. 9th Intersoc Energy Conversion Engng. Conf. 1974.

4.34 Gordon, H.T. A hybrid solar-assisted heat pump system for residential applications. Solar Heating & Cooling of Buildings, Vol.3. American Section of International Solar Energy Society, Floria, 1976.

4.35 Ruiter, J.P., Leentvaar, G. and Zeylstra, A.H. Tumbler dryer with heat pump. Elektrotechneik, Vol.58, No.4, pp 224-229, April 1978.

4.36 Ehringer, H. et al (Editors). The Second Energy R&D Programme. Energy Conservation. (1977-1983). Survey of Results, 2nd Edition, Report EUR 8661EN, CEC Brussels 1986.

4.37 Ehringer, H et al (Editors). Proceedings of Four Contractors' Meetings on Heat Pumps. Report EUR 8077EN, CEC Brussels, 1982.

4.38 Lloyd, A.S. Heat pump water heating systems. Heating, Piping & Air Conditioning, Vol.55, Pt.5, pp 83-86, 91-94, May 1983.

4.39 Paulick, W. I.C. Engine driven heat pump with a heat power up to 20 kW. In: Energy Saving in Buildings. Proc. CEC Seminar, The Hague, 14-16 Nov. 1983, pp 281-287. D. Reidel, Dordrecht, 1984.

CHAPTER 5

Heat Pumps in Commercial and Municipal Buildings

The heat pump used in the home, be it for space heating and air conditioning, or on domestic appliances, generally has a small heat output and particularly as far as Europe is concerned, has not been as succesful as was originally predicted in the 1970's. As shown in Chapter 5, however, many systems are available on the market, and developments having considerable long term potential are actively being persued.

In this Chapter, where heat pumps applied to commercial and municipal buildings are discussed, the situation is somewhat different. Systems for heating and cooling duties in large buildings such as office blocks have been available worldwide for many years, and have been extensively applied, in competition with other large scale air conditioning systems. Much pioneering work has been carried out over the years in Europe in this area, and two systems discussed in more detail later which fall into this category are the water-air heat pumps developed by Temperature Ltd in the United Kingdom and the units developed by Sulzer for application in swimming pools and other sports complexes.

In this Chapter it is proposed to describe a number of the larger heat pumps available on the market, and the economics of their operation. Applications in general buildings, swimming pools, and computer rooms will be discussed. The use of heat pumps in district heating schemes is described, and absorption cycle, as well as vapour compression cycle, units receive attention.

5.1 General Air Conditioning Plant

Before proceeding to discuss in detail the heat pump as applied to buildings, it is worth dwelling for a short time on the general features and applications of air conditioning equipment. Air conditioning, or control of the climate within a building, is not necessarily associated with heat pumps, although of course conventional refrigeration equipment is routinely used.

(Comparisons will be made later in this Chapter between the performance of a heat pump system and other air conditioning systems in commercial buildings).

It is beyond the scope of this book to describe in detail the many types of air conditioning systems available, but the concept of air conditioning has been universally adopted in developed countries where ambient temperatures and/or humidities are such that comfort conditions need artificial regulation. It is therefore no surprise that the development and application of comprehensive air conditioning systems has flourished in the USA, but has been, until comparatively recently, somewhat neglected in the United Kingdom. The human being, of course, is not the only subject which requires 'air conditioning' for efficient operation. As will be shown in Chapter 6, some industrial processes, particularly where high humidity is a problem, can benefit from the use of dehumidification equipment, and the heat generated by a large computer must also be removed from its vicinity. (It should be noted that heat sources such as these, while requiring cooling, can also provide useful heat for cooler parts of a building by the use of some form of heat recovery. This will be illustrated later).

A useful resumé still relevant today of the role of air conitioning in offices was presented some years ago in 'The Times' survey (Ref.5.1). The basis of an air conditioning system rests on the ability to provide controlled ventilation, the ventilating air being either heated or cooled, as required. The basic requirements for human comfort conditions in offices, for example, are as follows:

(i) The working area should be comfortably cool

(ii) Air movement should be adequate but without noticeable draughts. The velocity of the air should be increased in hot and/or humid conditions.

(iii) Relatively humidity should not exceed 70 per cent, and by preference should be much lower.

(iv) The room structure, e.g. walls, should be warmer than the contained air.

(v) The air at head level should not be noticeably warmer than that close to the floor, and there should not be excessive radiation at head level.

(vi) The air should be odour-free.

In the United Kingdom office workers who are predominantly sedentary find room ambient temperatures of 20-22°C acceptable. Air movement for a typical large office is of the order of 0.5 m^3/min, this being the rate of fresh air provided per person. Air change rate for the room may be between 1 and 1.5 per hour. The calculation of the heat balance must of course take into account heat emitted by the human body, typically 100W when in the resting state, and heat given out by lights and other fittings.

Humidifiers are incorporated to raise the humidity of internal air, particularly during the heating season. In the cooling season, dehumidification may be required. Air conditioning systems commonly incorporate some form of filtration - these may be electrostatic filters in which dust particles are electrically charged and then pass through a duct containing a number of parallel plates, alternatively earthed and held at a high voltage. The charged particles are attracted towards the earthed plates, where they are deposited.

Odours are commonly removed by increasing the ventilation rate. Alternatively activated carbon filters can be employed.

One major feature of air conditioning systems designed today is the possibilities offered in terms of ease of fitting to buildings, both old and new. This has been brought about by the advent of 'packaged' air conditioning systems. It is not necessary to indulge in severe structural modifications to a building to incorporate such a system. Packaged air conditioning equipment is available in a wide range of sizes, covering the large liquid chillers with capacities of several thousand tonnes of refrigeration down to single room air conditioning units. Such systems, whatever their size, do incorporate factory-matched refrigeration cycle and air moving equipment, as illustrated in Fig.5.1. In such a circuit incorporating a compressor, the evaporator is directly used to cool the fresh air being supplied to the conditioned space. A fan is provided for forced circulation of the air, and filtration is also incorporated. Note also in this case that a separate heating system, not linked to the refrigeration cycle, is used for space heating as required. The condenser of the refrigeration cycle is either air or water cooled, and in the system illustrated no facility is provided for heat recovery.

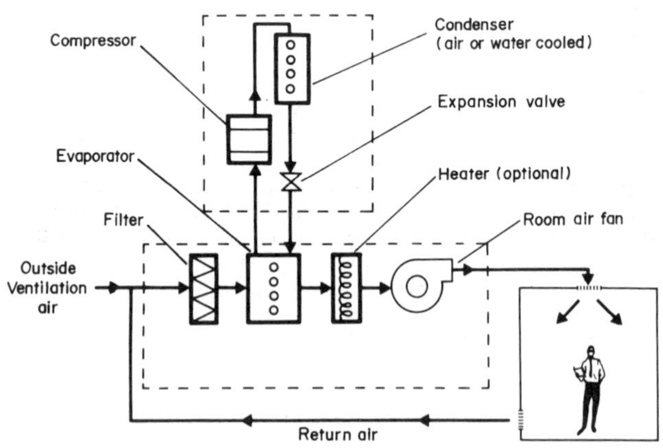

Fig.5.1 Layout of a Typical Air Conditioning System

However, a factory-assembled package of this type leads to reduced site work, and there is greater flexibility in selection and design, and application procedures have also been considerably eased and accelerated. The units are factory set and engineered, and factory-made assemblies of one or more finished cabinets designed to be installed on site are available. This leaves only electrical and, where relevant, water connections to be made at the final location.

The heat reclaim concept (see also Chapter 6, Sections 6.1 and 6.2) increases considerably the attractiveness of such an air conditioning system, bringing it into line with what we understand from previous Chapters to be the heat pump concept. The unused heat rejected to a cooling tower, or in any other manner, from the condenser illustrated in Fig.5.1, may be recovered to heat areas where previously some form of supplementary heating was needed. In this way heat recovered from occupants, as well as from lights and other equipment, may be usefully re-employed.

This leads us directly to the heat reclaim system as applied by many manufacturers to air conditioning plant in buildings.

5.2 Refrigeration Plant Heat Reclaim Systems

In non-domestic buildings, it is rare to find heat pump systems which operate solely in the 'heating only' mode. The major use of compressors in offices and the like has historically built up around the need to provide cooling, either for climatic reasons or to alleviate problems created by the presence of personnel, heat generating equipment, and perhaps lighting. The computer suite in an office building is one obvious area where a cooling load is certain to occur, as mentioned in the previous Section. This need for cooling has not always been associated with the realization that the refrigeration plant incorporated for the cooling duty, by its nature as a heat pump in the broadest meaning of the term, may also be able to provide a heating duty, (see also Chapter 6). Although this heat rejected at the condenser of the refrigeration plant, may be of a comparatively low grade, it is nevertheless often of use, and leads to substantial overall energy savings.

The technique of refrigeration plant heat reclaim is the simplest, in terms of its concept, to follow, although it is surpassed technically and economically by a decentralised air conditioning system which will be described in some detail later in this Chapter.

As will have become obvious to the reader, the operation of a heat pump (or refrigerator), in terms of the power absorbed related to that delivered at the condenser, is a strong function of the temperature difference between the evaporation and condensation processes. This may, as done by Bowen (Ref.5.2), be related to the economic condenser water leaving temperatures if the heat from refrigeration plant is to be utilised with benefit to the user. Fig.5.2 shows the power absorbed per ton of refrigeration duty, as a function of this water temperature, and with R22 as the working fluid Bowen recommends that

TABLE 5.1 Comparison of Air Conditioning Systems

What the User wants: What the Owner wants:	... with (a)	What he gets ... ... with (b)	... with (c)
	Central Refrigeration with 4-Pipe Induction, Fan Coil, or VAV* system.	De-centralised al-electric room conditioners with integral air-cooled condenser.	De-centralised heat room units on closed-circuit water system with boiler and cooling tower.
Vary room temperature quickly and over wide range.	POOR. Response to change in thermostat setting often slow and limited in range.	MODERATE. Response often slow, and temperatures swing erratic.	GOOD. Immediate response. Room temperature between 16-17°C normaly obtainable.
On-Off Control	POOR. Only fan-coil unit can be switched off and the coil remains active, with residual heat.	GOOD.	GOOD.
Always available.	POOR. At low or nil load standing costs are high. Repairs often mean large-scale shut-down.	GOOD.	FAIR. Water system to be kept in operation (but this is usually trouble-free).
Simple,quick maintenance.	POOR. House Engineer or Maintenance contract usually necessary. Chiller parts often on long delivery	FAIR. Removal difficult and often involves building work. Breakdown does not affect other units.	GOOD. Major unit repairs can be done at workshop. Units are reasonably easy to handle. Specialists expertise not needed.
Cheap to run	POOR. Can be energy wasters if simultaneous heating and cooling. Heat recovery is expensive in capital cost.	POOR. High Electrical consumption. Low efficiency on cooling. Expensive electric heat.	GOOD. Savings of 25-5Ø per cent compared to other two.
Easy to alter or extend.	POOR. Can be very costly and difficult unless carefully planned for at outset.	GOOD. But needs outside wall.	FAIR. Central plant and pipework may need alterations.
Easy to sub-let.	POOR. Difficult to divide control and to apportion costs according to usage, to added to rent on m^2 basis.	GOOD. Major running and repair costs can be borne by tenant, who has own control of operation.	GOOD. Major running and repair costs can be borne by tenant, who has own control of operation.

* Variable Air Volume.

an economical power-capacity ratio can be achieved if the water outlet temperature is of the order of 41 to 43°C, (for a conventional packaged chiller unit). In such a case the power absorbed by the compressor will not be much higher than that of a conventional 'cooling only' air conditioning system, while at the same time providing useful heat rather than rejecting it directly to the atmosphere.

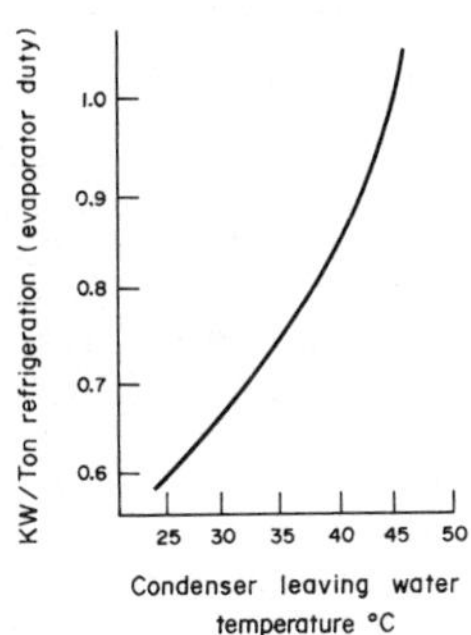

Fig.5.2 Effect of Condenser Leaving Water Temperature on Power Requirements of Refrigeration Plant

The most common application area of such a concept is illustrated in Fig.5.3. This shows how heat initially rejected to atmosphere via the refrigerator, which is used to cool the core of the building, may now be used to heat rooms around the perimeter, where heat losses via windows, doors etc will be greater.

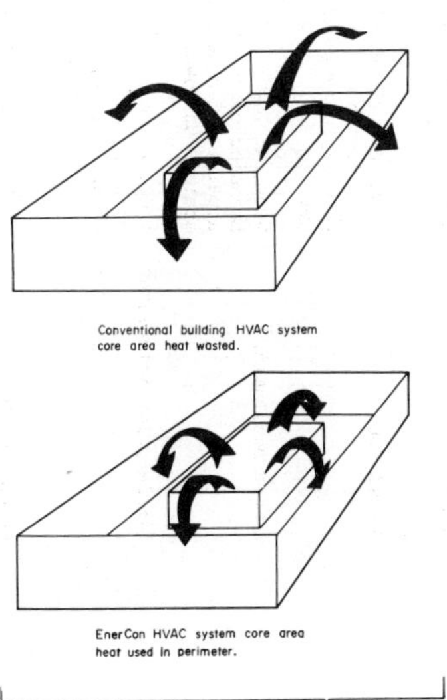

Fig.5.3 Schematic Illustrating How the Heat Resulting from the Cooling of a Building Core may be Effectively Used for Perimeter Space Heating

The layout of the refrigeration equipment used to effect the use of reject heat is illustrated in two forms in Figs. 5.4 and 5.5. In both cases the cooling coil would pick up heat from the building core, or any other location where a surplus of heat exists, transfer the heat via a pumped water loop to the evaporator, where the refrigerant would take up this energy for transport via the compressor to the condenser. The higher grade heat available there would be transferred via a second water circuit to the heating coil. During summer operation the system would dispose of the reclaimed heat from the condenser cooling water via a cooling tower, using either a 'double bundle' condenser or two separate condensers. The double bundle unit is illustrated in Fig.5.6, and a packaged chiller employing separate condensers for heat reclaim and cooling tower duties is shown in the sketch in Fig.5.7. During winter, when the combined heating and cooling function will be continuously required, a low grade heat source will be provided by the double bundle condenser or the second condenser on the package shown in Fig.5.7. The heat will be supplied by local fan coil units or by heat exchangers in ducted air systems. The cooling tower may of course still be used during such operation should any excess heat require to be dissipated to atmosphere. The arrangement shown in Fig.5.4, with a closed circuit cooling tower and a single condensing heat exchanger, operates such that when a heating duty is required, condenser cooling water bypasses the cooling tower and proceeds to the heating coil. (Note that the cooling tower need not necessarily be used to dissipate all the reject heat in the warmer months - where this heat can be used to supplement service water heating, or even in an industrial process, as discussed in the next Chapter, even greater cost benefits can result).

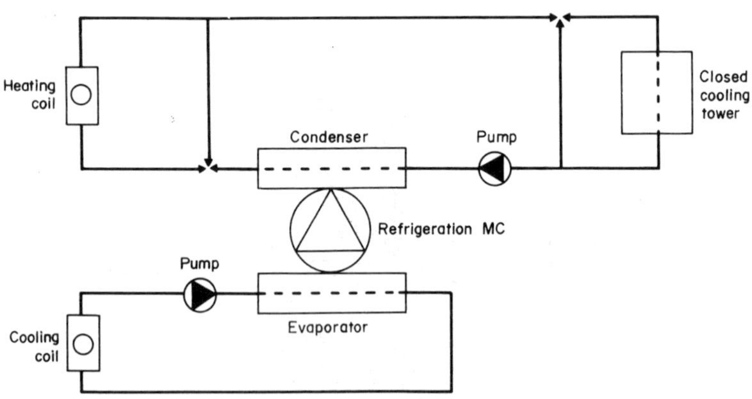

Fig.5.4 Heat Reclaim System Employing a Closed Circuit Cooling Tower

Two examples of the use of such a heat reclaim system may be quoted. An office in the United Kingdom utilizes two Dunham-Bush heat reclaim screw compressor packaged water chillers, with double bundle-type condensers. A conventional boiler and cooling tower are employed. The building is split into two sections, each with its own air conditioning plant, (Ref. 5.2). Provision is made, however, to use each plant to maintain conditions in both sections, should maintenance or breakdown create such a need. The individual offices are heated (or cooled) using fan coil units located under the windows, each module having a return air thermostat and control valve so that local temperatures may be individually controlled.

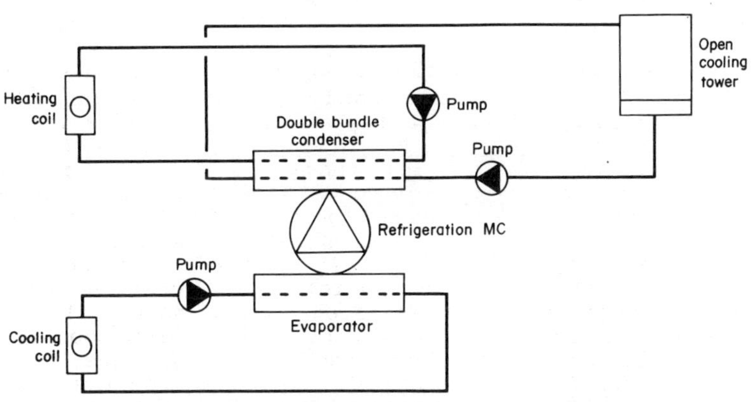

Fig.5.5 An Alternative to the System Shown in Fig.5.4 - Use of a 'Double Bundle' Condenser

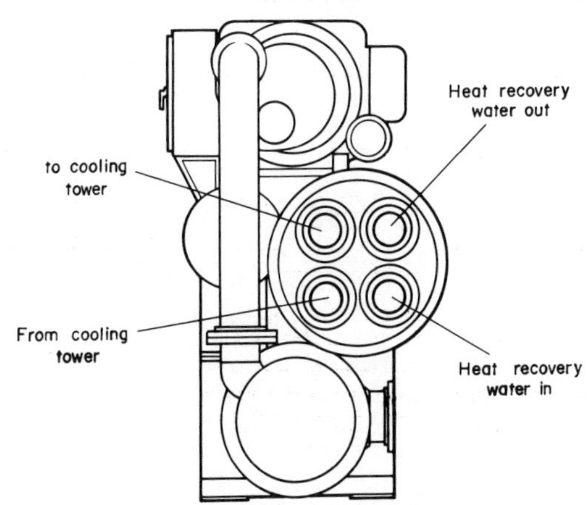

Fig.5.6 End View of a Package Chiller Unit Using a 'Double Bundle' Condenser

In this installation, the return air from the cooled core of the building is collected and return to the central air handling plant, but exhaust air from office and toilet areas is passed over coils so that the heat in it may be recovered directly for water heating. The building is maintained at a temperature of 10°C overnight, and control systems are incorporated, together with boost facilities, to enable warm-up at the start of the working day to be effectively implemented.

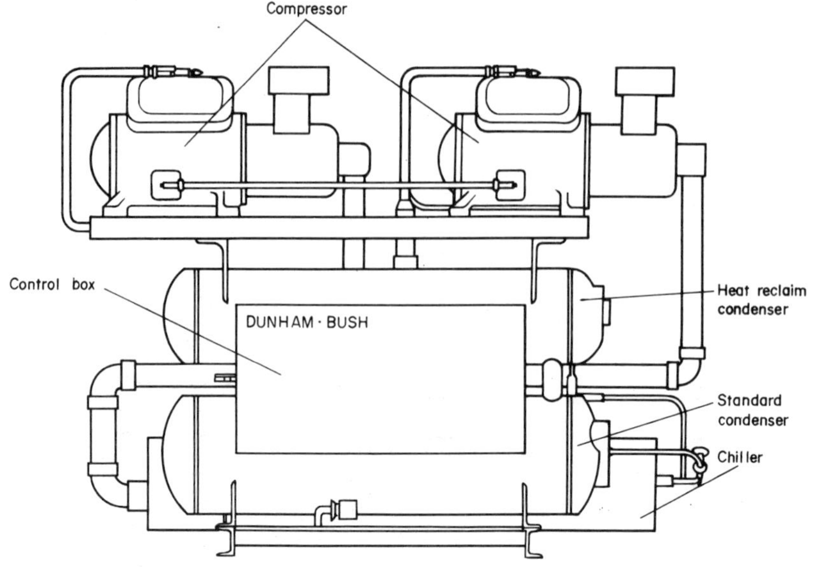

Fig.5.7 A Dunham-Bush Chiller which uses a Separate Condenser for the Heat Reclaim Role

Each of the two refrigeration plants, when operating in the heat reclaim mode, has a duty of 805 kW, when cooling water, passing through the evaporator at a rate of 34.1 litre/s, from 11.5°C to 5°C. Condenser water is heated from 30°C to 39°C, the flow rate being 27.3 litre/s. The power consumption at the full load condition is 240 kW.

The second example of a heat reclaim installation, based on equipment manufactured by Carrier, but installed in the United Kingdom, introduces the concept of recovering heat generated by lighting in offices. The building, completed in 1969/70, is the headquarters of the Merseyside and North Wales Electricity Board (MANWEB), and possesses the first large scale fully integrated heat-from-light air conditioning system in Britain, (Ref.5.3). The air conditioning system is designed to maintain an internal temperature of 21°C, with external temperatures varying between 26.6°C and -4°C. (It should be noted that removal of heat from the vicinity of light fittings can increase their lighting efficiency, in the MANWEB case by about 13 per cent).

A schematic of the layout of the air conditioning, and heat reclaim system used in the MANWEB headquarters is given in Fig.6.8. The luminaires (a term describing a complete light fitting) are cooled by air extracted across them from the space being conditioned. In addition, heat is recovered from the occupants and any heat generating plant, in the following proportions:

Lights	45%
Plant & Machinery	40%
Occupants	15%

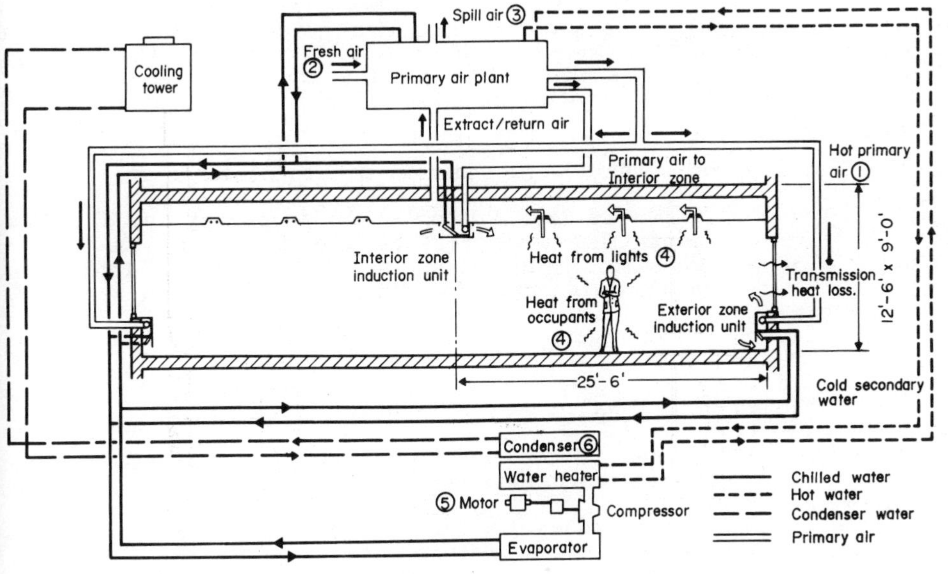

Fig.5.8 The Heat Reclaim System, Based Partly on using Heat Generated by Lighting, in the MANWEB Building

The air thus heated passes into the ceiling void and thence to a vertical duct through the core of the building to the central air handling plant. The refrigeration plant, two 300 ton Carrier centrifugal compressor units, are each capable of meeting the design heating load (space heating only) of 790 kW. The condenser inlet and outlet water temperatures in this installation are respectively 49°C and 54°C, the refrigerant evaporating temperature being of the order of 15°C. Heat exchangers are used to recover heat from the warm exhausted air, giving up heat to the chilled water circuit, thence feeding the evaporator. If an excess of heat is generated at the condenser, part of this may be dissipated to atmosphere by passing it to the cooling tower, as shown. Individual room units are provided, which can be adjusted for local heat control. These are also linked to the chilled water circuit so that heat recovered from these when they are operating in the space cooling mode can be applied to heating the building periphery, via the refrigeration plant.

No boiler plant was required in this building, and, incidentally when at night and during weekends a thermal store is used as a heat source when, of course, heat from occupants and lighting will not be available.

The heating and cooling of a large building complex in Switzerland, is yet another example of the use of 'double bundle' condensers. Scheduled for completion shortly, the Centre Geneve is a building complex in which are located administrative and commercial organisations, shops, restaurants and residential apartments. The heat pump installation provides heating and cooling, and can operate in the dual mode simultaneously. The heat source is the river Rhone, and this also serves as the heat sink in the cooling mode. Total cooling duty is 33 per cent greater than the maximum heat output capability, (Ref.5.4).

The installation comprises the following main components:

(i) Three Sulzer 'Unitop' heat pumps employing twin centrifugal compressors and double bundle condensers. Condenser heating duty is 1365 kW and cooling duty is 1800 kW.

(ii) Water accumulators - hot water storage capacity $200m^3$, chilled water capacity $130m^3$.

(iii) River water pumping station.

(iv) Standby Diesel generator sub-station with waste heat recovery capability.

(v) Boiler house for back-up operation - e.g. during low water temperature conditions in the river.

(vi) River water heat exchanger for directly chilling water for the complex. This can be used when the river water temperature falls below 10°C. The resulting rise in river water temperature locally created by this duty is used to benefit the heat pump when in its heating mode. This is visible in Fig.5.9.

Interestingly, the twin compressor layout is selected so that each can operate at its optimum efficiency at the different temperature lifts involved - one providing a heating duty and the other providing chilling.

For heating duties, the maximum demand is such that one heat pump is available always as standby. Hot water accumulators are incorporated, however, which enable the system to benefit from reduced rate electricity at night. As shown in Fig.5.10, the heat pump will only operate at full load if the river temperature exceeds 5°C. When the river temperature falls to between 4°C and 5°C, capacity is reduced by 50 per cent to eliminate freezing around the evaporators. A further fall in river water temperature leads to the sensors cutting out the heat pumps completely, and the heat supply is maintained solely by the boiler. Similar control is exercised on the condenser, where a 3-way valve maintains the condenser inlet temperature at 40°C in order to achieve a high efficiency. If the temperature exceeds a preset limit, the heat pumps are switched off.

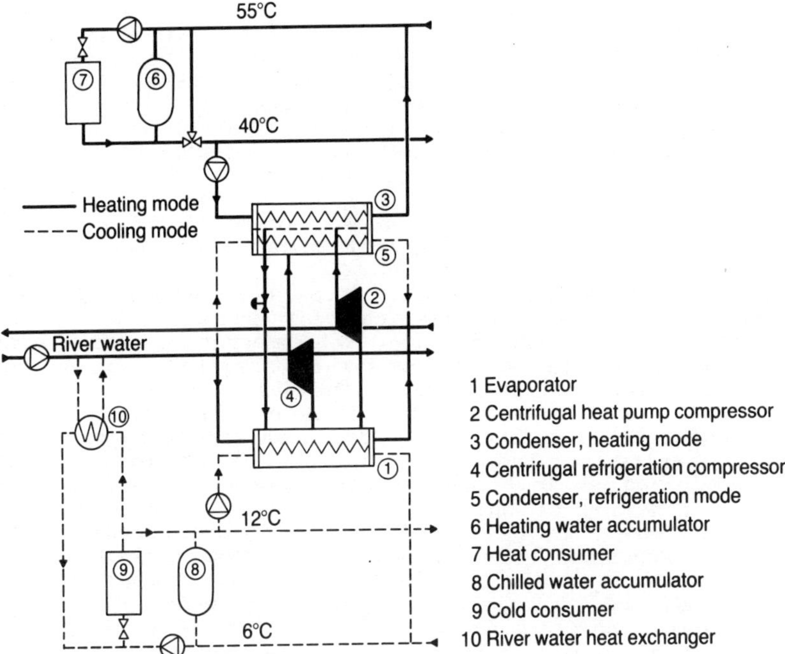

Fig.5.9 Centre Genēve Centrifugal Heat Pump/ Refrigeration Unit Operating Schematic

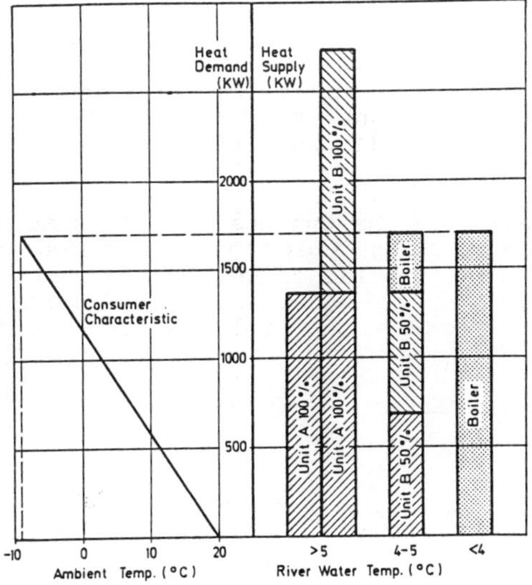

Fig.5.10 Centre Genēve Heat Pump Operating Mode

5.3 Decentralised Air Conditioning Systems

The system of heat reclaim from condensers on central air conditioning plant, often used in conjunction with individually controlled room units in the form of fan-coil heat exchangers, is one approach to the implementation of energy conservation measures in large air conditioned buildings. An alternative system which, we may argue, offers a number of advantages over the 'central' packaged chillers just described, is the use of individual heat pump units throughout a building, again based on the circulation of water around the building, with air as the final room conditioning medium. Such a system may be called 'decentralised air conditioning'.

Typical of the systems of this type on the market are the Enercon System, made by American Air Filter (AAF) and the Versatemp System, manufactured by Temperature Ltd, and the data given below is based largely on assistance given to the authors by these two companies. The basic circuit of the systems differs little from that of the central air conditioning plant, but instead of one large refrigerator/heat pump, a number of small units located at the points to be conditioned are employed, as shown in Fig.5.11. The Enercon layout illustrated utilises a series of unitary water-to-air reversible air conditioners, which provide either heating or cooling. These units are connected to a non-refrigerated central water system which is maintained within an approximate temperature range of 15 to 32°C by a supplementary water heater and a heat rejector (typically an evaporative water cooler or cooling tower). Each air conditioning unit linked to the water circuit consists of a complete refrigeration/heat pump unit, a fan for circulating room air, and controls for reversibility, and is connected to the water circuit as shown in Fig.5.12. It can be seen that the water serves as a heat sink to accept heat from an air conditioner in the cooling cycle, and to provide heat for an air conditioner to extract on the heating cycle.

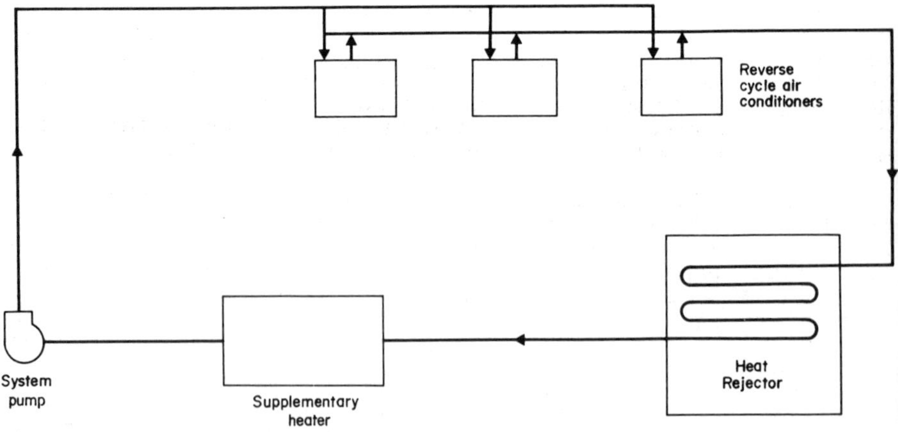

Fig.5.11 Layout of a Decentralised Air Conditioning System - Enercon. (Courtesy AAF Limited)

Supplementary heating is generally only required with such a system when most of the units are operating in the heating mode and the weather is cold. Heat input to the water circuit may be by means of a boiler, electric immersion heater, solar energy, or a waste heat source. The amount of heat required is reduced whenever one or more units is operating in the cooling mode. In moderate weather, units serving the shaded sides of a building are often heating the space, while those on the sunny side are providing cooling. When approximately 30 per cent of the units in operation are providing a cooling duty, they add sufficient heat to the water cicuit to obviate the need for any building heat rejection or addition.

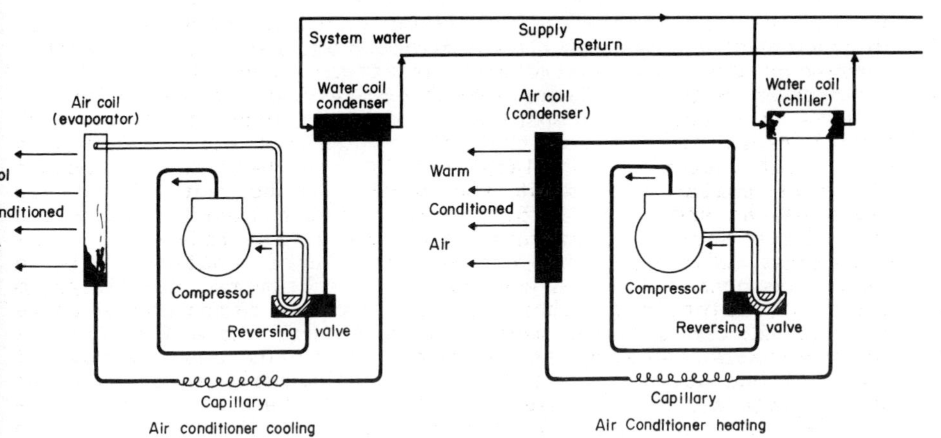

Fig.5.12 The Enercon Air Conditioners Connected to the Water Circuit, Illustrating Heating and cooling modes (Courtesy of AAF Limited)

As with the heat reclaim system described in Section 5.2, office buildings, for example, with high internal heat gains from people, lighting, computers etc., and possibly high rates of insulation may require local cooling throughout the year. Heat recovered from these areas can be transferred to the water circuit and thence to units on the periphery which in the winter months are likely to be operated in the heating mode.

The decentralised systems may also be used to advantage in buildings which might require cooling during daytime but heating at night. If during daytime the water circuit is raised to its maximum permissible temperature while still effecting cooling - of the order of 32°C - the heat in this circuit, provided that it is not rejected by the cooling tower to the atmosphere, may serve as a heat source during at least part of the heating cycle, before any form of supplementary heat needs to be added. This becomes necessary only when the water circuit temperature drops to about 15°C. The air conditioners thus frequently start the day with cool water for most efficient cooling, finishing with warm water for efficient overnight heating. The performance of the AAF Enercon system, as applied to a 7 storey office building, operating on a 24 hour cycle, is illustrated in Fig.5.13. The upper illustration shows the building heat requirements, this including heat losses which occur throughout the 24 hour period, and the heat demand arising from the need to heat ventilation air, arising during the main period of occupancy. In Fig.5.13(b), the internal heat gains which may be used as heat sources for the Enercon system are illustrated. This assumes that some plant in the building will be operating throughout the 24 hour period, and added to this are heat gains due to lighting (operating at a much reduced level between 10 p.m. and 7.45 a.m.) and the heat input from personnel, effective over a 9 hour period in each cycle. It is then possible to present the performance of the Enercon system, in terms of the amount of stored heat (stored in the water circuit) contributed towards the air conditioning demand, and this is given in Fig.5.13(c). The effect on the water temperature is shown, the temperature reducing from a peak at 10 p.m., when the contribution from lighting effectively ceases, starting to pick up again as the demand for air conditioning around midday reflects the heat input from personnel, lights, and plant.

A number of typical units, comprising compressor, fan coil, water loop, controls, and the container, are illustrated in Figs.5.14 - 5.16. Fig.5.14 shows a ceiling-mounted HW unit manufactured by AAF. Visible are, from left to right, the compressor, control box, and coil for air heating/cooling. The coil would incorporate a filter across its face, and in this package all connections, including those for the water circuit, are located on one side of the support structure. The control unit may be slid out of the package for repair and/or maintenance, without the need to disassemble or remove the complete package. In Fig.5.15, a cut-away view of a floor-mounted Versa Temp unit, manufactured by Temperature Ltd, is shown. The electrical control circuits, and the manual controls are located above the compressor unit, and to their right are the refrigerant-air heat exchanger, and the centrifugal fan and its motor. Provision is incorporated for the removal of any condensation which may occur on the coil, via a tray, leading to a pipe which projects from the bottom left side of the housing. Visible in the rear view of the partly-dismantled Versa Temp VM unit is the water coil, (Fig.5.16).

In order to give the reader an idea of the scale of these units, the cabinet (floor mounted) for the Versa Temp VM series ranges in length from 956 to 1280 mm, the maximum height being 743 mm. Weights range from 63 to 96 kg. With regard to the performance of the units, the smallest unit made by Temperature Ltd, the VM 150, has a rated cooling capacity of 1.76 kW, and a heating capacity of 2.42 kW. Total electrical input in this latter mode is 0.72 kW (fan and compressor). The large unit in the range, the VM 380, has a heating duty of 5.54 kW, for comparison. The total electrical power requirement is then 2.02 kW.

The manufacturers of decentralised equipment claim, justifiably in the view of the authors, that their products offer significant advantages over the central packaged refrigeration system, or self-contained decentralised electric units. Table 5.1

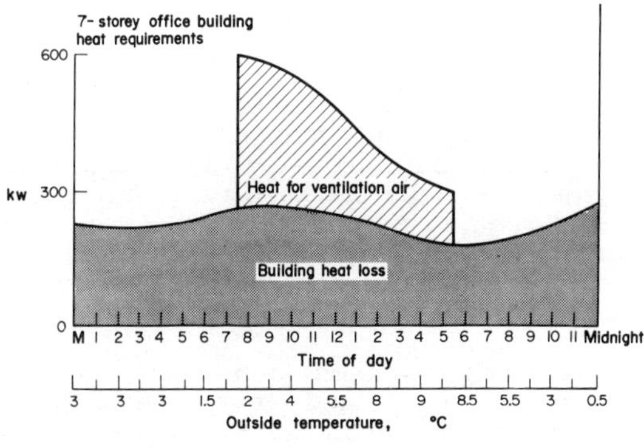

a

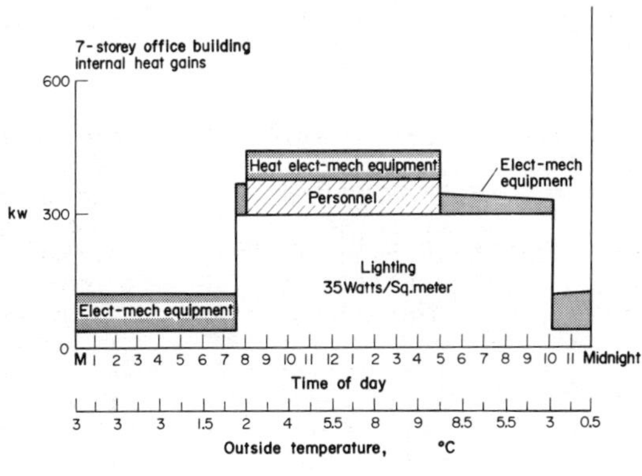

b

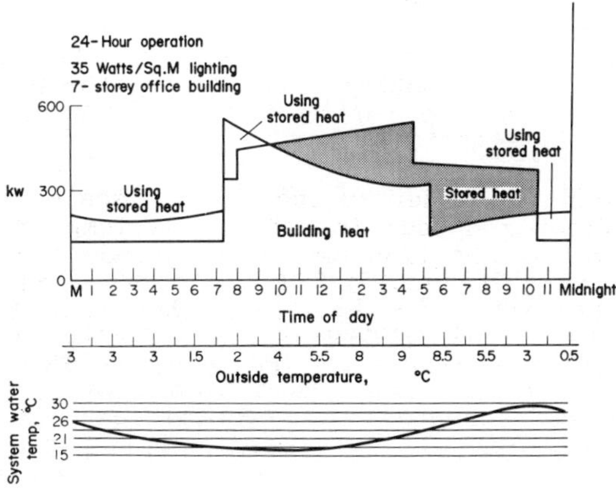

c

Fig.5.13 Enercon on a 24 Hour Cycle in an Office Building, using its Heat Storage Capability (Courtesy AAF Limited)

showed how the three types of air conditioning systems measure up to a number of basic parameters. In the article from which the Table was taken, it was pointed out that noise and humidity are omitted from the list of parameters. Noise from room units is more difficult to contain than that produced by a central air conditioning plant, but architectural encasement and other acoustic insulation can make noise levels most acceptable with the former. With regard to humidity, RH can vary, according to some studies, between 25 and 75 per cent, provided that temperature levels remain in the comfort zone, without being detectable by humans.

Decentralised heat pump systems can show advantages both in terms of lower capital cost and lower running costs. It is of course possible to fit heat recovery devices such as run-around coils and heat pipe heat exchangers to central air conditioning plant ductwork, and this will reduce operating costs, at the expense of increased capital expenditure, (although if designed in from the start, heat recovery equipment can lead to a reduction in size of the primary refrigeration plant). Decentralised units have advantages from the point of view of ease of installation, provision for extension as and when required, and also ease of repair. A single unit can be replaced without major system upheavals.

5.4 Combining Central and Decentralised Systems

As discussed above, the decentralised heat pump system, typified by the products of Temperature Ltd, and AAF, are particularly suited to the conditioning of single small offices in large buildings, where simultaneous heating and cooling may be required. However, in a building where there are also large spaces to be heated or cooled, such as a ballroom in an hotel, a central air conditioning system can fulfill the requirements in a more satisfactory manner. In such a case, AAF Ltd have combined a reverse cycle roof mounted packaged unit with the Enercon individual reverse cycle air conditioners described in Section 5.3, as shown in Fig.5.17. In this figure the water circuit of the roof-mounted unit is shown, linking up the room conditioners. Fig.5.18, which shows a cut-away view of the packaged unit, illustrates in addition the air coil (on the left). The central components are the centrifugal fans and drive motor, and on the right is the compressor and the water coils.

It is interesting to note in Fig.5.17 that a storage tank has been included in the circuit. In addition to being used for storing heat generated during a cooling operation for later use when heating is required, it may also, suggest the manufacturers, be used for storage of heat from solar collectors, heat rejected from refrigerated food facilities, waste heat from industrial processes, etc. Domestic hot water may also be preheated by this storage tank with the addition of appropriate heat exchanger coils.

5.5 District Heating

The use of low grade naturally-occurring or industrial heat sources to serve district heating schemes has become, during the last few years, a major business opportunity for large

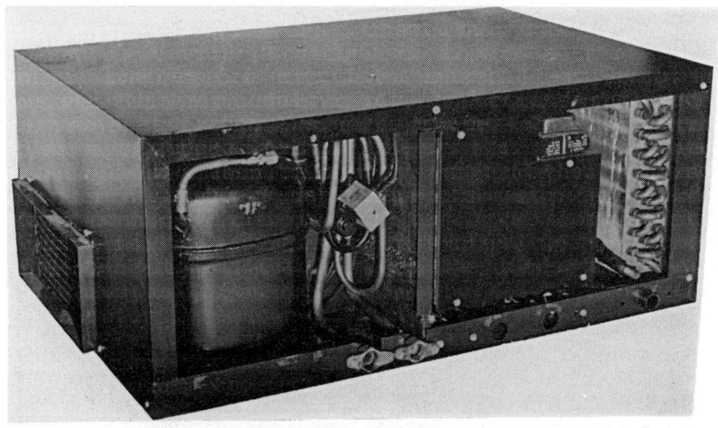

Fig.5.14 A Ceiling-Mounted Enercon Unit Showing Some of the Major Components. (Courtesy AAF Ltd)

Fig.5.15 A Cut-Away, View of the VersaTemp Unit - Details of the Components are Given in the Text. (Courtesy Temperature Limited)

Fig.5.16 Rear View of a Partly-Dismantled VersaTemp Unit, Showing the Water Coil. (Courtesy Temperature Limited)

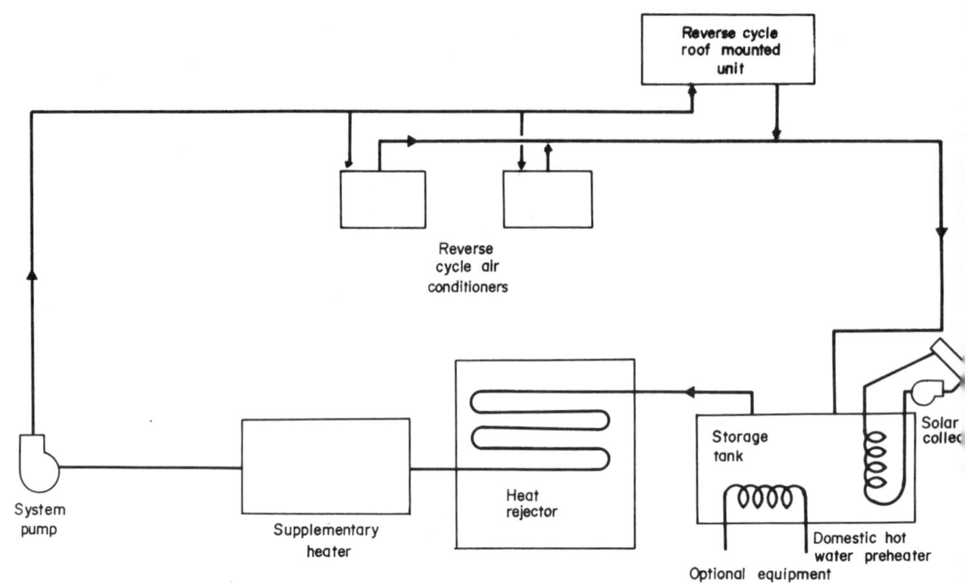

Fig.5.17 Enercon Units Combined with a Roof-Mounted Reverse Cycle Air Conditioner. (Courtesy AAF Limited)

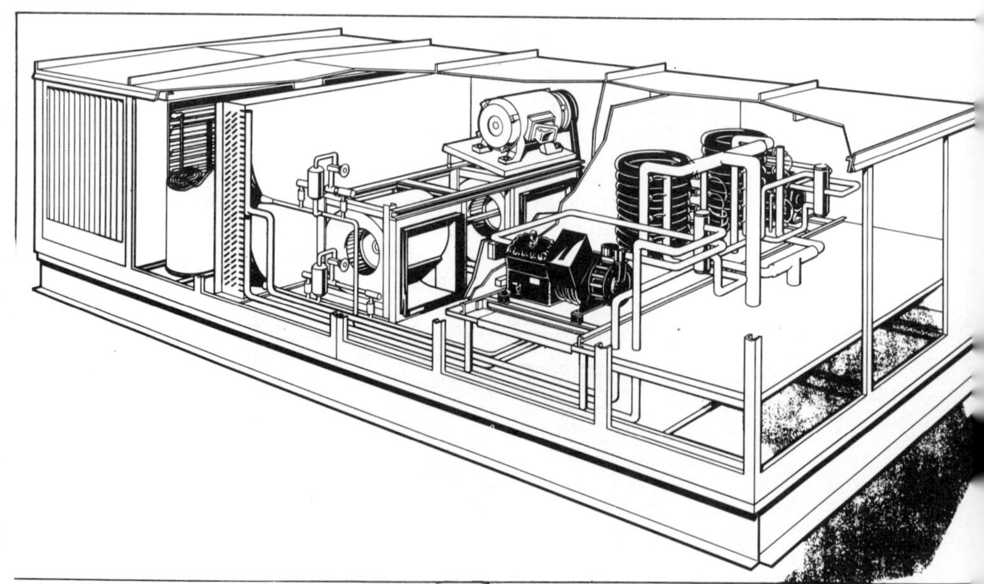

Fig.5.18 Layout of the Rooftop-Mounted AAF System, Showing Air and Water Coils

heat pumps. Much of the pioneering work in this area has taken place in Sweden, and most large Swedish towns now have at least one heat pump serving a district heating network. The heat pumps are generally in the MW heat output range, recent statistics (Ref.5.5) indicating that over 1000 heat pumps in Sweden are rated at 5 MW and above.

The largest of these installations to date is that supplied by Asea Stal to the Stockholm districts of Solna and Sundbyberg. With a total output of 120 MW, catered for by 3 40 MW heat pumps capable of producing 2.8 GWh of heat per day, the cost of fuel has been reduced by about 60 per cent, saving of the order of £6 million per annum. This gives a payback of around 4 years on an investment of £26 million. An additional benefit arising from the heat pump installation (which is electrically driven - made attractive by the availability of low cost electricity) is the reduction in local atmospheric pollution.

A smaller unit, based on the use of sewage water as a heat source, has been tested in Sweden for district heating applications.

Sewage has been for many years an attractive heat source for heat pumps, but exploitation has been hampered by fears concerning evaporator fouling. By utilizing treated sewage water and a special design of evaporator, a segment of which is shown in Fig.5.19, fouling problems were to some extent avoided. Four evaporator segments were used, the sewage water being sprayed over the outside of the tubes, while thermosyphon circulation ensures refrigerant flow.

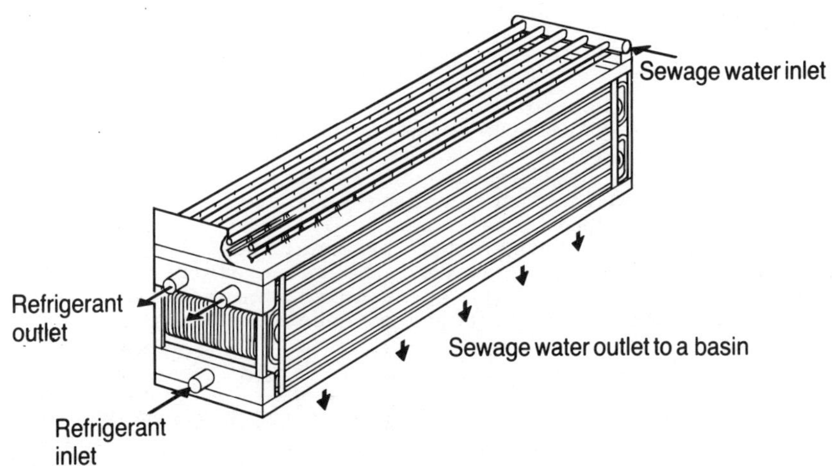

Fig.5.19 Evaporator Segments. The Evaporator Consists of Four Segments Like This. The Sewage Water is Sprinkled Over the Tubes. The Refrigerant is Circulated Through the Tubes by Thermo-syphon Circulation

The heat pump, the specification for which is given in Table 5.2, aids the performance of a number of conventional boilers serving the district heating scheme by preheating the return water from the district heating scheme before it re-enters the boiler complex, (Ref.5.6, Fig.5.20).

A number of operating problems were encountered following start-up of the system in June 1981. As listed in Table 5.3, these centred around difficulties in heat pump compressor capacity control brought about by the higher-than anticipated district heating return water temperature. This meant that the compressor was continuously operating below its design point. This problem has now been overcome.

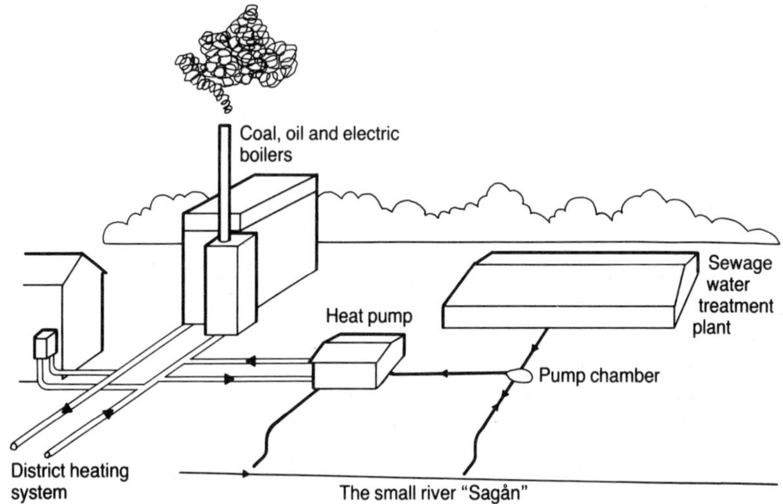

Fig.5.20 Schematic View of the Heat Generation Plant and the Sewage Water Treatment at Sala

During 1983, high utilization (93%) was achieved, and energy savings achieved during that period indicated a payback period of 6.5 years for the installation. Interestingly, no fouling has occurred on the evaporator external surfaces during the 2.5 years of operation. However, the water distribution system suffered from clogging of the spray holes, and regular (typically monthly) cleaning was found to be necessary.

5.6 Ground Source Systems

The use of the ground as a heat source for heat pumps has already received attention in Chapter 5, where some of its advantages and disadvantages have been explored. Ground source heat pumps have been installed in many locations, covering a variety of heat outputs. The two examples given below, located in West Germany and Denmark, are examples employing internal combustion engines as the compressor drives.

TABLE 5.2 Specification of Sala Heat Pump

Heat Pump Capacity:	
Sewage water temp.	8°C
District heating return temperature	45°C
Rated capacity	3.3 MW
COP	2.7
Compressor/Drive:	
Compressor type	Screw
Evaporating conditions	-5°C to +5°C
Condensing conditions	+55°C to +75°C
Motor type	Electric squirrel cage
Rating	1.4 MW
Evaporator:	
Type	Horizontal tubes + spray
Source inlet temperature	8°C to 16°C
Source temperature drop	6°C approx.
Source flow	300 m³/h
Material	Galvanized carbon steel
Condenser:	
Type	Shell & Coil
Sink Temperature	48°C to 65°C
Sink flow	300 m³/h
Manufacturer:	
Stal-Refrigeration, Sweden	
Cost: 5 million Sek	

5.6.1 Heating a local authority building In this example ground was selected as the heat source for a gas engine driven vapour compression cycle heat pump to serve local authority offices in Warendorf, West Germany. The total heat demand of the offices amounted to 1.8 MW, of which 650 kW was for space heating via ventilation air, and the balance, 1.15 MW, was supplied via hot water. The size of the heat pump system was selected to meet the hot water heating demand. It was calculated that it would be possible to meet 97% of the annual demand of the entire plant with the gas engine driven heat pumps. Boilers were provided to make up the balance, and the total heat demand of the building is summarized in Table 5.4, (Ref.5.7).

TABLE 5.3 Operating Experience with Sala Heat Pump

June 1981	Commencement of heat pump operation
Winter 1981-82	Vibration problems with slide valve used for compressor capacity control. Caused fractures in pipes and valves.
February 1982	Shutdown for 3 weeks for modifications Subcooler and oil cooling system installed, to improve performance at high water return temperatures.
Summer 1982	Vibration problems again, due to operation of heat pump below maximum capacity.
July 1982	Capacity control locked at maximum setting.
September 1982	Motor installed to operate compressor inlet valve, as means for controlling capacity. Mean COP in 1982 was 2.5, and total heat delivered 19.5 GW.h. Operating time 6450 hours.
February 1983	New compressor installed to enable modified capacity control system to function. Resulted in COP of 2 at minimum capacity during summer etc. Acceptable to purchaser.
March 1983	New compressor had higher start-up torque. Led to frequent stops during start-up periods Necessary to bypass low pressure trip switch. Mean COP in 1983 was 2.6, total heat delivered was 25.6 GW.h., and total operating hours reached 8156, representing 93 per cent availability.

The heat generation plant consists of two gas engine driven heat pumps together with three gas boilers, which cover the peak load and domestic hot water supplies. An earth source heat exchanger is installed, as illustrated in Fig.5.21, and it can be seen that both heat pumps operate on a common evaporator and condenser. The plant was developed by VEW, together with a number of other suppliers as listed in Table 5.5. Table 5.5 also gives full data on the design performance of the heat pump.

Two heat exchangers are used to extract heat from the soil, the tube material being polyethylene. The tubes are laid at depths of 0.8m and 1.8m, and a total tube length of 23000 metre is used. A brine solution is circulated to extract the heat from the soil and supply it to the heat pump evaporator. The specific heat absorption design value is 49 W/m^2 of soil area. Circulating pumps for the brine are located in the basement of the building, while the heat pump evaporator is located in the attic.

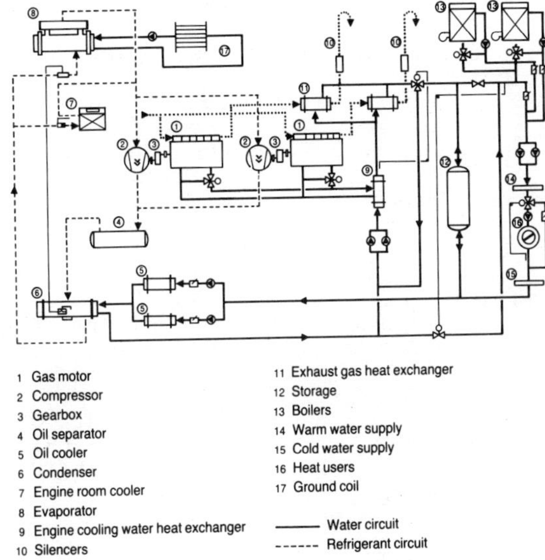

Fig.5.21 Connection Diagram of the Heat Generation Plant

TABLE 5.4 Maximum Heat Requirements and Annual Heat Demand of Warendorf Offices

Demand Type	Maximum Demand in any hour (kW)	Annual Demand (MWh)
District House	918	1,492
Police Wing	186	358
Ventilating Air Heating	640	207
Hot water preparation	82	88
TOTALS	1,825	2,145

TABLE 5.5 Warendorf Heat Pump Data

Design Heat Output	1,150 kW
Number of Heat Pumps	2
Compressor:	
Type	Screw
Manufacurer/Model	Mycom 160 LUDM
Power regulation	Slide valves
Drive power	121.6 kW
Working fluid	R12
Cooling capacity	218.5 kW
Speed	4,031 rpm
Engine:	
Type	6 cylinder Otto
Manufacturer/Model	Waukesha F817 GU
Maximum Power	143 kW at 1800 rpm
Continuous rating	130 kW
Evaporator:	
Type	Flooded
Manufacturer	Neunert
Duty	440 kW
Brine flow	86 m^3/h
Evaporating Temperature	-10°C
Brine inlet Temperature	0°C
Brine outlet Temperature	-5°C
Material	Steel
Condenser:	
Manufacturer	Neunert
Duty	680 kW
Water flow	98.9 m^3/h
Condensing temperature	55°C
Material	Steel
Engine Cooling Water:	
Duty	270 kW
Engine Exhaust Gas:	
Duty	200 kW
Engine Oil Cooler:	
Duty	78 kW

TABLE 5.6 Costs of Warendorf Heating Plant

Item	Cost
Planning & Execution	232,714.10 DM
Gas Engine Heat Pumps	922,063.51 DM
Ground Coils	473,538.03 DM
Gas Boiler	217,388.58 DM
TOTAL (incl. 13% VAT)	1,845,704.22 DM

Within the heat pump complex, heating water flows through two heating circuits at different temperature levels. In the low temperature circuit heat is supplied by the oil coolers, the condenser, and from other minor outputs. A proportion of the heating water is led through the high temperature circuit where heat is absorbed from the engine water cooling jacket and the exhaust gas heat exchanger. The design point is such that at an ambient of -12°C, the heat absorbed in these two circuits is completely utilized by the building heat distribution system.

A buffer store is incorporated in the system for storing heat at high temperature, and is charged when the demand is insufficient to utilize the full output of the heat pumps. The store thermostats can also respond to demands in excess of that supplied by the heat pump, at which time the gas boiler plant is brought on line.

Commissioning of the heat pump plant commenced in December 1982, and the plant was handed over on March 22 1982. A number of operational troubles were noted during the first few months of running, and these included a failure of the control slide valve regulation of the screw compressor, problems with motor speed regulation and ignition, and frequent oil leaks from the gas engine. More serious problems described below led to further delays in full operation of the system.

During erection of the plant, there were difficulties with a company which was contracted to build a car park in the area taken up by the ground source heat exchangers. Of particular concern was the degree of compression of the soil which might occur when the car park was in use, and the degree of compression permitted without distorting the heat exchanger. This was resolved by tests on a separate site, which showed that problems of this type were unlikely to occur.

As the gas engine heat pump is installed in the roof space above the fifth floor of the office building, the control of noise and vibration is therefore important. The room for technical monitoring equipment was isolated from the rest of the building structure, and all tube passages were equipped with rubber compensators. The heat pumps are mounted on spring elements, with care being taken to avoid matching the exciter frequency. Ventilation and deaeration plants and the exhaust heat exchanger were equipped with sound absorbers so that the maximum permissible sound level of 45 dB (A) at night measured 1 metre from the inlet and outlets to the unit, was not exceeded.

The safety of the overall system was subject to approval by the Technical Control Union based in Essen. This organisation insisted that the exhaust gas heat exchangers were regarded as self contained heat generation units, and it was therefore necessary to incorporate safety circuits dedicated to these heat exchangers. An additional expansion vessel was installed, and thermostats were located in positions where the entire plant could be switched off if the water temperature reached 100°C. It was pointed out that these were new requirements for gas engine heat pump plants.

Unlike a conventional heating system, the heat pump can be particular sensitive to off-design flow conditions. It was found particularly important in this installation to be able to co-ordinate the water volume flows of the heat distribution system to those of the heat pump plant. Problems resulted in this area which did not permit the gas engine heat pump to operate continuously in the heating mode and reliance had to be placed on other heat inputs. It was originally conceived that a constant temperature difference will be maintained with variations in the volume flow on the heat distribution side, and on this basis the heat pump would be able to respond to changes in heat demand. However, in practice, it was found that the volume flow and the temperature difference varied and this affected the heat extraction rate from the heat pump. In situations where the heating water volume flow was high, the buffer store was discharged too rapidly and was unable to perform the regulating function for which it was originally designed.

Two serious difficulties occurred at about this time. Firstly, during preparation of the ground for the car park, damage to the earth source heat exchanger occurred several times. In addition the temperature sensors and the lines leading to the temperature sensors were damaged, and this led to increases in cost and considerable delay. Secondly, corrosion problems have occurred which have necessitated shutting down of the heat pump. Both the sound absorbers and the exhaust gas heat exchangers showed signs of corrosion which was the subject of a metallurgical examination and analysis of the engine oil. It was concluded that hydrochloric acid had been formed and this had led to the corrosion.

It was concluded that the installation of a heat pump complex such as that at Warendorf necessitated close co-operation between all groups taking part in the construction of the building, and in spite of detailed planning, additional expenditure which has not been taken into consideration is highly likely to be incurred. (The costs of this particular heat pump, based on calculations done towards the end of commissioning, are given in Table 5.6).

5.6.2 A Diesel engine-driven unit for a school The decision was taken in Denmark in January 1978 to investigate how natural energy resources could be used with a view to replacing or at least in part supplementing district heating schemes. The particular application selected was a high school in Sonderborg in Southern Jutland. The overall scheme was handed over to a firm of consulting engineers who subsequently supervised the complete installation. A feasibility study led to the conclusion that a ground source heat pump would be most satisfactory, with diesel engines as the compressor drives. Based on studies of the heat loading in the ground, the ground coils were designed for a maximum heat removal rate of 25 W per metre length. As can be seen from Table 5.7, a total ground coil length of approximately 9000 metres was planned, the heat output of the heat pump being 400 kW, (Ref.5.8).

The heat pump itself comprised two Diesel engine driven compressors, this enabling half of the heating requirement at least to be met during maintenance of one of the heat pump units. The specification of the heat pump is given in Table 5.8. The principle of the installation is that all of the central heating radiators and one of the two ventilation plant heating coils are connected to the heat pump. Should the heat demand exceed the heat pump capacity, the extra requirement is supplied by the second heating coil which remains connected to the local district heating mains. A further heat input is available via solar gain through the glass roof between two of the buildings. The heating circuit is sized for a mean water temperature of 40°C, and a 10°C maximum differential between supply and return temperatures. The volume of water circulating is $34m^3/h$.

The system uses Danfoss thermostatic radiator valves, and injection of liquid refrigerant into the brine heat exchanger is regulated by a thermostatic expansion valve. High and low pressure control is applied to the compressor, and the oil differential pressure control is monitored. Capacity regulation of the heat pump is controlled by three electronic dead-zone thermostats. One of these controls the capacity cut-in and cut-out in relation to the temperature in the heating plant's supply line, while another undertakes the same duty in relation to the temperature in the return line. The third thermostat functions in a safety role. Two stages of capacity change are practised. One increase in capacity occurs when the temperature in the supply line falls below 40°C, while a further capacity input results from a drop below 34°C. If the supply temperature exceeds 45°C the safety thermostat cuts out heat pump capacity but allows the district heating scheme to take over provision of the resulting deficit in heat requirement. A programmable controller permits the loading and unloading of compressor cylinders to be synchronised with changes in engine speed.

The heat pump system was commissioned in November 1979 and has proved to be highly successful during the trial period. A significant number of continuous temperature reading points are located both in the ground and external to the ground coil. Oil consumption has been recorded, together with the heat absorbed from the ground and the heat supply to the buildings. The preliminary data available indicated that the assumptions made concerning the ability of the soil to regenerate its heat capacity during the summer were too pessimistic. It was found that the soil temperature returned to its original level approximately 3 months after the system was switched off in the spring.

TABLE 5.7 Data on Danish Ground Coils

Available Ground Area	18,000 m^2
Total Pipe Length	9,000 m (approx)
Pipe Diameter	40 mm
Pipe Depth	1.5 m
Circulation Brine Volume	57 m^3/h
Assumed Soil Temp at Maximum Load	3°C
Brine Temp at Maximum Load	-6°C in, -10°C out
Frost Protection	20% calcium chloride

TABLE 5.8 Heat Pump Specification

Supplier of Package:	Brd. Gram A/S
Plant Output:	400 kW
Engine:	Diesel Type D-226-6 (MWM, Mannheim)
Engine Size:	2 x 64 kW
Compressor:	Reciprocating Type HC-6-100 (Brd. Gram, Denmark)
Refrigeant:	R22
Condenser:	Water in 35°C, out 45°C
Evaporator:	Brine in -6°C, out -10°C

TABLE 5.9 Energy Consumption of Ground Source Heat Pump

Monitoring Period	Nov 1979 - May 1980
Heat from District Heating	50,000 kW.h
Heat from Heat Pumps	952,000 kW.h
Total Heat to Buildings	1,002,000 kW.h
Diesel Engines Oil Consumption (equivalent to	64,000 litres 650,240 kW.h)
Circulating Pumps - Electricity Consumption	26,000 kW.h
COP	1.44

The measured system COP has been calculated to be approximately 1.4, and thus the system has generated about 60 per cent more heat for a given oil consumption than that which would have been used by a conventional oil fired central heating system, Table 5.9.

As a result of this project, encouragement was given to further installations using ground as the heat source, but it was pointed out that the planned use of natural gas in Denmark would lead to a greater use of gas engine driven units, and provision was made for converting the Diesel engines at Sonderborg to gas at a future date.

TABLE 5.10 Capital Costs of Ground Source Heat Pump

Plant	Cost (Dkr)
Ground Coils	449,000
Heat Pump	533,000
Pipework	195,000
Oil Tank and Flue	84,000
Automatic Controls	25,000
Civil Engineering	20,000
Connection Charge for District Heating (saved)	184,000
Net Extra Cost (ex. VAT)	1,122,000

TABLE 5.11 Operating Costs of Ground Source Heat Pump, (for Period as per Table 5.9)

	100°C District Heating	Heat Pump + District Heating
Heat to Buildings from Distrct Heating	200,000 Dkr	10,000 Dkr
Diesel Engine Oil	-	140,000 Dkr
Circulating Pumps (Electric)	6,000 Dkr	12,000 Dkr
Fixed Costs for District Heating	90,000 Dkr	30,000 Dkr
Heat Pump Maintenance	-	10,000 Dkr
Total (incl. 22% VAT)	296,000 Dkr	202,000 Dkr
Saving with Heat Pump		94,000 Dkr

(Annual saving estimated to be 117,500 Dkr, and 8.2 Dkr = $US 1)

5.7 Heat Pumps in Swimming Pools

The use of heat pumps for energy conservation in swimming pools has become a major application area in recent years, particularly in Europe, including the United Kingdom. The applications to which the heat pump might be put in swimming pools, and for that matter in other sports complexes, are many and varied. It may be used in most swimming pool installations to recover heat from humid exhaust air, for heating of indoor swimming pool complexes. It may also be used, in conjunction with ground water sources of one form or another, to directly heat indoor or outdoor swimming pools. The heat pump may be combined with heat storage in such applications. For indoor swimming pools, the provision of hot water for additional services, over and above those for pool heating, has been demonstrated. Of even greater benefit to the operator is the possibility of using a heat pump in a leisure centre where both cooling and heating duties on a substantial scale are required, as can occur, for example, in a complex where an ice rink and a heated swimming pool exist. All of these examples will be illustrated below.

5.7.1 Heat pumps for exhaust air heat recovery A typical public indoor swimming pool is a large user of energy, particularly if it is located in the cooler climates of North America or Europe, for example. The heat losses, expressed as a percentage of the total loss, may be approximately represented as shown in Fig.5.22 as a function of ambient air temperature. Figures quoted for the annual energy use of typical public indoor pools (Ref.5.9) suggest values of 5000-10,000 kW.h per square metre of water surface. A water temperature of approximately 30°C is required, and the air temperature is normally maintained at a temperature slightly higher than this. Ventilation rates of between 4 and 20 air changes per hour are also required.

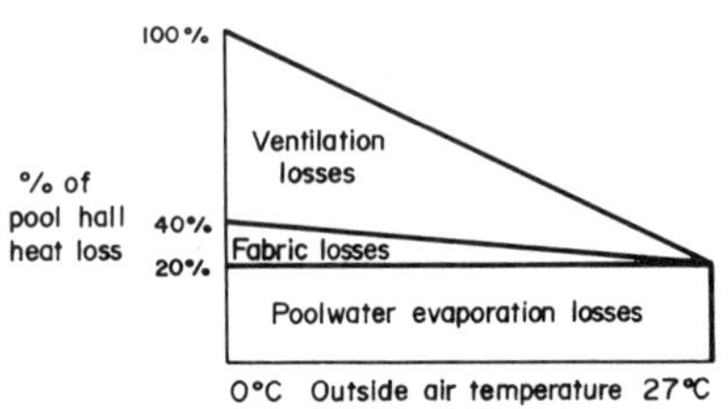

Fig.5.22 Typical Proportions of Heat Losses from an Indoor Swimming Pool

Conventional heat recovery systems, such as rotating regenerators, can be applied to the exhaust air from swimming pool buildings, the heat recovered by these devices being used to preheat ambient incoming air thus saving energy (Ref.5.10). The use of such heat exchangers in swimming pools is becoming a matter of routine. However, this only recovers a proportion of the heat contained in the air. Due to high humidity, the moisture content of the exhaust air is high, and most of the conventional heat recovery systems based on heat exchangers recover predominantly sensible heat. Recuperative heat exchangers are capable of condensing out a proportion of this moisture only, sometimes a comparatively small proportion. The recovery of latent heat can be much improved by using heat pumps, in many cases in conjunction with conventional heat recovery systems.

Typical of a heat pump installation using heat available from within an indoor swimming pool complex is that at Northgate Arena, Chester, in the United Kingdom. The two swimming pools form part of a larger indoor leisure centre, but the greater part of the energy requirements of the building, which has a design heating load of 2 MW, arise from the swimming pool areas, (Ref.5.11). As mentioned earlier, high ventilation rates are necessary in indoor swimming pools. In the case of Northgate Arena, the fresh air input to the complex is of the order of 46 m^3/sec, of which 21 m^3/sec is supplied to the swimming pool hall. High ventilation rates in this case minimise condensation in the pool and changing room areas, and also reduce the odour problem associated with the use of chlorine as a sterilization agent. In addition, the balance of the 2 MW is made up from pool water heating, the provision of hot water for showers and other domestic uses, and for heating of associated office accommodation. Nearly three quarters of the total heat load is for ventilation, of which that for the swimming pools is more than half.

Cost benefit studies carried out on a number of heat recovery systems showed that in this application the use of run-around coils in the ventilation ducts, together with a heat pump system would be the most economically attractive arrangement, and a flow diagram of the scheme is illustrated in Fig.5.23. Extract air is taken across the coil leading to the heat pump evaporator where it is cooled, and a considerable amount of latent heat is recovered, before it passes across one half of the run-around coil, where it is cooled by a further 4°C, as shown in Fig. 5.24. The fresh air is introduced across the second half of the run-around coil before being further heated by the output from the heat pump condenser. The overall heat balance indicated that the run-around coil recovers about 400 kW, and the heat pump slightly in excess of 1 MW, leaving a comparatively small proportion of the heat demand to be made up using conventional boiler plant. Gas fired boilers are used in conjunction with heat exchangers in series with the low temperature heating circuit from the heat pump, although the whole preheating requirements of the air for the swimming pool halls can be met using the heat recovery systems, leaving a surplus for other duties.

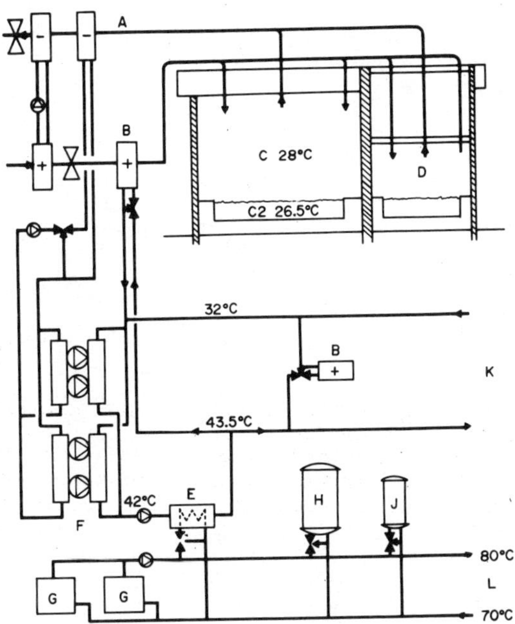

Fig.5.23 The Flow System of the Heat Pump and Run-around Coil Heat Recovery System at Northgate Arena Leisure Centre

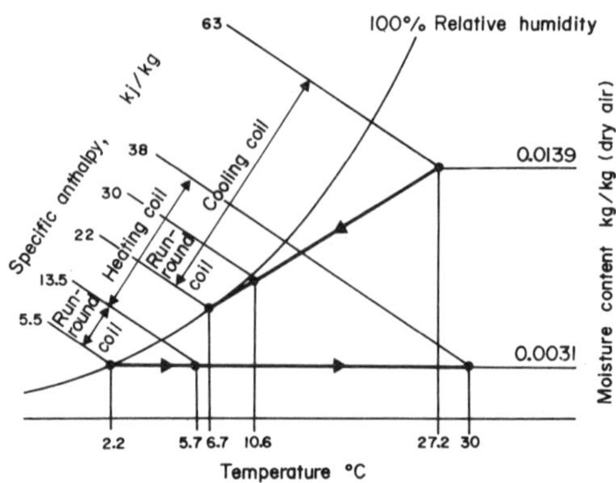

Fig.5.24 Psychometric Chart Showing Exhaust and Fresh Air Conditions in the Swimming Pool Area

Selection of the heat pump was based on the fact that it was believed prudent to operate as near as possible to normal air conditioning design temperatures, as standard packaged water chiller units could then be used. Two packaged plants were procured, each having twin reciprocating compressors. Both plants are operated in parallel, control being from the chilled water flow temperature.

Boiler capacity was determined by the need to heat the pools and associated buildings and services from cold. From initial operational experience of the systems, including the heat pump, which has a COP of slightly in excess of 4, it has been noted that the amount of evaporation from the pool surfaces and surrounding wetted area is significantly lower than that indicated by calculation. As a result the amount of heat available for recovery in the exhaust air ducts, particularly as far as the heat pump is concerned, is smewhat lower than anticipated. As a result, more supplementary heating is required for ventilation air introduced into the building, although this is to some degree offset by the lower heat requirement of the swimming pool water itself.

Recent data on the Northgate Arena heat pump (Ref.5.9) shows how effective the energy conservation measures have been. The performance is given in Table 5.12, and it can be seen that a simple payback of 2.2 years has been achieved. Pool data are given in Table 5.13.

5.7.2 Combining exhaust air heat recovery with other heat pump systems The use of heat pumps in indoor swimming pools is not restricted solely to air-air operation. Sulzer, who probably have more experience of varied applications of heat pumps in swimming pools than any other manufacturer, have combined a number of heat pumps, each fulfilling a different function, in indoor swimming pools. Typical of these is the installation at Lindenberg, Allgau, illustrated in flow sheet form in Fig.5.25. This indoor pool has a water surface area of 315.5 m², an air temperature maintained at between 30 and 32°C, and a water temperature 2°C below the air temperature.

TABLE 5.12 Energy Performance

		Standard Equivalent Pool (GJ/a)		Standard equivalent Pool (£/a)	
Typical Pool	Fuel Electricity				
	Total	22,000	20 GJ/m²	90,000	£83/m²
This Pool	Fuel	11,000		33,000	
	Electricity	3,000		30,000	
	Total	14,000	12.8 GJ/m²	63,000	£58/m²
	Energy Saving per year	8,000 GJ (36%)		£27,000 (30%)	

Installation cost £60,000 (over standard equivalent pool). Simple payback 2.2 years

Current energy cost per bather £0.15

TABLE 5.13 Pool data

Water temperature	28°C
Air temperature	29°C
Water volume	1,360 m³
Poolwater turnover	1.3 and 6 hours
Relative humidity	55%
Airchange	4-6 per hour
Air dilution (Winter)	50-100%
Disinfection	gas chlorine
Heating	2 gas boilers
Heater rating	860 kW

In this system, which may be largely understood by following the legend below the figure, a semi-hermetic reciprocating compressor-driven air to air heat pump (a) having a capacity of 228 kW, uses exhaust air from the indoor swimming pool as the heat source, in a similar manner to that described in the previous section. The fresh incoming air is passed over the condenser, which is followed by a conventionally heated coil which is able to provide supplementary heating as and when necessary. In cases where the space heating demand is less than the output of the heat pump, a secondary water-cooled condenser diverts this heat for swimming pool water heating.

A second heat pump (B), one of the standard Sulzer 'Liquifrigor' models, is used for heating the swimming pool water. These heat pumps also employ reciprocating compressors, and can provide heating capacities of up to 582 kW. The unit incorporated in the system illustrated in Fig.5.25 has an output of 258 kW, and is a liquid-liquid system, using as the heat source warm waste water from a nearby industrial plant. As well as heating the swimming pool water for the bathers, a proportion of the heat is stored for use by yet another heat pump. This is also of the Liquifrigor type, is a water-water unit, and has a heating capacity of 151 kW. Two condensers are used in parallel, providing heating of shower and other service water, and for floor heating. It is interesting to note that even the splash water collected around the edges of the pool is used as a heat source for heat pump (C).

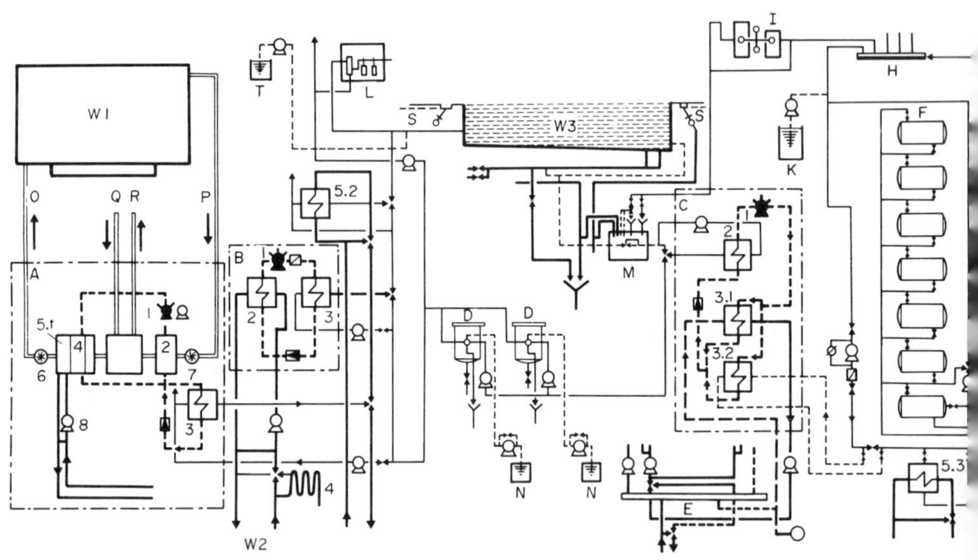

Fig.5.25 One and Two Stage Heat Pump Installation in an Indoor Swimming Pool. (Courtesy Sulzer Bros Ltd)

Refrigerant line
Air duct
Water line
Heat pump operation

Water line
Water Line

A Air/air heat pump
1 Compressor
2 Air Cooler (evaporator)
3 Shell and tube for removing surplus heat to swimming pool
4 R22 condenser for heating air
5.1 Booster radiator for air
6 Supply air fan
7 Return air fan

C Heat Pump: Heating of shower and utility water, floor heating
1 Compressor
2 Water cooler (evaporator)
3.1 Shell and tube condenser for floor heating
3.2 Shell and tube condenser for heating shower and utility water.

D Pool-water filter installation
E Heating distributor
F Warm water Storage tank
G Warm water distributor manifold
H Cold water distributor manifold
I Softener installation
K Phosphate proportioning

8 Heating circulation pump

5.3 Counterflow unit (PWW)

L Chlorine proportioning
M Splash-water collector tank
N Flocculant proportioning
O Incoming air to pool building

B LIQUIFRIGOR heat pump for heating the swimming pool
1 Compressor
2 Water cooler (evaporator)
3 Shell and tube condenser
4 Heating of outdoor side walk
5.2 Counterflow unit

W_1 = heat source: humid exhaut air from building for drying
W_2 = heat source: heated cooling water for pool heating and heat storage for floor heating and shower water
W_3 = heat source: stored heat from W_1

P Return air from building
Q Fresh air
R Exhaust air
S Floor heating
T pH correction

PWW = Pump warm-water heating

5.7.3 General comparison of swimming pool enengy efficiency measures The UK Energy Efficiency Office has, via the Demonstration Scheme and other opportunities, obtained comparative data on the effectiveness of a variety of energy efficiency measures applied to swimming pools. These include relatively simple activities, such as the provision of pool covers and associated 'good housekeeping', and conventional heat exchangers in addition to a number of heat pump examples. The comparison is summarised in Table 5.14, (Ref.5.9).

Included in the table is the heat pump at Farnborough Recreation Centre, illustrated in Fig.5.26. This gas engine-driven unit cost £150,000 and has a payback period of 4-6 years.

5.8 The Absorption Cycle Heat Pump in Buildings

A considerable effort is being made to introduce absorption cycle heat pumps into large scale commercial space heating. In addition, much research is being directed at developing domestic absorption cycle heat pumps which, operating with an air source, will be able to serve a wet central heating system. In both cases it is common to find that the heat pump generator is served by natural gas, and while commercially available systems have been introduced onto the market, they have been, in general, reported to be expensive and unreliable. The British Gas Corporation purchased three absorption cycle heat pumps claimed to be suitable for the small commercial market. These units have been installed in three of the British Gas Regional buildings for the provision of space heating, and a report (Ref.5.12) described operating data recorded during part of the 1983/4 season. Useful experience was also gained in the installation and operation of these types of heat pump systems.

Fig.5.26 View of Heat Pump at Farnborough Recreation Centre

The absorption cycle heat pump selected used the ammonia/water working fluid pair, and the principal parameters are given in Table 5.15. The design of the system was such that the units would only operate down to air temperatures of 2°C, and were unable to deliver water at temperatures above 50°C because of the high pressures which would result on the ammonia side. Approximately 0.82 kW was required in the form of an electrical input in order to power the solution pump, combustion fan and evaporator fan. It was pointed out by British Gas that this would amount to about 15 per cent of the total running costs of the system.

Before the three heat pumps were installed in buildings for monitoring, laboratory tests were carried out on one of the 25 kW units. It was found that combustion efficiency was 81 per cent, and this compares favourably with a conventional boiler. Recovery of heat from the flue gases was effected by diverting them over the evaporator. It was proposed that it would be better to use the flue gas via an economizer for water heating, and it is obviously of benefit to recover heat in this manner in absorption cycle heat pumps. Measurements of COP were made for a range of ambient air temperatures varying between 2°C and 15°C. The maximum COP achieved was 1.02, and the heat output varied from 15.7 kW to 19.5 kW, rising with the ambient temperature. Start up of the heat pump was such that it was found to take approximately 25 minutes for the system to achieve a steady state, delivering water at an acceptable temperature. During this period of warm up, the Coefficient of Performance was also adversely affected. At low ambient temperatures the COP was only 0.8. A number of other tests were carried out at half load, prior to the systems being installed in the three buildings.

TABLE 5.14 Case Study Summary - Descending Order of Simple Pyaback Period

Technique	Installation Cost (£)	Payback (years)
Pool covers and energy management programme	7,000	0.5
Variable ventilatioin + economiser	10,500	1.0
Thermal wheels + CHP	64,000	1.8
Electric heat pump heat recovery following improved conservation/ insulation	32,000[1]	2.1
Electric heat pump + run-around coil heat recovery	60,000[1]	2.2
Small electric heat pump + cover	2,000[1]	2.5
Variable flow ventilation, covers, high efficiency gas burners and CHP	74,000	2.6
Run-around coil and controls, conservation and low energy lighting	60,000	2.6
Ozone treatment + heat pump heat recovery	40,000[1]	3.3
Reverse flow circulation + ambient	2,500	4.2
Gas heat pump heat recovery	150,000	4.6
Run-around coil, variable ventilation, covers, solar heating + CHP	55,000[2]	5.0
Electric heat pump heat recovery	180,000	5.8
Solar heating	2,000	8.3
Thermal wheel and heat pump	-	not applicable
Plate heat exchanger, cover and insulation	27,000	uneconomic
Solar heating and insulation + heat pump heat recovery and controls	70,000	not applicable
Heat dump/heat reclaim	-	not applicable
Run-around coil and variable ventilation	-	not available
Low cost construction, controls and covers	160,000 (total pool)	not applicable

[1] Heat pump only

[2] CHP equipment with coil heat recovery

NB Payback periods have been calculated wherever possible for all the energy-saving measures installed in a pool. Indications are given, both in the Case Study Summary and in the individual Case Studies, where this is not the case. All costs and paybacks are based on the best information available.

TABLE 5.15 25 kW Absorption Cycle Unit Data

Type:	Air-water
Rated Heat Output:	25 kW
Primary Water Flowrate:	2.2 m^3/h
Water Pressure Drop:	0.17 bar
Maximum Delivery Temp:	50°C
Air Flow Rate:	6000 m^3/h
Minimum Air Temp:	2°C
Noise Levels:	55 DB at 5m
Dimensiions:	1.51m long x 0.95m wide x 1.42m high
Weight:	695 kg
Electricity Consumption:	0.82 kW

One heat pump was installed at Watson House, the British Gas domestic research centre. This was directed at providing heating for an office area, with a water delivery temperature of 50°C and a differential of 10°C. A back up boiler was also provided, this being relied on to provide the heat demands at ambient temperatures below 2°C. Switch over to the boiler at this minimum temperature is effected by an outside thermostat, and an interlock relay is incorporated to make sure that the heat pump and its circulating pump are unable to operate at this time.

The second unit is in Livesey Hall, where it provides space heating via fan convector units. The heat pump is used in conjunction with a condensing boiler, which takes over the heating duty when the ambient falls below 5°C. The heating system operates on two zones, the water flow temperature to each zone being controlled in a similar way to that at Watson House. An external temperature sensor and a water temperature sensor are linked to an electronic controller which provides the heating control based on a specified temperature schedule. Again, the flow through the heat pump is stopped when the minimum operating temperature specified is reached.

The third unit is located at the North Thames Energy Management Centre. It provides heat to a lecture theatre by means of a warm air distribution system. Ventilation heat recovery is also employed, this being able to provide up to 8 kW of the required 25 kW heat demand. The heat pump in this case is also an air to water unit, although the water coil from the heat pump condenser is passed to the air circuit where it gives up its heat via a finned tube bundle. It is not possible to directly condense in the finned heat exchanger because of regulations governing the use of ammonia.

All three heat pumps were installed in 1983, and were monitored during the relatively mild heating period of 1983/84. A typical heat pump operating cycle during a cold day is illustrated in Fig.5.27. The heat pump commences operating at full load in the morning because of the comparatively low overnight temperatures, and the effect of the thermostatic radiator valves can be seen on the heat pump cycling. Operation of these valves affects the return water temperature, leading to an increase which switches off the heat pump. If the room temperature increases sufficiently, it was found that the return water temperature did not fall below the thermostat setting, preventing the heat pump from firing. Although this did not happen in the case of the cold day illustrated, on a warmer day, it was found that the heat pump did not operate for a period of seven hours in the middle of the day when the ambient was rising quite rapidly. The variation in COP during the day can be seen in Fig.5.28.

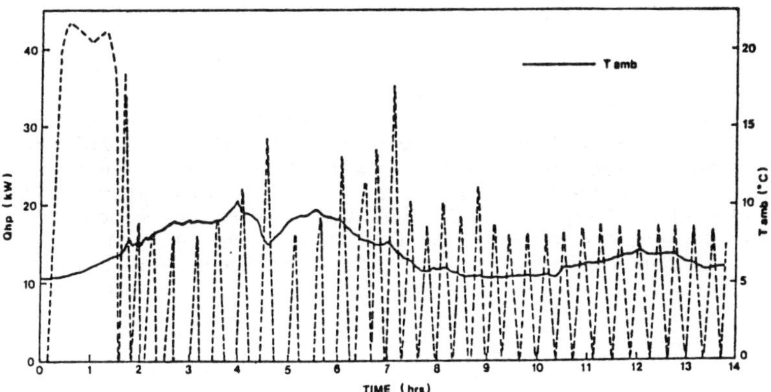

Fig.5.27 Heat Pump Operting Cycle - Cold Day

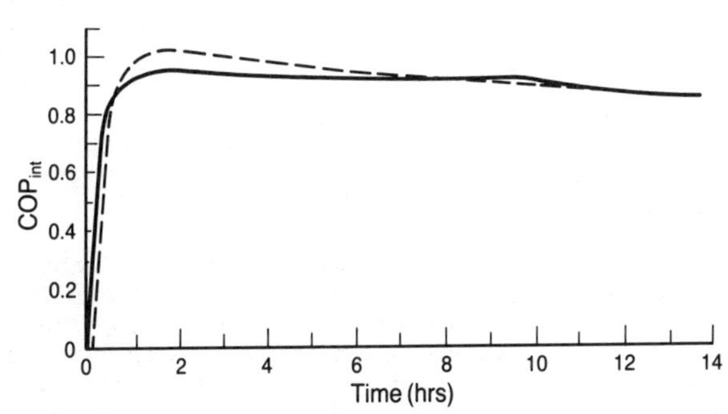

Fig.5.28 Daily Variation of Absorption System COP

Although the units had not been operated for a sufficiently long time to enable estimates of annual fuel savings to be made, some useful overall conclusions were reached. It was found that the total installed cost of the three heat pumps which were the subject of the monitoring exercise was approximately four times the basic cost of the heat pump. Although the cost of the larger radiators needed because of the lower delivery temperature can lead to higher costs for air-to-water systems, the extra ductwork involved in warm air systems can be set against this. If one compares this to the equivalent boiler system, the total heat pump installation cost will be approximately 2.5 times as great.

British Gas found that the reliability of the systems was high, the only failure being a pump drive belt. Noise levels were found to be not significantly different from those of conventional commercial boilers, although it was pointed out that in view of this fact, the domestic absorption cycle heat pump may create noise difficulties. The use of a back up boiler was considered a realistic solution to the problem created by the restriction on the minimum operating temperature of the heat pump evaporator, and while this would be acceptable for commercial systems, it is unlikely to be a viable solution in the domestic context. The cycling of the system can lead to reductions in overall coefficient of performance, and this has an influence on the economic viability. It was recommended that cycle rates of less than 0.5 cycles per hour should be the aim of any control system. It was appreciated that this could involve the use of a thermal store.

As a result of these monitoring exercises, it was recommended that further development on the heat pump design be carried out to improve performance and reduce capital cost.

REFERENCES

5.1 Dorman, R. Offices: better atmosphere aids comfort. Review of Industrial Heating, Ventilating and Air Conditioning, The Times, London, 21 Oct. 1974.

5.2 Bowen, J.L. Why you should consider heat recovery. Building Serv. Eng., Vol.44, pp.A20,22,24,25, Nov. 1976.

5.3 Anon. Heat from light in Chester electricity headquarters. Heating and Ventilating Engineer, pp.116-119, Sept. 1969.

5.4 Hess, J.H. and Moser, P. Large centrifugal heat pumps integrated into energy systems and economic aspects. Paper D2, Proc. 2nd Int. Con. on Large Scale Applications of Heat Pumps, York, 25-27 Sept. 1984. BHRA, Cranfield, 1984.

5.5 Sacks, T. Sweden turns to sewage and icy lakes for winter warmth. Electrical Review, Vol .219, No.4, 5/12 Sept. 1986.

5.6 Lindstrom, H.O. Experiences with a 3.3 MW heat pump using sewage water as heat source. Paper C2, Ibid.

5.7 Gunia, J. Compressor type heat pump, with the ground as heat source, for heating a local authority building. Energy Saving in Buildings, Proc. CEC Conference, The Hague, 14-16 Nov. 1983, pp.242-249. D. Reidel, Dordrecht, Holland, 1984.

5.8 Barmwater, P.A. Danish heat pump system. Australian Refrigeration, Air Cond. & Heating, Vol.37, Pt.6, pp. 42-44, June 1983.

5.9 Butson, J. et al. Energy Efficiency Technology for Swimming Pools. Energy Technology Series Report No.3, Energy Efficiency Office, London, Jan. 1985.

5.10 Reay, D.A. Heat Recovery Systems: A Directory of Equipment. E & FN Spon, London, 1979.

5.11 Shave, R.E.J. Northgate Area. The Consulting Engineer, pp. 38-39, April 1978.

5.12 Searle, M., Sharma, V. and Evans, R. Operating experience of three gas fired absorption heat pump installations. Proc. Conf. on Directly Fired Heat Pumps, Bristol, 19-21 Sept. 1984.

CHAPTER 6

Heat Pump Applications in Industry

This Chapter describes in some detail the application of heat pumps in industry, an area which to date has seen much activity in research and development effort, but a rather limited amount of practical implementation.

As will be shown later, the use of heat pumps in industry has been regularly proposed, and in some cases practiced, over the past 30 years, and other processes capable of upgrading waste heat have also been routinely used over a long period. Oliver Lyle's book, The Efficient Use of Steam, is most revealing in this respect (Ref.6.1). It has, however, remained until recently a comparatively dormant area in terms of growth, but there are a large number of industrial applications receiving attention now that systems are available. Some of these will be discussed in detail below.

In particular, Section 6.4 includes details of the development of a heat pump designed to operate in an industrial environment at a condensing temperature of 120°C, and examples of the economic implications of its use.

6.1 Broad Areas of Application

The application of heat pumps in industry, (and this covers areas such as power generation, as well as the manufacturing and processing industries), may be categorised in the following way.

6.1.1 Heat recovery from refrigeration Firstly, there are a very large number of application areas where refrigeration is regarded as the primary function of the heat pump/refrigeration cycle. However, as is obvious, any refrigeration duty necessitates rejecting heat from the condenser at a temperature higher than ambient. Conventionally this heat may be rejected via a cooling tower, or by means of an air or water cooled condenser directly to atmosphere or waste. Such heat may alternatively be recovered and used for space heating, water preheating, or some other function. Because the heat is often of a comparatively low grade, its use may be restricted to areas conveniently located with respect to the condenser, such as warehouses and factory buildings, rather than offices. One of the main advantages of such an installation, however, is the fact that replacement of the conventional condenser arrangement with one suitable for heat recovery is comparatively inexpensive, and a rapid return on the investment may be anticipated. It must be borne in mind however that the coefficient of performance will not be optimised for the heating mode, although because of the need to install plant for refrigeration duty, producing some heating at a relatively low additional cost, makes this point less significant.

6.1.2 Drying, evaporation and boiling processes A second major application area, considerably more diverse than heat recovery from refrigeration equipment, both in terms of process use and the equipment available to fulfill the function, is associated with drying, evaporation and boiling.

The potential of conventional vapour compression cycle heat pumps in drying is beginning to be realized. In this role the evaporator recovers heat from the humid exhaust air, at the same time cooling this air, as shown in Fig.6.1. This cooling process leads to condensation of some of the moisture, drying the air sufficiently for at least a proportion of it to be recirculated, where it and any make-up air required is heated using the condenser section of the heat pump. It may be argued that on many dryers, particularly those involving only the evaporation of water, the heat pump could well surplant conventional gas-gas waste heat recovery systems in cases where the economics are attractive. (Note that this concept is identical to that applied to the domestic tumble dryer described in Chapter 4).

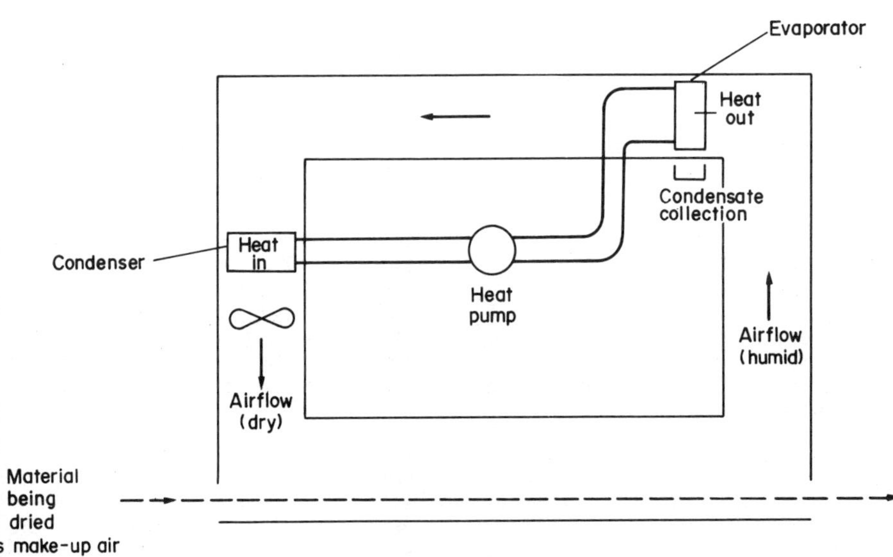

Fig.6.1 Schematic of a Vapour Compression Cycle Heat Pump in a Drying Process

More unconventional types of heat pumps may be used in this category of applications. As proposed for dryers, the high temperature recompression cycle uses superheated steam as the heat transfer medium. In a practical arrangement the exhaust air from the dryer would be split into two streams, one stream being compressed adiabatically, thus raising its temperature and pressure. At the higher pressure the latent heat of condensation is transferred via a heat exchanger at high temperature to the second stream. Heated and dried in this manner, the second stream may be recirculated through the dryer. The first stream is rejected as condensate.

The use of steam in heat pump cycles can overcome some of the problems associated with the more conventional refrigerants normally used, when high evaporating and condensing temperatures are required. Reporting in 1984 on the latest results of the long term development of a steam recompression cycle heat pump dryer, the Electricity Council Research Centre at Capenhurst in the UK presented an economic assessment of a system providing a dryer inlet temperature of 160°C (Ref.6.2). This type of heat pump dryer, illustrated in Fig.6.2, was compared with a hot air dryer having an evaporative efficiency of 56 per cent. Use of the heat pump system would, it was claimed, result in annual savings of £350,000, based on 5000 hours per annum operation. The additional capital cost of such a system would be £481,000, thus giving a very attractive return.

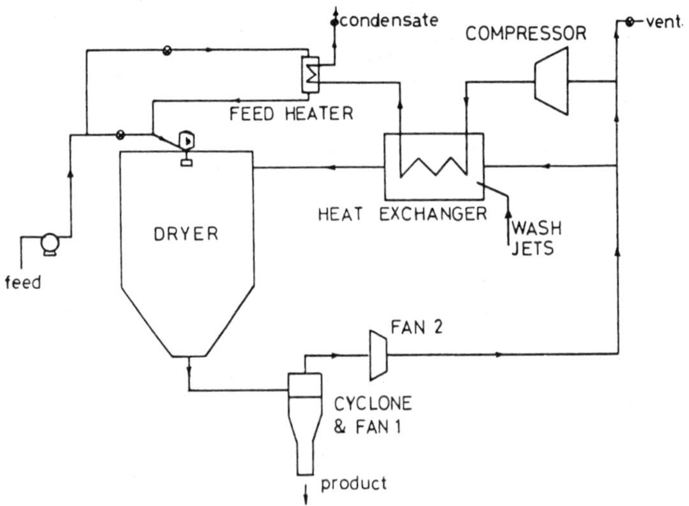

Fig.6.2 Steam Recompression Dryer Pilot Plant Developed at Capenhurst by ECRC

Evaporation and distillation processes make use of a third form of heat pump, that based on mechanical vapour recompression (MVR). MVR is similar in concept to high temperature recompression, but in this case all the vapour flow is passed through the compressor, thence to the heat exchanger in the evaporation vessel, as illustrated in Fig.6.3. The implications of this in evaporation and concentration processes are discussed in more detail later.

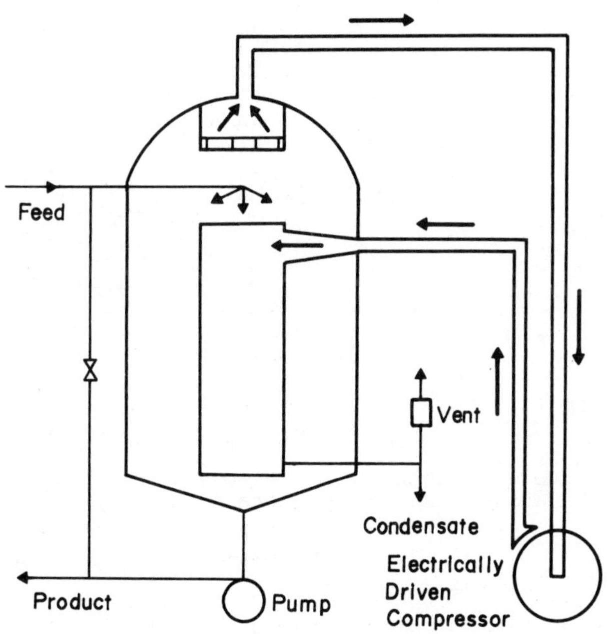

Fig.6.3 Layout of Typical Mechanical Vapour Recompression Cycle Unit

6.1.3 Heat recovery from liquid effluents This is one of the main areas for industrial heat pump applications, utilizing both electric and gas or diesel engine-driven compressor systems normally operating on the conventional vapour compression cycle, and more recently absorption cycle units. Typical application areas are shown in Fig.6.4 (based on data describing the Westinghouse Templifier industrial heat pump, a system marketed in the 1970's and early 1980's. The range of industries where liquid effluents are available at temperatures of between 10 and 60°C is large, and not included in this figure is the potential application of preheating boiler feedwater, universally acceptable in all of these industries. (The figure also includes some drying applications, which of course require modified condenser and evaporator configurations).

It is in the field of heat recovery from liquid effluents that the heat pump is of particular interest in the industrial context, because, unlike many gaseous effluents which are at relatively high temperatures, the low grade heat cannot be always reused by applying conventional heat exchangers. Of the liquid-liquid heat exchangers available, the plate heat exchanger is the most compact and efficient, with regeneration efficiencies

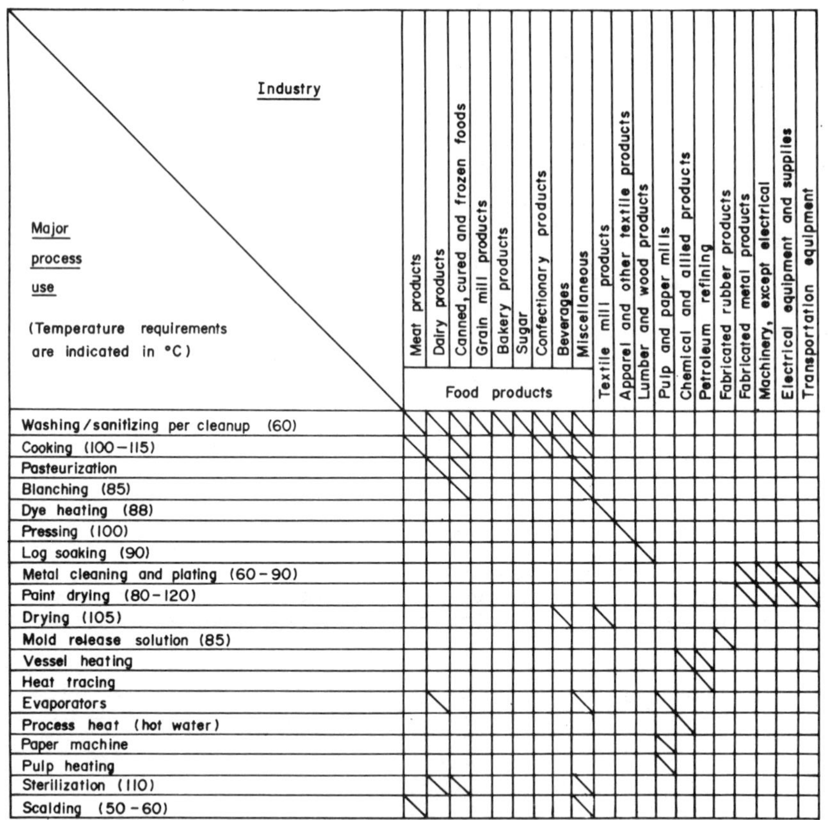

Fig.6.4 Potential Applications and Temperature Requirements of a Templifier Heat Pump

in excess of 94 per cent being economically viable (Refs. 6.3, 6.4). However, this recovered heat may be often at an insufficiently high temperature for reuse, whereas using, for example, a heat wheel on a high temperature gaseous exhaust enables high grade heat to be recovered for reuse without the need for severe upgrading. A role can be found, of course, for both a plate heat exchanger and a heat pump in the same system, where this can effectively increase the COP by reducing heat source-sink temperature differences. It can also be used to reduce the size of heat pump required for a given duty, by putting a proportion of the load through the heat exchanger.

The liquid-liquid heat pump can be applied to higher temperature processes of the type described in Fig.6.4 in a similar manner to the system briefly described in Section 6.1.1. Where both a cooling and a heating duty exist together, easing the load on a cooling tower or a separate refrigeration process can lead to benefits at both the condenser and evaporator 'end' of the heat pump.

6.1.4 Other application areas The application areas broadly described above, which serve as an introduction to this Chapter, are largely characterised by the type of heat source and/or sink. Refrigeration or chilling, with heat recovery for space heating, humid exhaust air or other vapours, and liquid effluent. This essentially covers the potential heat sources and sinks in industrial processes, with the exception of hot exhaust gases which in most instances can benefit very well from the application of conventional heat exchangers. An example will be given later of such an application. This particular case is all the more interesting because the locations of heat source and heat sink are some distance apart, making the use of ducting of the waste heat possibly prohibitively expensive.

No mention has been made yet of the use of sources outside the process environment which could be considered for some process applications. In general, however, industrial processes where heat recovery is of value, use heat at such a temperature and in such a quantity that the location of an evaporator in a river, the ground, or even in the ambient air, would be impractical from an engineering point of view and uneconomic. The degree of upgrading required would be such that only low COP's would be achieved. Although of course by using the evaporator of a refrigerator as a heat source, one may be achieving low COP's, the refrigeration duty in such instances is the 'raison d'être' for the compressor assembly, and cannot be directly compared with a 'heating only' air or natural water source heat pump.

Two other heat sources which have not yet been mentioned will, however, be considered in more detail later. The use of geothermal heat, in the form of hot water, as a heat source has been proposed in an assessment of heat pumps for industrial use (Ref.6.5), and heat sources in power stations are also of interest in this context (Ref.6.6).

It is now proposed to detail these broad application areas, giving data on the types of equipment used, particularly where this differs significantly from heat pump systems described earlier, the performance and economics, and potential areas of wider application. Where appropriate case histories of installations illustrating applications will be given.

6.2 Heat Recovery from Refrigeration and Cooling Water Systems

The potential for using the condenser heat output of refrigeration plant has long been realized, and has been practiced in many fields. In Chapter 4 reference was made to a system which could cool the food store in a home, at the same time providing space heating for rooms. Chapter 5 cites examples of the use of rejected heat from refrigeration plant in an ice rink being used to warm the water and environment of a swimming pool within the same complex, as well as the routine use of heat recovery from refrigeration in air conditioning plant.

As in this latter case, it is to Sulzer that one may turn to find an early example of an industrial process refrigerator which also provides useful heat. The chemical industry has for many decades been the user of large scale refrigeration plant. A massive refrigeration installation constructed in Basle in 1947, for example, produces 240 tonne of ice per day, and also handles the cooling of brine to -10°C, with a capacity of 270 kW, plus a further 480 kW for water cooling. This installation serves simultaneously as a heat pump to deliver 2320 kW to heat general purpose water from 18°C to 72°C (Ref. 6.7).

The layout of such chillers is similar to those shown for applications in the commercial sector, described in the last Chapter.

Another early application area where heat recovered from refrigeration plant could be usefully employed is in milk coolers. Work in the United States in 1952 (Ref.6.8) on the conversion of farm milk coolers which conventionally used air-cooled condensers to fulfilling a dual role as milk coolers and water heaters indicated useful savings when electric heating was used on the farm for heating of churn washing water etc. With the modified arrangement, a condenser coil placed within an insulated water tank could be used to preheat mains water before it passed to the conventional water heater. Theoretically the daily heat loss from a 4 churn milk cooler was of the order of 6 kW, the cooler reducing the temperature of the milk from 32°C to 4.5°C. If all this rejected heat could be employed for water heating, it was calculated that approximately 250 litres of water could be heated from 16°C to 46°C in each 24 hour period, providing a useful input, as the water may be used for cleaning processes at temperatures of between 65°C and 70°C.

A prototype installation, using R22 as the refrigerant, with a compressor power of approximately 250 W, was assembled using a water tank of 225 litres capacity heated electrically, and a tank of half this capacity served by the refrigerant condenser. The results of the tests are shown in Fig.6.5. On average better than 50 per cent of the electrical energy was saved by placing the refrigerator condenser in the preheat tank.

The heat source for such heat pumps need not, of course, originate solely from refrigeration circuits. The use of ambient temperature water to cool equipment and processes also results in some cases in sources of interest to heat pump users. Examples of this are given later.

One may look more generally at the effect of putting the heat from refrigeration plant to use, before moving on to more specific examples. The consulting firm of W.S. Atkins, responsible for many large refrigerating schemes, cite examples in the food industry, where regularly both cooling and heating are required in the same building. Here heat from the refrigeration plant could be put to use for heating process water, the additional heat required to raise it to its final operating temperature being obtained from the boiler. Based on W.S. Atkins figures, Table 6.1 shows the operating costs of basic refrigeration plant, using a shell and tube condenser, with that of a system incorporating a condenser where the heat is recovered for process use. In the conventional system the costs associated with the cooling tower are very significant. Cooling towers are used to reduce the temperature of the water used to take away the rejected heat at the condenser, so that the water may be recirculated.

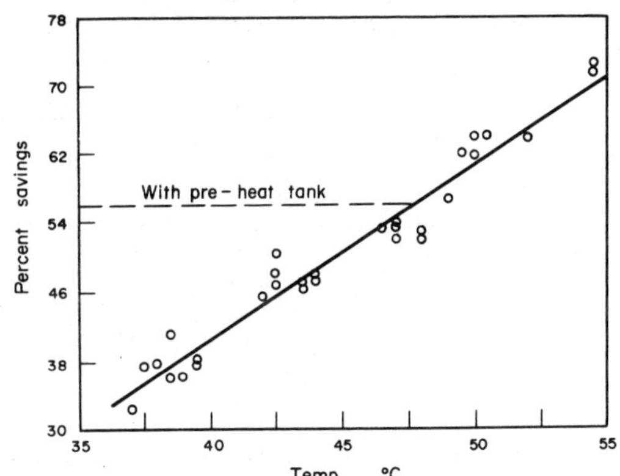

Fig.6.5 Savings Achieved by Recovering Heat from a Milk Cooler, as a Function of Water Inlet Temperature to the Tank Containing the Condenser

TABLE 6.1 Operating Costs of Refrigeration Equipment

(Operating Costs given in £Sterling)

	Conventional System	Heat Recovery Systems		
Condensing Temperature (°C)	35	47.5	35	22.5
Electricity Consumption	50000	61300	46900	33300
Value of heat recovered	0	-58400	-55200	-52000
Water (cooling towers)	5000	500	750	1000
Maintenance	7500	1300	1300	1300
TOTAL (£/annum)	62500	4700	-6250	-17000

W.S.Atkins point out, as is evident from Table 6.1, that the power costs are less when the heat is recovered at a lower temperature. This is analogous to the COP performance of a heat pump as a function of increasing temperature difference between evaporator and condenser. In the case illustrated, the capital expended in modifying the plant to recover the heat was recovered in a little over one year.

6.2.1 Heat pumps in the plastics industry One of the classic applications of refrigeration heat recovery is in the field of plastics injection moulding, typified by the installation at the Link 51 factory in the United Kingdom. The annual energy savings due to this installation amounted in 1976 to £15000, and the layout is illustrated in Fig.6.6, (Ref.6.9).

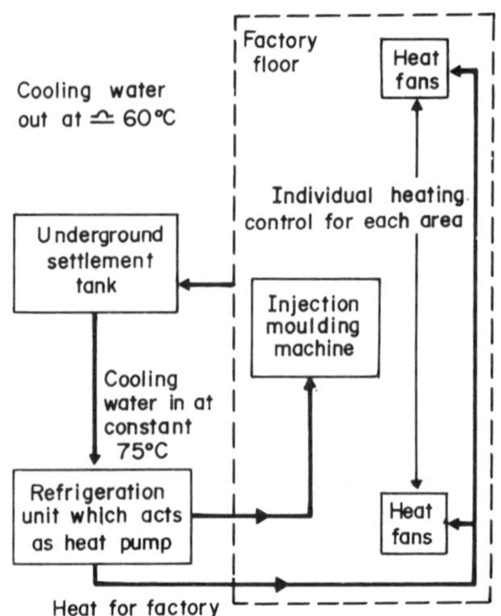

Fig.6.6 A Heat Pump Functioning as a Water Chiller and Space Heater in a Plastics Factory

Cooling water from the injection moulding machines is continuously recycled via a hot-well and a Prestcold Central heat pump. The heat pump, which replaces a conventional cooling tower, reduces the temperature of the cooling water from about 11°C as it leaves the injection moulders to a steady 7.2°C at the rate of up to 1140 l/min. The heat extracted from the cooling water is transferred to the space heating system, consisting of fourteen horizontal fan-type air heaters.

A major benefit of using the heat pump system was a 5 per cent reduction in cycle time and a corresponding increase in productivity. This was because the heat pump provided cooling water to the injection moulding machines at the correct low temperature all the year round, whereas temperature variations would be common with a cooling tower.

There was also a significant saving in fuel costs because oil or gas is not needed to heat the 3160 m² factory. When the machines are working the heat saved from the process, plus the heat generated by the compressor plant, amounts to about 325 kW - which provides adequate space heating under the coldest conditions. Provision was made for space heating when the machines were off-line, or just prior to start-up after weekends. Two 140 kW electric flow heaters were installed in the hot-well and could be switched on prior to work recommencing or during injection moulding machine maintenance so as to keep the temperature of the hot-well adequate for the space heating requirement. This made optimum use of energy since any low grade heat in the hot well was not wasted but was brought up to operating temperature by the immersion heater.

A further installation is in France, where heat recovery from extruding machines is being used for factory space heating.

In this particular instance a water-to-water heat pump manufactured by Cliref is used to recover heat from five extrusion machines used for coating cables. The extrusion machines have a total installed electrical load of 500 kW. The units utilise a total energy input of 115 x 10^3 MJ (electrical energy input) per annum while providing workshops environmental heating amounting to approximately 400 x 10^3 MJ, (Ref.6.10).

In a second example in the United Kingdom, Revell, Injection Moulders, who produce plastic hobby kits moulded in polystyrene, had a requirement for both heating and cooling. Chilling is required to cool the water being supplied to the injection moulding machines, this water being used to cool the moulds and thus hasten the solidification of the plastic. 22 kW of chilling requirement exists, and a total of 108 MJ/h at 0°C is recovered at the evaporator. This, together with energy from the compressor, is delivered via the condenser to provide air heating for the moulding shop during the winter heating season, (Refs.6.10, 6.11).

An injection moulding plant in Wisconsin, USA, also effectively utilizes heat pumps. The factory area totals approximately 20,500 m^2, with the injection moulding machines being arranged in four separate groups. Each group of machines is provided with a ring-main cooling water system, this being served in turn by a number of chillers, the number of units operating being determined by the temperature duty requirement. The total of four groups had 2, 3, 7 and 13 heat pumps providing a total chilling capacity of approximately 400 tons of refrigeration (1.4 MW). The quantity of heat recovered totalled 7.85 GJ/h, and this was used to maintain an area of approximately 19500 m^2 devoted to manufacturing and storage at a temperature of 21°C. Using this heat recovered via the heat pump chiller system, 78 per cent of the factory's heating requirement was met.

As both heating and cooling are required in the types of factories under consideration, it is useful to make a comparison of the economics of heating and cooling methods. Data available compare electric heating, as illustrated in Table 6.2, and three different water cooling methods, as detailed in Table 6.3. It is obvious from Table 6.2 that the heat pump is at the higher end of the capital cost range of equipment available, and it therefore could well restrict its applications to capital-intensive industries or processes which operate for in excess of seven or eight hours per day. Injection moulding is one of these industries.

As far as the cooling is concerned, heat needs to be rejected from the process and all production machines are equipped with water cooling. A number of different cooling methods are available as detailed in Table 6.3, and these have different capital and running costs. There is also some difference in the cooling effects which they produce, according to their type. Different temperature levels may be achieved, and different variations in temperature can occur. It is believed that the heat pump can provide a number of economic benefits which would not result from the use of other systems, as it can provide constant temperatures throughout the year. The ability to maintain constant process temperatures means that consistant production quality will be achieved, leading to fewer rejected components. In addition, the ability to vary the duty to give the most appropriate temperature for the material being processed will lead to optimum production rates being achieved.

TABLE 6.2 Cost Comparison of Electric Heating Methods (Ref.6.10)

	Approximate Capital Cost £/kW of Mains input	Efficiency of Application	Approximate Cost £/kW Applied
Metal sheathed elements (liquid heating)	7-10	0.95	7-10
Cartridge elements) Band (barrel) heaters)	12-15	0.75	15-20
Package steam boiler	40-50	0.98	40-50
Heat pump/chiller	300	2.5	120
Dielectric	500	0.7	700
Microwave	1000	0.6	1700

TABLE 6.3 Temperature and Specific Cost Comparison of Different Methods of Water Cooling Indicative (Ref.6.10)

Method	Summer Water Temp Achieved (°C and Variability)	Specific Costs £/100 000 kcal/h extracted	
		Capital (installed)	Running
Direct (e.g. town mains)	10 - depends on water source	Negligible	2.2 (10°C rise) 4.4 (5°C rise)
Cooling tower	25 - depends on bulb temp	900 (1 000 000 kcal/h)) 2500 (100 000 kcal/h)) 4000 (25 000 kcal/h))	0.15-0.2
Air cooled Heat pump/ chiller	0 - 10 Typically constant to ± 1% within this range	12000 (1000,000 kcal/h)) central system) 17000 (25 000 kcal/h)) 40000 (2500 kcal/h)))	1.2

The use of vacuum pump cooling water as a heat pump heat source provided some useful data on operating experience. This heat pump system was studied as part of an exercise funded by the International Energy Agency (IEA), under the auspices of the Advanced Heat Pump Programme, and is available as a detailed case history (Ref.6.12). The installation is located in an electronics component manufacturing plant, and the heat pump system was commissioned in March 1980. The reporting period covers the subsequent 18 months of operation.

The layout of the system is shown in Figs.6.7 and 6.8. The heat source is cooling water from a set of four vacuum pumps, the heat dissipated from the pumps totalling 51 kW. The cooling water flow rate is 9 m^3/h. Performance of the vacuum pumps is influenced by the cooling water temperature, and in this application the cooling water inlet temperature was generally maintained below 20°C. While the number of vacuum pumps in operation can vary, the average load is 75 per cent of the above maximum value.

Following its passage through the vacuum pumps, the cooling water enters a reservoir where it is cooled by the heat pump evaporator (or by the addition of cold water). The air-cooled condenser preheats incoming air for an assembly region. The design conditions are given in Table 6.4.

The operating hours of the vacuum pumps average 2800 per annum, while a space heating demand exists for 2000 hours per annum. The space heating demand far exceeds the heat output of the heat pump system, and thus in the economic analysis it was assumed that the heat pump would function for the full 2000 hours per annum.

The original specification had a somewhat lower condenser duty (26.83 kW) than that given in Table 6.4. In addition, it is important to point out that by cooling the vacuum pump water, and recycling it, significant savings in water charges could be effected. As may be seen below, these savings were the main justification for the project, the actual running cost of the heat pump being more than the value of the energy (natural gas) saved.

Savings due to decrease in well water consumption	:	100,397 BF[1]
Savings in natural gas	:	28,440 BF
Heat Pump running cost	:	41,950 BF
Net annual saving	:	86,887 BF

[1]Costed in 1979. Current, £1≃ BF 60 (Jan 1987)

The predicted payback period was 2.93 years, and while this was outside the range normally accepted by this company (2 years maximum), it was believed that natural gas prices in Belgium would rise at a faster rate than those for electricity. A unique situation arose also because it was anticipated that the use of well water by industry would in future be severely limited, or even prohibited. The water savings were, therefore, particularly significant, and fouling problems sometimes associated with the use of large quantities of fresh water were minimized.

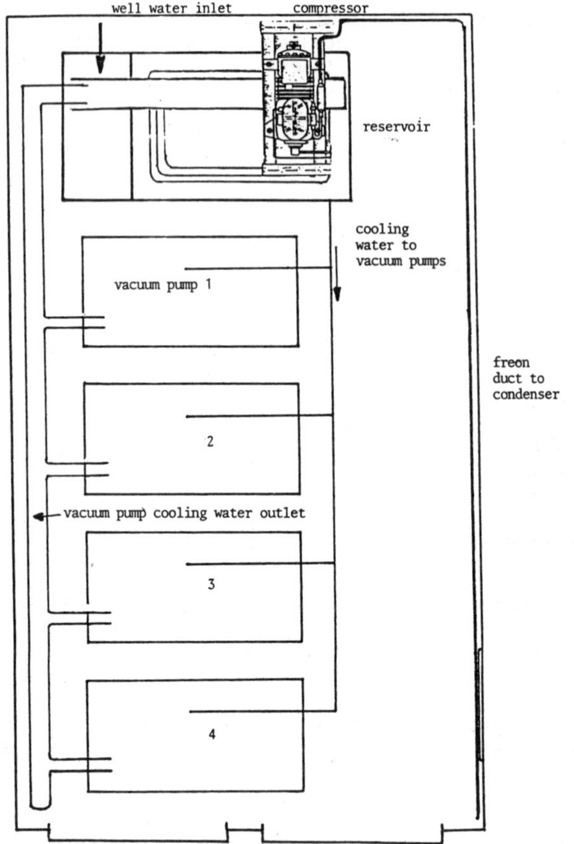

Fig.6.7 Heat Pump Layout in Belgium Plastics Factory

Problems shortly following commissioning of the heat pump system centred on leakages of refrigerant, which were soon cured. Monitoring of the heat pump over a 1 week period 2 years after installation revealed the following:

(i) the control system regulating the flow of cold well water did not operate properly, the thermostat setting being too high.

(ii) Poor condenser performance was evident. It was found that severe fouling of the external finned surfaces had occurred, and obviously no presentative maintenance had been carried out. The corresponding compressor power of 10 kW was significantly greater than the design absorbed power.

(iii) The condenser was located 13m from the compressor/ evaporator assembly. An insulated copper tube of 17 mm bore was used to carry refrigerant vapour to the condenser. It was discovered that during typical operating conditions, the pressure drop between the compressor and condenser was unacceptably high at 2 bar. This also partially explained the high compressor power consumption.

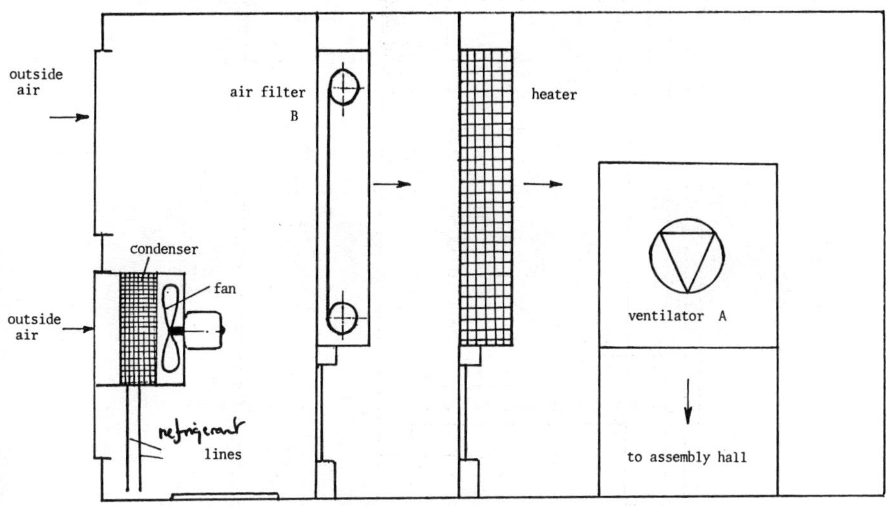

Fig.6.8 Heat Pump Layout - Local detail in Belgian Plastics Factory

It was found that condenser heat output was also increased, and in fact the increased electricity cost was almost offset by the increased savings in well water and natural gas, as illustrated in Table 6.5.

The use of combined heating and cooling will be dealt with further in Section 6.4, principally in the context of liquid-liquid duties. The reader is also referred to Chapter 5, where a discussion on the Dunham Bush, and other, heat reclaim systems is given.

However, it is worth emphasising here that any process using refrigerators to provide cooling, chilling or freezing conditions can directly reclaim heat from the condenser cooling water or air. As pointed out by Braham (Ref.6.13), it is also practical to use a refrigerator which is cycling intermittently for a constant supply of warm water or air for factory heating, when a thermal buffer tank is installed between the process heat recovery and the environmental space heating system.

Some cooling was carried out by the heat pump shown in Fig.6.9, which was driven by a steam turbine. Installed as a UK Energy Efficiency Office Demonstration Project at Marfleet Refining Company, Hull, the system heated boiler feedwater to 80°C. The turbine recovered energy which would otherwise have been expanded via a valve.

Fig.6.9 Steam Turbine-Driven Heat Pump for Feedwater Heating

Appropriate to Sections 6.2 and 6.3, some data are given in Tables 6.6 and 6.7 on the working fluids used in heat pumps and refrigerators for these duties. This gives a comparison of the vapour pressures of the various refrigerants, at evaporator and condenser temperatures, for a variety of applications ranging from low temperature food freezing to high temperature heat pumps (see also Section 6.4), (Ref.6.14).

Note that ammonia has too high a vapour pressure at heat pump temperatures to be of use in such systems. Pressure is not the only criterion which is used in determining the correct working fluid to be used in a particular application. The amount of heat which can be transported by the refrigerant is a strong function of its latent heat, or enthalpy, values of which are listed in Table 6.7. The specific volume of the working fluid determines the displacement of the compressor needed to deal with the flow of fluid necessary to meet the duty. Critical pressure and temperature influence the limiting conditions beyond which the fluid cannot be successfully used, and the miscibility of the working fluid with the compressor lubricating oil is a significant factor in the selection of fluids.

As well as the above properties, such considerations as cost have to be taken into account, although the market for most of the commercially available refrigerants is sufficiently large and competitive to keep prices within reasonable bounds. In addition, as the secondary circuit is in theory a closed system, loss of refrigerant should be minimal, and the quantity required in even a large system will not be a significant proportion of the capital or running costs.

TABLE 6.4 Design Conditions of Belgian Electronics Factory Heat Pump

Condenser:	
Air inlet temperature	10°C
Air outlet temperature	25°C
R12 inlet temperature	63°C
R12 outlet temperature	35°C
Air flow	2.27 m^3/s
Heat supplied	40.7 kW
Evaporator:	
Water inlet temperature	25°C
Water outlet temperature	20°C
R12 inlet temperature	10°C
R12 outlet temperature	18°C
Saturation temperature	10°C
Pressure	4.2×10^5 Pa
Water flow rate	5.85 m^3/h
Heat extracted	34 kW
Compressor:	
Pressure ratio	2.02
Suction flow rate	43.1 m^3/h
Isentropic efficiency	50%
Power of motor	6.7 kW
Discharge temperature	63°C
Heat Pump COP	5.44

TABLE 6.5 Comparison of Design and Actual Costs of Belgian Heat Pump

	Design	Actual
Water Savings	180,150 BF	204,200 BF
Natural Gas Savings	69,636 BF	85,500 BF
Electricity Cost	71,786 BF	111,700 BF
Net Savings	178,000 BF	178,000 BF

6.3 Drying, Evaporation and Boiling Processes

The field of drying, evaporation and boiling covers a large range of industrial processes, and the heat pump in its conventional form is in competition with other heat recovery systems (Ref.6.15) and a variety of other compression cycles. The cycles available, and examples of their application in the above fields, will now be discussed.

TABLE 6.6 Saturation Pressures with Various Refrigerants at Conditions Varying from Applications in Low Temperature Refrigeration to High Temperature Heat Pumps

Typical Application	Temperatures Evap.	Cond.	Refrigeration	Pressures Evap. kg/cm abs	Cond. kg/cm abs	Remarks
Food Freezing	-40°C	+35°C	R12	0.65	8.6	Usually two
			R22	1.07	13.8	stage
			NH3	0.73	13.8	compression
Food Storage	-20°C	+35°C	R11	0.16	1.5	Low pressure
			R114	0.38	3.0	refrigerants
			R12	1.54	8.6	
			R22	2.50	13.8	High pressure
			NH3	1.94	13.8	refrigerants
	-10°C	+35°C	R11	0.26	1.5	
			R114	0.60	3.0	
			R12	2.24	8.6	
			R22	3.60	13.8	
			NH3	2.96	13.8	
Water Chilling	+1°C	+35°C	R11	0.43	1.5	
			R114	0.94	3.0	
			R12	3.26	8.6	
			R22	5.25	13.8	
			NH3	4.56	13.8	
	+1°C	+50°C	R11	0.43	2.4	High condensing
			R114	0.94	4.6	temp. for use
			R12	3.26	12.4	of air cooled
			R22	5.25	20.0	condenser
			NH3	4.56	20.7	
Heat Pump	+25°C	+70°C	R11	1.05	4.2	
			R114	2.18	7.4	
			R12	6.6	19.0	
			R22	10.5	30.5	
			NH3	10.2	35.0	
	+25°C	+80°C	R11	1.05	5.3	
			R114	2.18	9.5	
			R12	6.6	23.2	
	+25°C	+90°C	R11	1.05	6.7	
			R114	2.18	12.3	
			R12	6.6	29.0	
	+25°C	+100°C	R11	1.05	8.3	(see also
			R114	2.18	14.4	Table 6.7)

TABLE 6.7 Theoretical Performance of Refrigerants at Typical Heat Pump Temperature Conditions

REFRIGERANT		11	113	114	12	31/114	12/31
Evaporator Pressure	kN/m^2	220	103	425	1190	758	1200
Condenser Pressure	kN/m^2	1000	560	1700	3950	2920	4140
Critical Temperature	^{o}C	198	214	146	112	142	118
Compression Ratio		4.45	5.3	3.95	3.35	3.86	3.4
Specific Volume	m^3/kg @ 50°C	0.0801	0.1298	0.0320	0.0146	0.0324	0.0170
Weight circulated	kg/sec	0.098	0.060	0.245	0.530	0.242	0.450
Net Refrigerating Effect	kJ/kg	114.5	86.4	48.5	36.6	118.8	65.0
Heat of Compression	kJ/kg	27.9	25.6	18.6	18.6	35.0	25.6
Capacity	kW	13.9	6.76	13.0	28.3	37.6	41.7
Coefficient of Performance		5.1	4.35	3.5	2.9	4.48	3.55

Evaporator 50°C }
Condenser 100°C } Assume compressor swept volume of 28.32 m3/hr (1000 cu ft/hr)

The net refrigerating effect is obtained by subtracting the enthalpy of the refrigerant at the condensing temperature from the enthalpy of the superheated vapour entering compressor at the evaporator temperature. The heat of compression is the difference between vapour enthalpies upstream and downstream of the compressor (assuming isentropic compressor efficiency).

6.3.1 Drying and dehumidification The application of heat pumps in the field of drying and dehumidification is an area of considerable research and development, and commercial systems for this purpose are available on the market. Some of the best work on the use of heat pumps in drying and dehumidification has been carried out at the Electricity Council Research Centre in England over a period of several years, where it is continuing. Hodgett, writing in The Chemical Engineer (Ref.6.16), has analysed the state of the art in dryer technology, and also calculated the amount of water removed in industrial processes in the United Kingdom. Shown in Table 6.8, these figures do not give the water removal from products in the chemical, pharmaceutical or food industries, and it is therefore estimated that in the UK between 20 and 30 million tonnes of water are evaporated by industry per annum. Taking a value of 30 million tonnes, the energy required to evaporate this would be 74 x 10^6 GJ, and, assuming an overall dryer efficiency of 50 per cent, the total energy consumption in these processes would be 148 x 10^6 GJ.

TABLE 6.8 Water Removed by Evaporation in Industry

Material	Annual Production tonnes x 10^6	Average Moisture Content Drop, %	Water Removed, tonnes x 10^6
Paper and Board	4.6	200	9.2
Bricks	15.7	15	2.4
Milk, dried	0.21	900	1.85
Milk, condensed	0.17	500	0.85
Gypsum	3.7	20	0.74
Plaster and plasterboard	2.3	45	1.0
Textiles	1.4	30	0.4
China clay	3.5	10	0.35
Fertilisers	4.0	3	0.12
Timber, softwoods	0.27	45	0.12
Timber, hardwoods	0.24	20	0.05
Dyestuffs	0.1	50	0.05
Vitrified china clay pipes	0.75	15	0.11
Tiles, pottery, and sanitary ware	1.0	15	0.15
Total for 14 materials	-	-	17.4

The efficiency of conventional convection dryers is a function of the amount of air recirculated, such a dryer being illustrated in Fig.6.10. The efficiency of such a dryer is described by the equation:

$$\eta = \frac{T_2 - T_3}{(T_2 - T_3) + (1 - W)\,(T_3 - T_1)}$$

Where T_1 is the ambient air temperature
T^1 is the temperature of the air entering the dryer
T^2 is the exhaust air temperature
W^3 is the ratio of recirculated air to total air flow

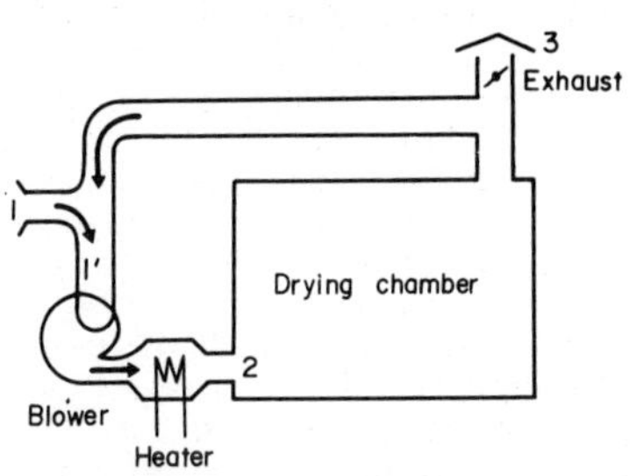

Fig.6.10 A Conventional Dryer Employing Partial Recirculation

Typical results of dryer efficiency as a function of the recirculation ratio W are shown in Fig.6.11. It is of course impossible in a conventional dryer to completely recirculate the exhaust air, as increases in humidity would quickly reduce its drying capacity. However, the heat pump may be used most effectively in dryers as a means of dehumidifying, or removing the moisture from, the exhaust air so that it may be recirculated in significantly greater quantities (as was illustrated in Chapter 4 in the discussion on the topic of the tumble dryer).

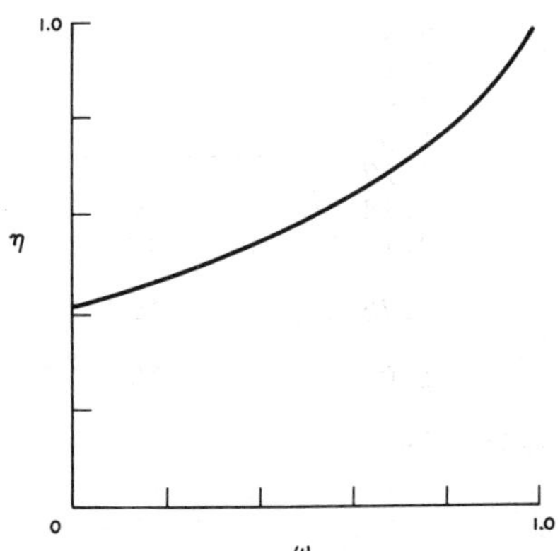

Fig.6.11 The Effect of Recirculation on the Efficiency of a Conventional Dryer

One may examine how this occurs with reference to the psychometric chart shown in Fig.6.12, and the heat pump dehumidifier in Fig.6.13. The exhaust air leaving the product being dried is passed over the evaporator coil of the heat pump, where it is cooled (path 1 or 2 on Fig.6.12). As a result of this cooling, some of the moisture contained in the exhaust air condenses out, and is drained away. The recirculated air which has been cooled then has to be raised in temperature before re-entering the product to be dried. This is done by passing the air over the condenser, where it picks up both latent and sensible heat recovered from the evaporator, and the heat associated with the work of compression.

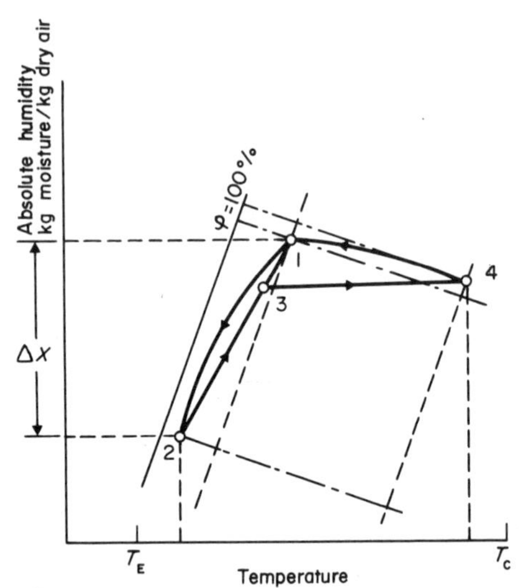

Fig.6.12 Cycle of a Heat Pump in a Dryer Effecting Moisture Removal. T_C is the Condensing Temperature and T_E the evaporating Temperature

The efficiency of the system is subject to the same considerations as any other heat pump application - the COP being a function of the difference between evaporating and condensing temperatures. A reduction in temperature difference is commonly achieved in the dryer by incorporating a plenum chamber where, at condition 3, chilled saturated air is mixed with unchilled air before passing over the condenser.

It is more usual to present the performance of a heat pump in a dryer in terms of the rate of specific moisture extraction, typically having the dimensions kg/kWh. The effect on the specific moisture extraction rate of increased dryer temperature and humidity is illustrated in Fig.6.14. This shows that the specific moisture extraction rate tends to increase both with increasing temperature and humidity. Extraction rates of up to 4 kg/kWh are possible, with an average figure of about 2.5 kg/kWh.

Taking into account the efficiency of generation and distribution of electricity, this being assumed to be 30 per cent, it has been calculated that a typical heat pump dehumdifier dryer will be more efficient in terms of primary energy consumption than a conventional steam heated dryer with a thermal efficiency of 75 per cent or a direct gas fired dryer having an efficiency of 58 per cent. It should be noted that in general a majority of dryers which do not employ high amounts of recirculation operate at efficiencies considerably less than these figures, (Ref.6.17).

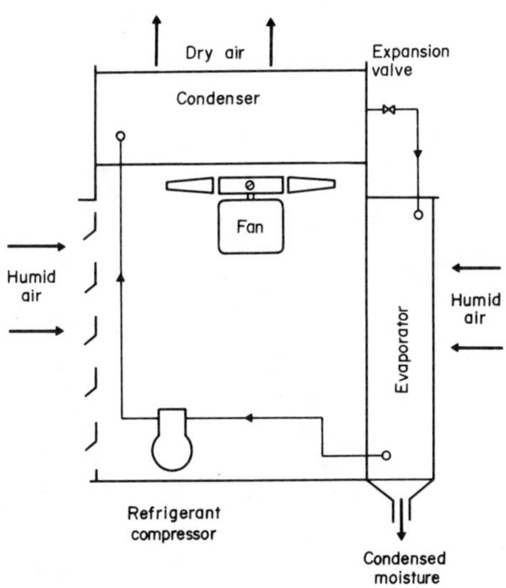

Fig.6.13 Heat Pump Dehumidifier

The initial application area for this type of dehumidifier was in timber seasoning, but more recently the textiles, ceramics, rubber, gypsum and confectionery industries, amongst others, have taken up the heat pump dehumidifier. In all of these cases the drying takes place at a temperature less than 50°C, but more recently a heat pump dehumidifier capable of operating at 80°C has been developed and is now in commercial use. The first application for this system was in the field of timber seasoning also.

Applications: One of the first applications of heat pumps in a dehumidification role was as a result of work by Sulzer (Ref.6.7). These were constructed in Germany in 1943 for the dehumidification of underground caverns. The heat accrueing from condensing the water vapour in the air was used to preheat incoming fresh air. Another early study, carried out in the United States in the 1950's, was associated with the application of heat pumps to grain drying, (Ref.6.18). Using a heat pump, the temperature and relative humidity of the dryer air could be closely controlled - an important aspect of many dryer applications - and the heat from the evaporator coil could also be recovered as described above.

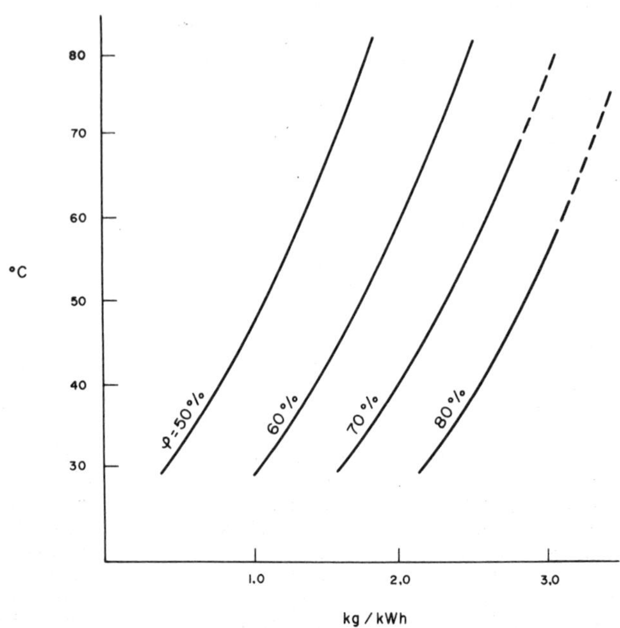

Fig.6.14 Rate of Specific Moisture Removal by dehumidification, as a Function of Humidity and Dryer Temperature

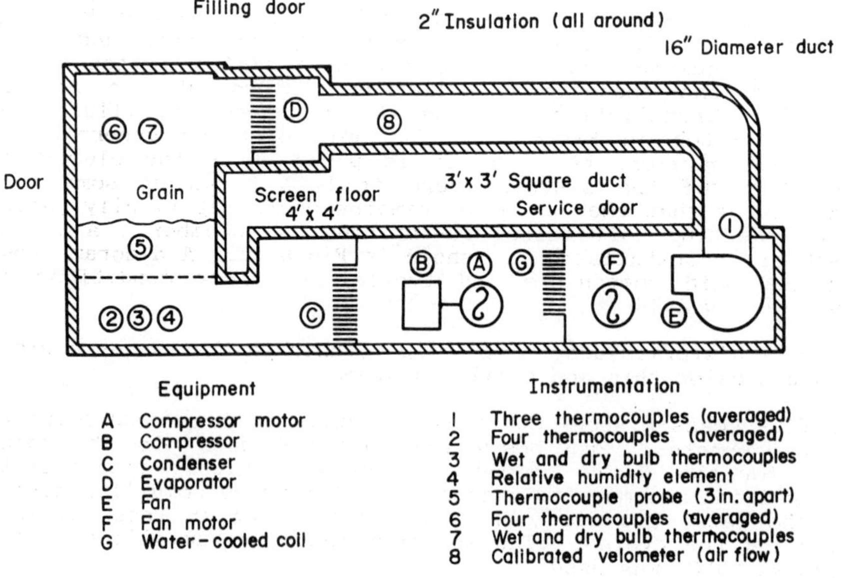

Fig.6.15 Experimental Heat Pump Grain Dryer Tested in the USA in 1950's

An experimental grain dryer was constucted, as shown in Fig. 6.15, and tests were carried out over a range of conditions, partly to verify some theoretical studies carried out as early as 1949 (Ref.6.19). The dryer contained a grain bin with a floor area of approximately 1.3m square. The heat pump used a 570W electric drive, with R-12 as the refrigerant. A centrifugal blower of 380W was used to circulate the air. A water-cooled coil (G) was installed for the purpose of controlling the maximum drying temperature.

The two control variables used in the study were air flow and drying air temperature, the air flow covering a range of 550 m^3/h to 2000 m^3/h, and the temperature of the air being between 43°C and 54°C. The tests were terminated when the grain had been dried to about 12 per cent moisture content. It was found that the power consumption in kilowatt-hours per kilogram of water evaporated agreed very closely with the theoretical values predicted by the work carried out in 1949, and the tests showed that the minimum cost occurred at an airflow of between 800 m^3/h and 1000 m^3/h. The value also varied with the drying air temperature, and expressed in terms of specific moisture extraction rate, a figure of 0.28 kWh/kg was achieved at 43°C and 0.27 kWh/kg at 54°C. It is interesting to note that these figures suggest high heat pump COP's, albeit at fairly low condensing temperatures, when compared with systems currently on the market. At the time this study was carried out, it was pointed out that the main disadvantage of the system was the additional capital costs incurred. It will be shown later that commercial systems recently introduced onto the market can result on returns on the investment in a matter of months.

One of the first applications of heat pumps to drying processes on a commercial scale, and one where control of temperature and humidity are of prime importance, is in timber kilning. Pioneers in this field were Westair Systems Ltd, and their range of equipment was developed over a period of ten years, in about one thousand installations throughout the world. A typical installation, showing the air path, is illustrated in the sketch in Fig.6.16. The unit operates generally as described earlier, but the air is passed over the electrical components of the system, where it both picks up some heat and ensures that the fan motor remains cool. It finally passes over a heating coil before re-entering the chamber. A photograph of an installation is shown in Fig.6.17. A diagram showing the main components and typical air flow conditions is given in Fig.6.18.

The case histories described below illustrate the energy savings possible using this and similar systems.

The first case history concerns a textile company manufacturing woollen berets. Previously fuel oil had been used to raise steam for the drying chambers, which were highly inefficient. The decision was made to change to heat pump dehumidifier drying and the hats, and all yarn required treating in these dryings following the dyeing. The low temperature dehumidifier, operating at 50°C, was used.

Batch drying times were of the order of 5.75 hours, and this necessitated occasional use of a third oven to overcome bottlenecks in the drying process. Trials showed that the heat pump system could reduce drying times and energy usage, and as a result the two principle ovens were upgraded by having thermal insulation to give a wall panel U value of 0.37 W/m²K. The system was totally enclosed by adding pneumatic seals to doors, ensuring a gas tight structure.

The effect of the change in the drying equipment is shown in Table 6.9. A comparison was made in 1980 and it can be seen that site energy use has been reduced by approximately 90 per cent, and drying costs have been reduced by between 70 and 75 per cent. The shorter drying time has also eased production hold ups, and it was also claimed that product quality was improved, although this is a subjective assessment. As in most of the heat pump dehumidify installations, the heat pumps are comparatively small, in this case unit capacities of 15 kW and 7.5 kW being employed. As in the case of the steam heated dryer, the drying temperature is maintained at 50°C, and a total payback period, taking into account additional installation etc has been estimated to be 2.4 years.

A second installation is in a dryer for processing gypsum moulds. The production of ceramic tableware is frequently implemented by pouding liquid clay into the gypsum mould. Following removal of the clay once it has hardened, the mould requires drying. A typical mould drying process involves reducing the moisture content from 30 per cent to less than 1 per cent, and again drying temperatures are less than 50°C, as higher temperatures can induce mould burn.

The mould dryers in this case were originally fired using natural gas, and their specific energy consumption was estimated at approximately 4,600 kJ/kg. Trials were first undertaken using a 3 kW packaged heat pump dehumidifier, and this demonstrated that high quality products could be dried using a significantly lower energy consumption. Data on the heat pump dehumidifier trials are given in Table 6.10 and the specific energy consumption for the heat pump system has been measured to be 1676 kJ/kg, this representing a 64 per cent reduction in site energy usage.

The ability to operate a heat pump dehumidifier at a delivery temperature of 80°C was first tested on a commercial scale at a saw mill at Boroughbridge, North Yorkshire. This operator had previous experience of heat pump dehumidifiers, albeit at the lower temperature of 50°C, but an increase in production capacity by the saw mill operator led to the decision to choose the temperature unit for reasons of low energy cost and high throughput.

Two 7.5 kW heat pumps are used, and detailed monitoring was carried out by the local electricity board. The results of the trials are given in Table 6.11, and while an attempt was made to obtain a comparison between the 50°C and 80°C units with identical timber loads, it was emphasised in the report that the chances of obtaining identical loads are remote, and the results should serve only as an indication of the efficiency of the two heat pump systems. Nevertheless, the high temperature system has an energy usage of almost 50 per cent less than that of the 50°C system. An 80°C heat pump dehumidifier was tested as part of a development programme and a comparison made with a direct oil fired system. Data on this are given in Table 6.12. It can be seen that for two identical loads, dried to the same moisture level, the heat pump dehumdifier used only 29 per cent of the energy used by the direct oil fired system. This can be translated into a cost saving of 39 per cent and a reduction in primary energy usage of 18 per cent.

It should be emphasised that all of the applications for this type of heat pump described are associated with batch processes. The use of heat pump dehumidifiers on continuous dryers is much less well established, although attempts are being made to develop a number of systems, including high temperature steam recompression units, for this use.

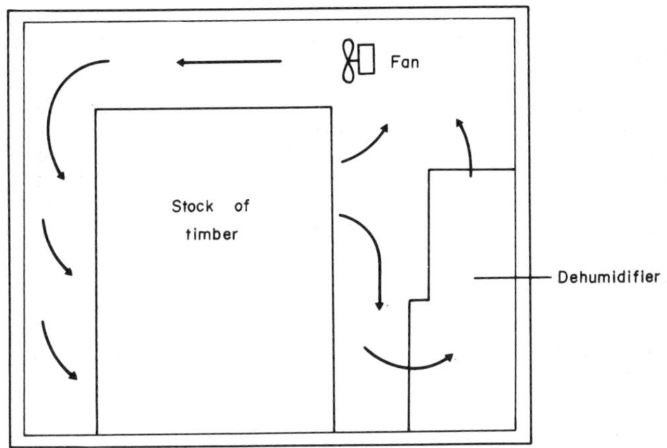

Fig.6.16 A Heat Pump Dehumidifier used for Timber Kilning

In drying, the basic refrigerant heat pump system may also be supplemented by other compression systems. In a study carried out in France some ten years ago (Ref.6.20) proposals were put forward concerning the use of water vapour compression in dryers, in one case in association with a conventional heat pump vapour compression cycle. Fig.6.19 shows the scheme for water vapour compression. Water vapour at atmospheric pressure is collected above the wet product and passed through a compressor. It is condensed at 5 bar (150°C) in a heat exchanger in the air stream leading to the product dryer. If the steam is not of a sufficiently high quality, a closed steam loop with supplementary heat exchangers is recommended. An efficiency of 1000 kJ/kg water evaporated (3.6 kg/kWh) is claimed. In the second concept, a superheated steam circuit is followed by a conventional heat pump using a Freon, as shown in Fig.6.20. The efficiency is of the order of 750 kJ/kg water evaporated (4.8 kg/kWh) for wet zone drying. This, as shown in Fig.6.20, is applied to a continuous drying process in a tunnel dryer, whereas most of the other applications described above are batch processes.

TABLE 6.9 Performance of Heat Pump used for Textile Drying

		Chamber One	Chamber Two
General	Product	Woollenberets	Hanks, cheeses, Ribbons
	Chamber dimensions	9.74m x 2.75m x 2.05m	9.1m x 2.4m x 2.05m
	Construction	Galvanised steel panels with high thermal bridging, poor insulation and door seals.	
	Modifications to construction	Insulation improved to U value of 0.37W/m²°C, all door seals improved.	
Oil Fired System	Oil usage	16.64 litres/hr or 599000 kJ/hr	13.54 litres/hr or 487700 kJ/hr
	Electrical power Consumption	2.29 kW or 8244 kJ/hr	5.59 kW or 20124 kJ/hr
	Cycle time	5.75 hrs	5.75 hrs
	Water removal	154 kg	54 kg
	Specific energy consumption		
	on site	24080 kJ/kg	54074 kJ/kg
	primary	31748 kJ/kg	74515 kJ/kg
Heat Pump Dehumidifier System	Unit capacity (nominal)	15 kW	7.5 kW
	Total power consumption (average)	18.37 kW or 66132 kJ/hr	9.88 kW or 35550 kJ/hr
	Cycle time	4.5 hrs	4.7 hrs
	Water removal	154 kg	54 kg
	Specific Energy Consumption		
	on site	2052 kJ/kg	3095 kJ/kg
	primary	7600 kJ/kg	11463 kJ/kg
Comparison	Energy usage reduction		
	on site	91%	90%
	primary	76%	85%
	Energy cost reduction (customer's estimate)	70%	75%
	Payback period (including cost of total conversion).	2.4 years at 1980 fuel price levels	

TABLE 6.10 Gypsum Mould Drying using a Heat Pump Dehumidifier

General	Product	Gypsum moulds
	Chamber dimensions	3m x 2m x 2.5m
	Construction	Standard cold store panels with an insulation value of 0.25W/m²°C
Gas Fired System	Specific energy consumption (estimated)	
	on site	4600 kJ/kg
	primary	7050 kJ/kg
Heat Pump Dehumidifier System	Heat pump capacity (nominal)	3 kW
	Product load (wet)	800 kg
	Water to remove	232 kg
	Cycle time	18 hrs
	Electricity consumed	108 kWh or 388800 kJ
	Specific energy consumption	
	on site	1676 kJ/kg
	primary	6207 kJ/kg
Comparison	Energy usage reduction	
	on site	64%
	primary	12%

Fig.6.17 Photo of a Westair TS2 Timber Seasoner in a Timber Kiln in Northumberland

TABLE 6.11 Timber Drying up to 80°C using a Heat Pump

Low Temperature 50°C, Heat Pump System	Specie	Ash
	Thickness	25 to 50 mm
	Kiln load	$24m^3$
	Moisture content reduction	30% to 13%
	Water to remove	103.7 kg/m^3
	Total energy consumption	4330 kWh
	Specific energy consumption	6300 kJ/kg
High Temperature 80°C, Heat Pump System	Specie	Ash
	Thickness	25 mm
	Kiln load	$24m^3$
	Moisture content reduction	32% to 13%
	Water to remove	115.9 kg/m^3
	Total energy consumption	2720 kWh
	Specific energy consumption	3520 kJ/kg

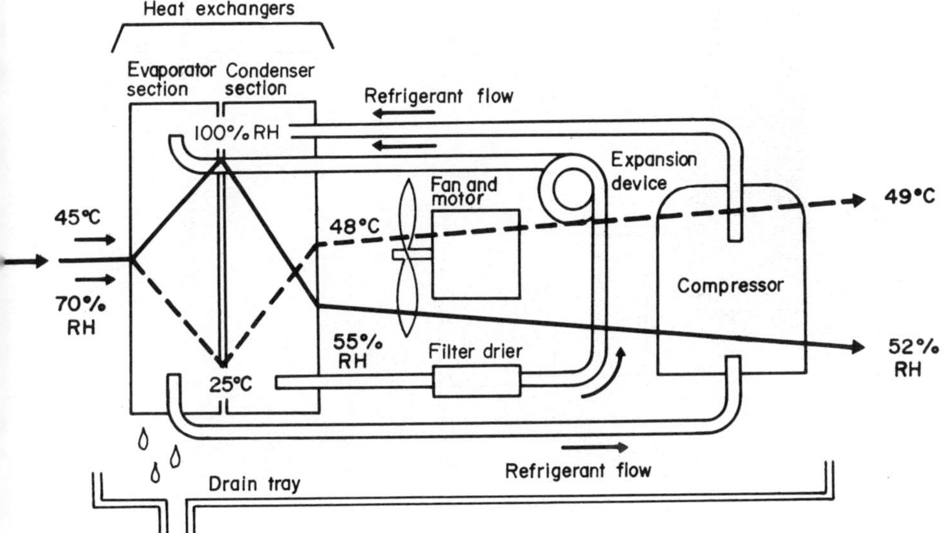

Fig.6.18 The Main Components of a Westair Timber Seasoner and Typical Air Flow Conditions

TABLE 6.12 Timber Drying: Comparison of Heat Pump and Heat and Vent System

Drying of 32 mm thick W/E Beech from 27% to 12.3% mc

Direct Oil Fired System	Kiln load	21m^3
	Oil used	27.5 litres/m^3 or 101 x 10^6 kJ/m^3
	Electricity used	29.3 kWh/m^3 or 0.11 x 10^6 kJ/m^3
	Water removed	91.1 kg/m^3
	Specific energy consumption	
	on site	12,294 kJ/kg
	primary	15,794 kJ/kg
High Temperature Heat Pump Dehumidifier	Kiln load	11.5m^3
	Electricity used	88 kWh/m^3 or 0.32 x 10^6 kJ/kg
	Water removed	91.1 kg/m^3
	Specific energy consumption	
	on site	3,477 kJ/kg
	primary	12,878 kJ/kg
Comparison	Energy usage reduction	
	on site	71%
	primary	18%

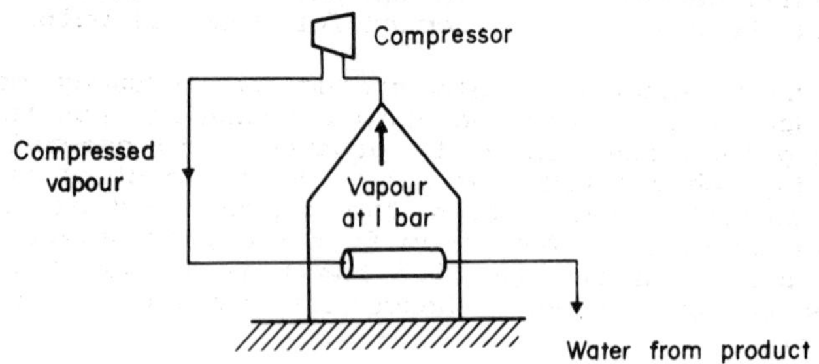

Fig.6.19 Drying with a High Pressure Steam Circuit

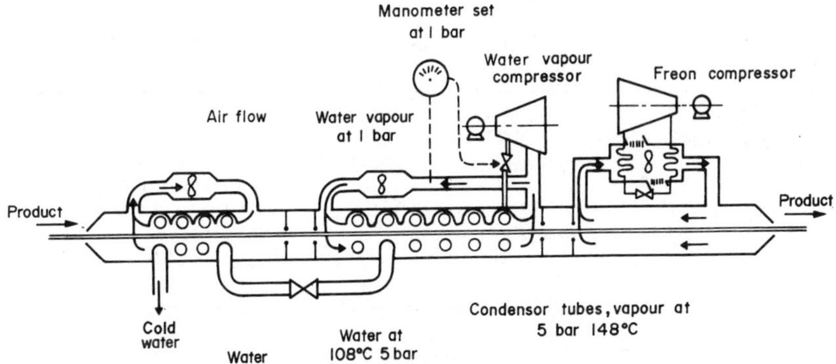

Fig.6.20 Convection Drying with a Superheated Steam Circuit and a Conventional Heat Pump

Application of Steam Compressors One of several demonstration projects involving heat pumps supported by the European Commission concerns the use of steam compressors to improve efficiencies in pulp production, and the operation is centred around the recovery of energy from the steam generated in the thermomechanical pulp production shop. The company where the installation is sited is Papeteries Beghin-Say, in Corbehem, France. The factory was set up in the 1920's where a sugar refinery had previously existed. At the time of reporting (Ref.6.21), it employed 1200 people and the output was 250,000 tonne/year of lightweight coated paper and 35,000 tonne/year of coated cardboard, representing a turnover of 1,300 million francs.

The production facilities at Corbehem comprise a chemical pulp grinding shop, a mechanical pulp shop, a thermomechanical pulp shop and four paper-making machines. The thermomechanical production of pulp is the process which has gained considerable ground throughout the world during the last decade. This is due to a number of factors, including three principle ones. A fall in the cost of raw materials has occurred owing to the use of sawmill wastes, the pulp has better mechanical characteristics, and there is a lower demand for chemical pulp.

However, thermomechanical pulp production consumes more electricity. Specific consumption varies depending upon the quality of the pulp that one wishes to produce. In general refiners consume 1,800 kWh per tonne of pulp. 90 per cent of this energy is converted into heat during the process and is dissipated in the form of steam. The steam from the pressurized preheater for the fibre separator (at 1.7 bars) is the most attractive as regards energy recovery, hence its use as the heat source in this unit.

The steam generated by the process contains various acids, wood, resins, a small quantity oif air and solid particles such as wood fibres or even small chips. Particular attention has been paid to the steam washing system in order to avoid clogging of the exchangers by fibres and resin, (Fig.6.21).

Process steam is extracted from the preheater via an automatic valve (VA2) operated in series with the existing automatic valve (VA1) via the pressure regulator (PC1) on the existing preheater. The steam is then fed into a cyclone washer (L) where solid impurities are removed continuously. The heavier particles are separated out on the internal walls which are sprayed continuously, while the finest particles are removed when the steam passes through a fine curtain of water in the upper part of the washer. The water, which is atomized by a free-flow vortex (PO) comes from the condensates arising in the exchanger (EC) and returning by gravity to the lower part of the cyclone washer.

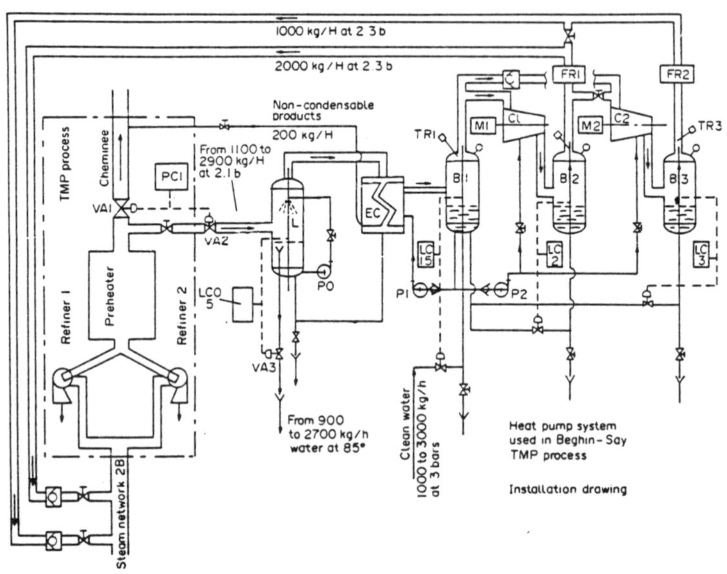

Fig.6.21 Installation Drawing of Heat Pump System used in Beghin-Say TMP Process

The exchangers acting as the heat pump evaporator consist of stainless steel plates arranged in such a way as to provide vertical ducts where the steam from the preheater condenses freely, while the boiler condensates evaporate in a cross-flow system.

Excess condensate is continuously pumped from the first separator (B1) onwards and returns to this at the same time as the evaporator steam. This device was included in order to increase heat transfer on the evaporator side. The clearance between the exchanger plates on the condenser side is enough to avoid blocking by fibres or chips should the cyclone washer (L) fail and to enable these to be washed easily if there is resin precipitation due to condensation. For this reason the welded plate assemblies on each exchanger can also be removed completely and quickly so that they can be washed easily if necessary.

This has never been necessary even after several months of operation.

Two steam compressors have been installed in such a way that they can operate either in parallel, or in series. Compressor choice fell upon steam systems because:

. water is cheap;

. water is easy to replace in open circuits, and the make-up of losses in closed circuits is relatively simple;

. Water vapour can condense at relatively high temperatures;

. water vapour is universally available in industry; it is easy to connect a water pump to a factory steam network, thus enabling heat recovered from one process to be used in another;

. the compressor is well-matched to low flow rates;

. the steam produced is saturated.

The biggest compressor (C1) is able to compress 1,500-2,500 kg/h and the temperature rise of the saturated steam between the inlet and outlet lies between 10 and 14°C. The installed electric power is 132 kW. The smallest compressor (C2) can compress between 700 and 1,300 kg/h and the temperature rise in the saturated steam betwen the inlet and outlet is 10-14°C. The installed electric power is 55 kW.

The system is highly flexible: each of the compressors may be shut down or started up instantly by means of a simple push button. In order to achieve this, the pump (P1) supplying the exchangers (EC) and the pump (P2) supplying the water rings for the compressors with condensate are kept permanently in service. The steam throughput perfectly matches that of the outlet process steam without affecting the pressure in the preheater, this being an essential characteristic of the process. For example, where the production capacity of the process is reduced, the pressure regulator (PC1) on the preheater will immediately detect that the demand of steam by the heat pump has become too high, which would tend to reduce the steam pressure in the preheater. The regulator will first close the automatic stack exhaust valve (VA1) and, if this is not enough, will begin to close the automatic exchanger inlet valve (VA2). Both pressure and temperature will then drop on either side of the exchangers and, of course, in the separator (B1) upstream of the compressors. Since the compressors must supply the factory network at constant pressure and temperature their flow rate will be reduced immediately, thus creating new, stable operating conditions.

If, on the other hand, process capacity increases, the preheater pressure regulator will open automatic valve (VA2) in the direction of the exchangers and begin to open automatic valve (VA1) towards the stack, thus creating new stable compressor operating conditions at an increased mass flow rate.

The design and development work was carried out by Papeteries Beghin-Say, in conjunction with Air-Industrie. The compressors were supplied by Nyrec, heat exchangers by Barriquand, separators and washer by Leneveu, and Mathieu was responsible for the control system. The system was commissioned in early 1981. Economic data are given in Table 6.13.

TABLE 6.13 Economics of Beghin-Say Installation

Amount of Steam Recovered	- 9230 t
Electricity Consumed	- 857,440 kW.h
Average Fuel Price	- FF 1,160/t
Average Electricity Price	- FF 0.218/kW.h
Savings (Nett)	- FF 543,000
Capital Cost (Marginal)	- FF 860,000
Payback Period	- 19 months

Absorption Cycle Heat Pumps for Drying While comparatively little work has been done on the application of absorption cycle heat pumps to drying processes, a number of concepts have been examined. An industrial absorption dryer has been studied by Gaz de France, (Ref.6.22). The operation of the absorption dryer is illustrated in Fig.6.22.

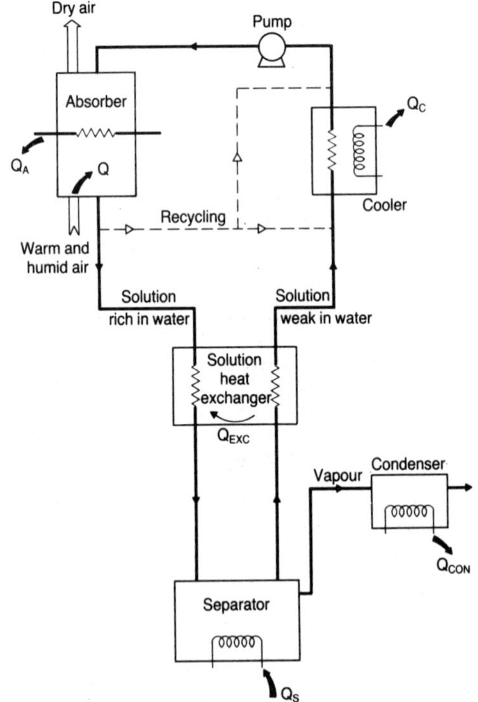

Fig.6.22 Operation of an Absorption Dryer

Hot humid air from the exhaust of the dryer is directed into the absorber. Here it comes into intimate contact with an absorbant, such as lithium bromide, and transfers some of the water vapour into the absorbant. The dry air is discharged from the absorber.

If a comparison is made with the operating cycle of a conventional absorption cycle heat pump (see Chapter 2), the drying unit differs in that it operates on an open cycle, as the condensate leaves the system. Also, the evaporator is eliminated.

To date only parametric studies have been carried out, examining the influence of solution temperature and pressure on COP and heat recovered.

6.3.2 Evaporation and boiling processes The application of heat pump systems to evaporation, particularly in concentrating liquids and boiling is typified by the mechanical vapour recompression process illustrated in Fig.6.3 and also described, as applied to timber dryers, in Section 6.3.1. As already mentioned, water is commonly the working fluid of the system - in evaporation processes this represents the vapour/condensate of the feed liquor. Two other main differences set the mechanical vapour recompression (MVR) system apart from the conventional heat pump. The heat pump works on an open cycle with both the condensate (water) and the concentrate being removed as appropriate, and further supplies of feed added to the system. Also, as shown in Fig.6.3, the evaporator and condenser are arranged as an integral component so that the latent heat of condensation can effect further evaporation from the liquor.

The use of MVR in evaporation processes, as in drying, has the same energy implications. Since the latent heat of vaporization required to evaporate the water from the feed is recovered when the vapour is recondensed, the only energy needed to process the concentration of the feed liquor is that required by the compressor. Provided that the temperature difference between the condensing vapour and the boiling point of the feed liquor can be kept to about 7°C, the energy input for compression, taking into account compressor overall efficiency, is of the order of 70 kJ/kg. A conventional single-effect steam heated evaporator would require an energy input of 2790 kJ/kg, and a six-effect system, of high efficiency, would still need an energy input of 465 kJ/kg.

It is worth pausing here to briefly describe the process of multiple effect evaporation. This process was invented by Rillieux in 1843 for use in cane sugar manufacture, and involves the reuse of latent heat for successive evaporations (without recompression). In a single effect evaporator the evaporate is rejected, but by adding a further stage of evaporation, this rejected vapour may now be used in a closed system to effect further evaporation. If the second vessel is to boil at atmospheric pressure, the steam supplied to it from the first vessel must be under pressure.

Assume that the second vessel, as shown in Fig.6.23, is open to the atmosphere. When steam enters the first vessel, boiling will occur there, but not in the second vessel. However, as

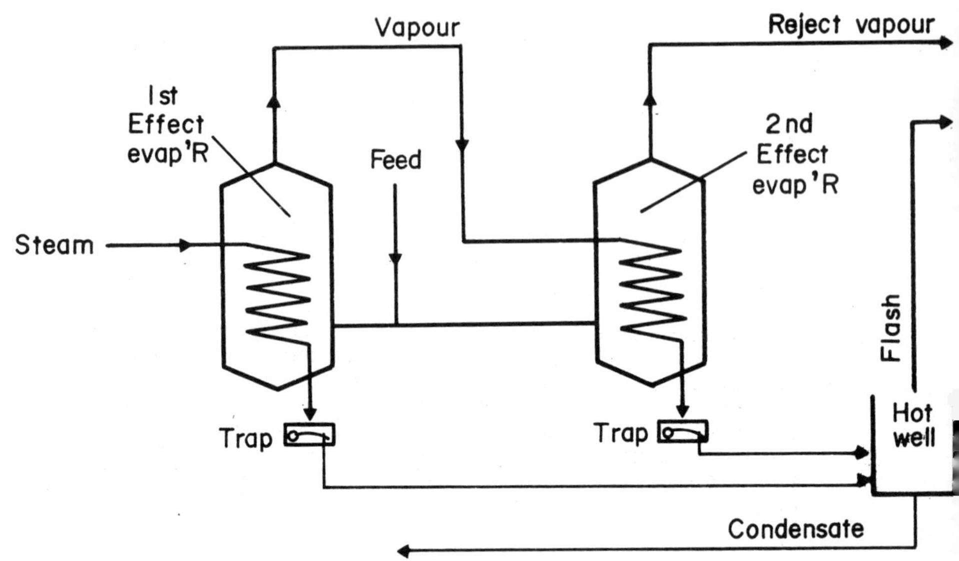

Fig.6.23 Double Effect Evaporator

the pressure in the first vessel increases, the temperature difference between the input steam and output vapour will decrease, reducing heat exchange. At the same time, a temperature and pressure difference is building up between the vapour and liquid in the second vessel, initially resulting in condensation of the vapour from the first effect vessel. The latent heat given up will eventually provide sufficient energy in the second vessel to effect evaporation. The number of stages used may be typically up to six. In the dairy industry, for example, as plant replacement is implemented, the number of effects used in evaporators will be gradually increased from the current two to three, to four or five, (Ref.6.23). This will be logically followed by the use of mechanical vapour recompression, particularly in larger plant. The effect of this on energy consumption for the evaporation of approximately 450,000 litres per hour of skim to 48 per cent total solids is given in Table 6.14.

It can be seen that on a heat supplied basis the total energy consumption is reduced by almost 80 per cent, and even taking into account conversion losses associated with the generation of electricity, MVR still represents a saving of over 40 per cent in primary fuel.

TABLE 6.14 Energy Requirements for Evaporation

	Electricity kWh/litre	Steam kg/litre	Total MJ/litre	
			Heat Supplied Basis	Primary Fuel Equivalent
4-Effect Evaporator	0.0027	0.16	0.49	0.58
Evaporator with MVR	0.0230	0.01	0.11	0.39

It is more commonly and successfully applied to the concentration of very dilute feeds. This is because of the need to operate the system with low temperature differences between the condensing vapour and the feed boiling point. As concentration occurs, elevation of the boiling point of the solution may be too great for continuing efficient operation of the MVR system. Another limitation is put on its use by its inability to cope with liquids of high viscosity. In a multiple effect evaporation system, however, the MVR unit may be applied to the first stage only, where the solution remains comparatively weak.

Highly viscous liquids are normally treated using a special type of evaporator, known as the agitated thin film type. These operate with a high temperature difference, typically about 80°C, to ensure a high heat flux and maximum utilisation of the heating surface, (ref.6.21). To increase temperature and pressure to this level so that recompressed vapour's can be reused in the same evaporators requires a compression ratio of about 10:1. The maximum attainable pressure ratio of single stage MVR systems is only of the order of 1.8:1, however.

The installation of commercial MVR systems has included units to concentrate pulp and paper wastes, the concentration of whisky effluents, and wide use in the chemical industry. A typical unit employing a vapour recompressor (or thermo-compressor) on the first stage is shown in Fig.6.24. The term thermo-compressor is included because MVR is not the only recompression system. It has become quite common to use steam jet recompression to upgrade the used vapour. The low pressure vapour may be boosted up to a slightly higher pressure by mixing it with high pressure steam in a suitable injector. The resulting mixed steam has a pressure higher than that of the original vapour (as in MVR), and can be efficient if conditions are appropriate. This necessitates that the steam or vapour should be at or about atmospheric pressure, and that only a small increase in pressure should be attempted. The main disadvantage of the system is that high pressure (e.g. 10 bar) steam must be used, (Ref.6.25). The reader is also advised to read this reference for details of the many other energy saving schemes which may be applied in evaporators.

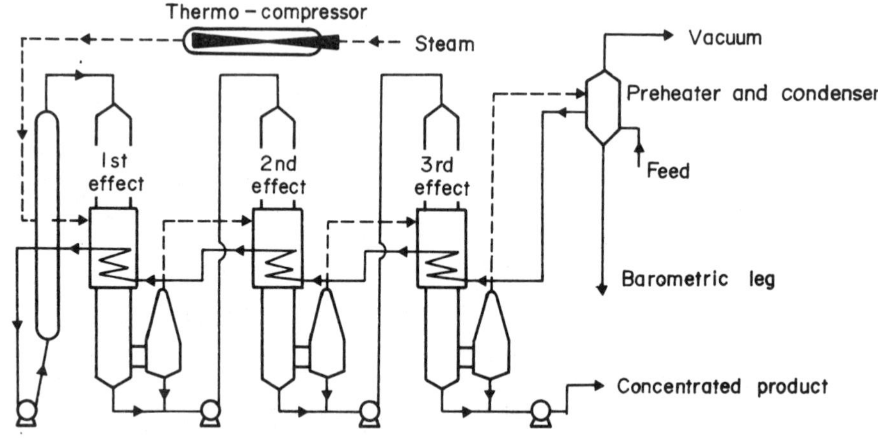

Fig.6.24 Typical Arrangement of Eva-Dry Three Stage Falling Film Evaporator with Vapour Recompression

The use of heat transformers (see Chapter 2) in evaporation duties has been proposed. One scheme investigated by Krupp in West Germany (Ref.6.26) also involved a thermocompressor. An aggressive aqueous solution is concentrated in a 4-effect evaporator. In order to remove 34.5 tonnes of water per hour, 10 tonnes of process steam at 130°C are needed. Vapour from the final evaporator stage (V4) in Fig.6.25, together with some vapour from the expansion trap (ET) is available at a rate of 8.7 tonnes/hour and at a temperature of 75°C. In addition, site cooling water can be provided at a temperature of 18°C and a flow rate of 350 m^3/h, and process steam can also be supplied at a pressure of 15 bar. Also described in US Patent 4,379,734, a solution based on the use of a heat transformer is to condense the low temperature vapour in the evaporator and stripper of the heat transformer. A heat ratio of 0.375 can be achieved by recovering the heat and delivering it in the form of low pressure steam (105°C-saturated) from the absorber. As this is at an insufficiently high temperature to provide a driving force for evaporation, a jet compressor is employed to boost the pressure and temperature to 2.7 bar and 130°C respectively. This can now be fed to the first evaporator stage.

The resulting total process steam consumption is 0.192 tonnes/tonne water evaporated. It is claimed that using conventional practice such a low energy consumption could only be achieved using a 7-effect evaporator, which would necessitate a three-fold increase in the amount of heating surface. Data on the process and heat transformer are given in Table 6.15.

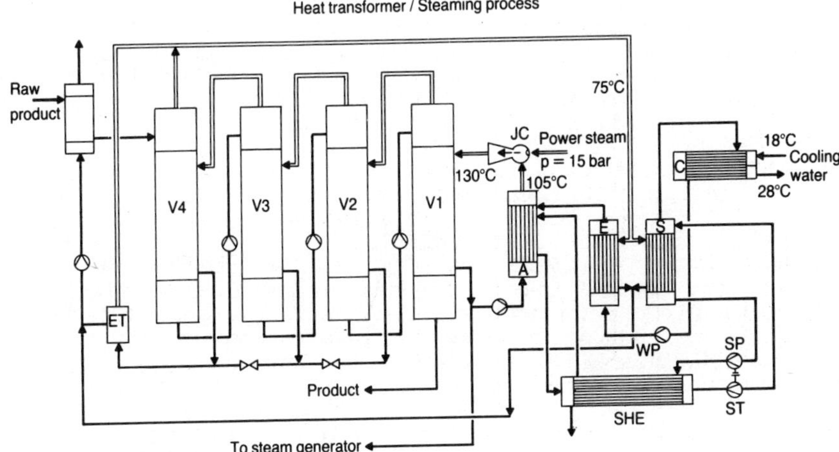

Fig.6.25 Connection Diagram of a Four-Stage Evaporator used in Combination with a Jet Compressor and a Heat Transformer

TABLE 6.15 Heat Transformer on Multiple Effect Evaporator

Mass flow of process steam	6.62 tonnes/h at 15 bar 10.0 tonnes/h at 2.7 bar
Heating temperature	130°C
Vapour volume (heat source)	8.7 tonnes/h
Vapour temperature	75°C
Vapour pressure	0.386 bar
Absorber rating	2.1 MW
Temperature from absorber	105°C
Heat ratio	0.375
Process specific steam consumption	0.192 kg/kg water
Process steam savings	33.7%
Heat transformer electricity requirement	24 kW
Cooling water requirement	335 m^3/h
Cooling water temperatures	18°C in 28°C out
Heat transformer cost	1.48 x 10^6 DM

The second major area of heat pump application in the context of this Section is in distillation. The primary difference between evaporation and distillation, which significantly affects the way in which the heat pump can be used, is the fact that the vapour emitted from a distillation column is rarely water. This implies that direct steam compression cannot be used - compressors capable of handling the product of the distillation column must be employed, or an indirect system using a conventional refrigerant can be incorporated.

Monsanto Chemicals was one company at the forefront of the study of the application of heat pumps in distillation in depth, (Ref.6.27). A conventional distillation column, illustrated in Fig.6.26, consists of, in addition to the column, a reboiler and condenser. The reboiler is normally steam heated, and the condenser coolant choice depends on the condensation temperatures required. One comparatively straightforward technique involving heat pumps for energy conservation in distillation is to upgrade heat recovered from the condenser for replacing the steam supply to the reboiler. This is implemented as shown in Fig.6.27. Thus the reboiler becomes the heat pump condenser. In order to minimize the work required by the compressor, the differential temperatures in reboiler and condenser should be kept as low as possible by good heat exchanger design. This must be traded off against the increased cost of large surface area heat exchangers, of course.

A second way in which heat pumps can be used in distillation plant is by employing the column fluid as a refrigerant, if its properties are sufficiently attractive. This eliminates the problems associated with the temperature differentials at the condenser and reboiler, although overall temperature remains important as far as the heat pump COP is concerned. Column configurations which could incorporate column fluid as the refrigerant are shown in Figs.6.28 and 6.29.

The study carried out by Monsanto also involved a limited economic assessment, from which a number of conclusions were drawn. It was found that heat pumps would not be competitive when condenser temperatures were in the range 35-110°C, if a combined heat and power plant was present on the site. The primary areas of application were in distillation plant where direct refrigeration or chilled water were needed for condenser duties, and in many new process plants.

Among the companies follwoing up such work with more detailed studies was British Petroleum (Ref.6.28). One scheme investigated involved using the distillatiion column vapour as the working fluid, as in Fig.6.28, with a conventional gas compressor providing the heat pump duty. Initial calculations indicated that at the time the heat pump would add something like £150,000 to the cost of the distillation column, representing 10 per cent of the investment. In return, it would save in excess of 25 per cent of the energy necessary to operate the plant. No radical changes in distillation column technology would be needed, the major modifications being restricted to permitting the column to operate at a higher pressure than normal, to keep down the size of the compressor. It was emphasised that careful design of the heat exchangers would be of importance, as the temperature rise attributable to the heat pump would only be about 20°C, thus restricting heat transfer due to the fairly low temperature differences involved.

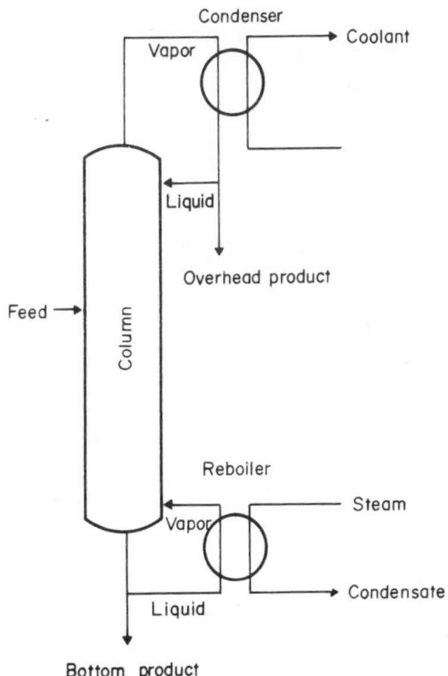

Fig.6.26 Layout of a Conventional Distillation Column

The technical problems can be overcome relatively easily - the main drawback in this, as in many other applications, has been the short period required for the return on the investment. Recovery of costs involving energy conservation equipment is regularly expected to take place in, at the most, two years in industrial processes.

Units developed by Krupp Industrietechnik (Ref.6.26) and already discussed in the context of evaporation, use water and ammonia as the working fluids. Typical performance data for units operating with a heat source of 80°C are given in Fig.6.30. The heat ratio E is defined as the ratio of the useful (high temperature) heat output to the waste heat input. The thermal efficiency η is defined as the ratio of the achieved value of E to that which is theoretically possible. Specified by some as a 'Coefficient of Performance', it is a measure of losses within the cycle, the heat exchanger efficiencies, and the presence or otherwise of components such as the solution heat exchanger, (analogous to the intercooler in a vapour compression cycle).

The case study given below is based on proposals put forward by the manufacturer. This deals with a heat transformer in a rectifying column, as illustrated in Fig.6.31. Total heat input requirement for the process is 5.5 MW, the temperature at the base of the column being 120°C to generate vapour from the top of the column at 80°C.

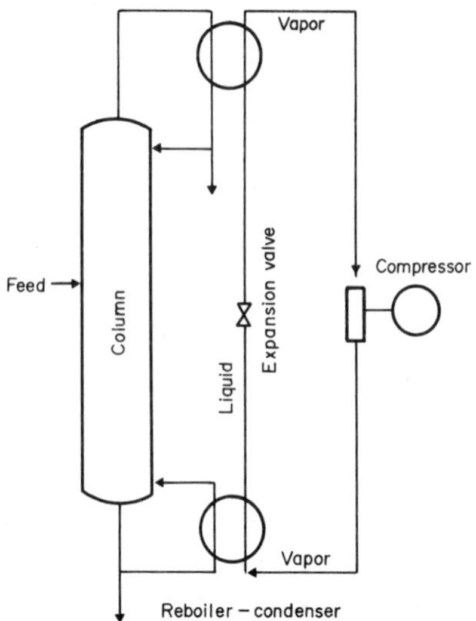

Fig.6.27 A Distillation Column Incorporating a Heat Pump using a Separate Refrigerant Circuit

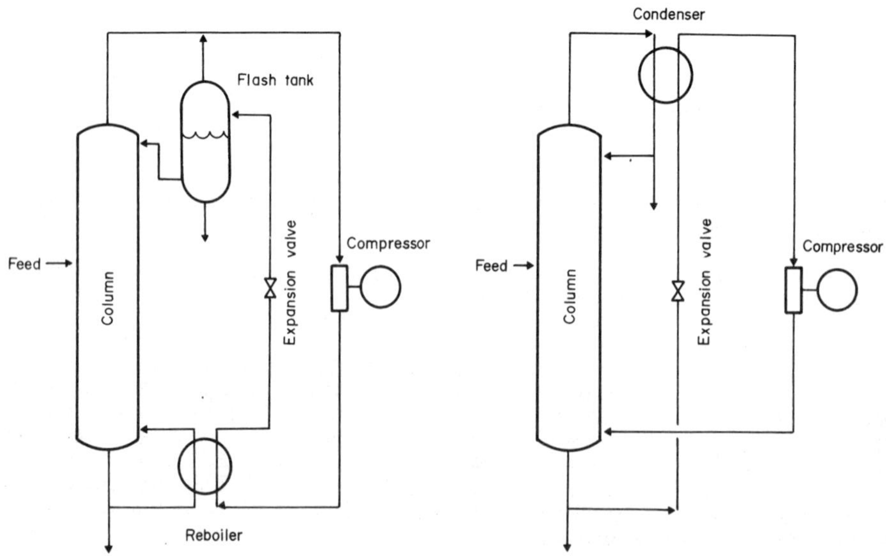

Fig.6.28 One Heat Pump system where the Column Fluid is used as the refrigerant

Fig.6.29 Use of the Column Fluid as Refrigerant can also Lead to Elimination of the Reboiler

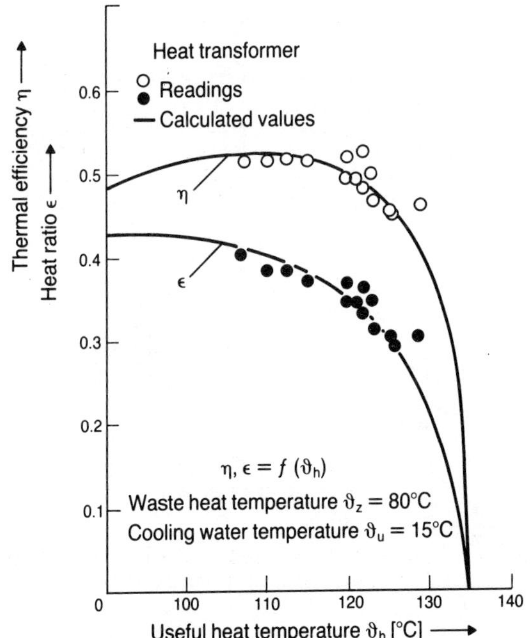

Fig.6.30 Heat Ratio and Thermal Efficiency of the Heat Transformer for Different Useful Temperatures at ϑ_z = 80°C and ϑ_u = 15°C

The vapour discharged from the top of the column is condensed in two locations, the stripper (S) and the evaporator (E). A proportion of the resulting condensate is returned to the column, while the remainder is used to preheat the working fluids in the heat transformer.

The heat transformer has an ambient air-cooled condenser with supplementary water sprays, designed to maintain heat transformer condensing temperatures within the range 17 to 27°C (varying with seasonal fluctuations). The resulting heat ratio is between 0.3 and 0.39, with a mean of 0.365. The absorber is thus able to deliver 2.0 MW of useful heat at 120°C. Operating for 7500 hours per year, the steam saving amounts to approximately 24,800 tonnes at conditions of 2.7 bar and 130°C. Other data are summarised in Table 6.16.

Based on the investment cost given (1.58) million DM), and a steam cost of £10/1000 kg, a payback period of the order of 2 years would appear possible in this instance.

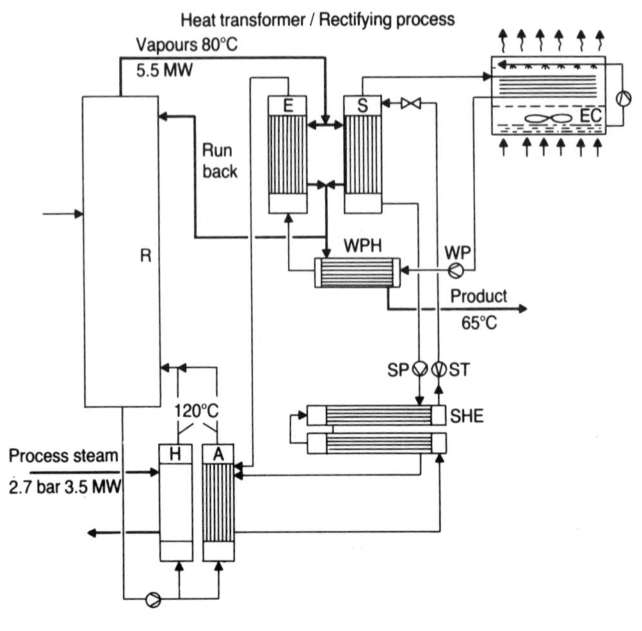

Fig.6.31 Connection Diagram of Rectifying Column in Combination with a Heat Transformer

TABLE 6.16 Main Parameters of Heat Transformer on Rectifying Column

Waste heat temperature	80°C
Temperature from absorber	120°C
Condenser temperature	17-27°C
Heat ratio (average)	0.365
Quantity of waste heat	5.5 MW
Quantity of useful heat delivered (average)	2.0 MW
Electricity consumption of unit	61 kW (pumps)
	32 kW (fan on condenser)
Cooling water consumption	7.1 m^3/h
Investment cost	1.58 x 10^6 DM

TABLE 6.17 Heat Pump on Vacuum Stripper

Heat Source		
Quantity of heat	-	0.94 MW
Source water flowrate	-	80.7 tonne/h
Source inlet temperature	-	60°C
Source outlet temperature	-	50°C
Heat Sink		
Heat output (hot water)	-	2.28 MW
Sink water flowrate	-	196 tonne/h
Water inlet temperature	-	80°C
Water outlet temperature	-	90°C
Generator Heat Source		
Heat source	-	Steam
Heat source temperature	-	183°C
Steam pressure	-	10 kg/cm(g)
Steam consumption	-	2230 kg/h

The use of a lithium bromide-water absorption cycle heat pump on a vacuum steam stripper has been analysed by Sanyo (Ref. 6.29).

The system flow chart of the stripper, incorporating the heat pump, is illustrated in Fig.6.32. The absorption heat pump is used to produce hot water at 90°C by using as a heat source waste steam at 75°C. The heat in the steam is first recovered in hot water, which is heated to 60°C by the steam. The generator of the heat pump is heated by higher temperature steam. The ultimate requirement is for low pressure steam for process use. Thus the hot water at 90°C is used via a flash tank to generate steam at 80°C.

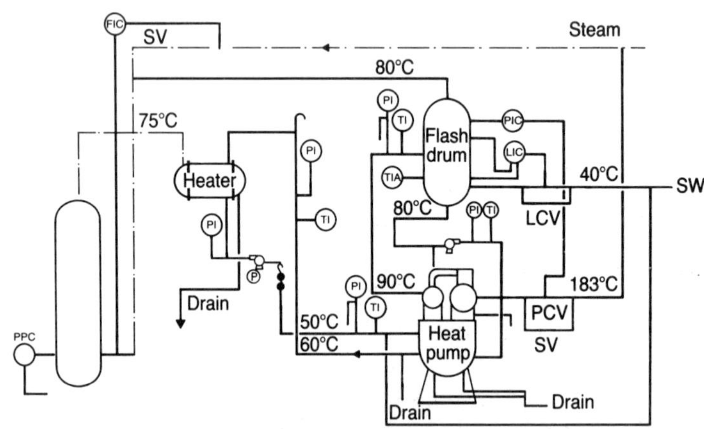

Fig.6.32 System Flow Chart of the Type I Absorption Heat Pump System

The amount of steam available from the flash vessel is 1.37 tonne/hour. At a steam cost of 5000 Yen/tonne, and based on a utilization of 7600 hours per year, the value of the savings is approximately 52 million Yen. With an electrical consumption of 14 kW, the net energy savings are approximately 50 million Yen. On this basis, the manufacturer claims a payback period of about 1 year.

The study of the use of vapour compression cycle heat pumps in distillation has not been neglected. Mixtures such as ethane and ethylene, propane and propylene, and in this particular case butane mixtures, have been studied with a view to applying heat pumps, and it has been predicted that something approaching 50 per cent of the energy consumption in the separation process could be achieved in a butane splitter, (Ref.6.30).

Two particular heat pump or recompression systems may be considered with regard to fractionation, these being over-head vapour recompression fractionation, and bottoms flash recompression fractionation. These are compared with the conventional fractionator shown in Fig.6.33, in Figs.6.34 and 6.35 respectively.

In the case of over-head vapour recompression, the vapours leaving the top of the fractionator are superheated prior to being compressed by a turbine-driven compressor. The reboiler heat is provided principally using hot isobutane, by condensing against the column bottoms in the reboiler/condenser. The isobutane product is chilled by seawater and refrigerant before boiler stored. Any additional reboiler heat required is provided by steam heating, while the bottoms product, normal butane, passes to a store via the feeds/bottoms heat exchanger. In the case of bottoms flash recompression, mixed butanes for the recompression system are pumped via the feed/bottoms heat exchanger to the fractionator. The vapourised bottoms material is superheated by steam before being compressed, again in a turbine-driven compressor system, and returned to the fractionator as reboiled vapour. The heat is also recovered from the column overheads, prior to this being cooled following a similar route to that in Fig.6.34.

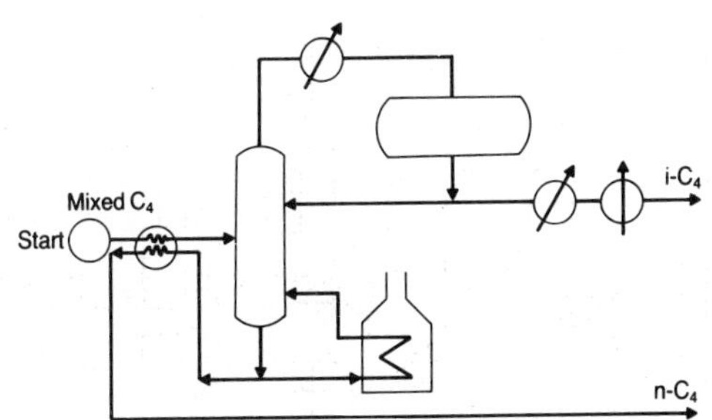

Fig.6.33 Conventional Fractionator

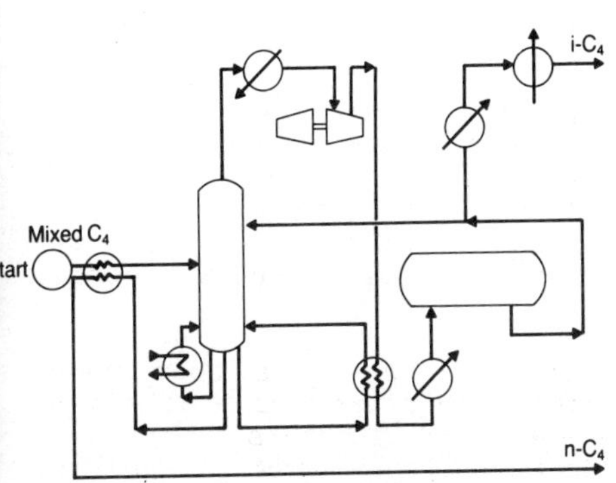

Fig.6.34 Heat Pump Configuration - Overhead Compression

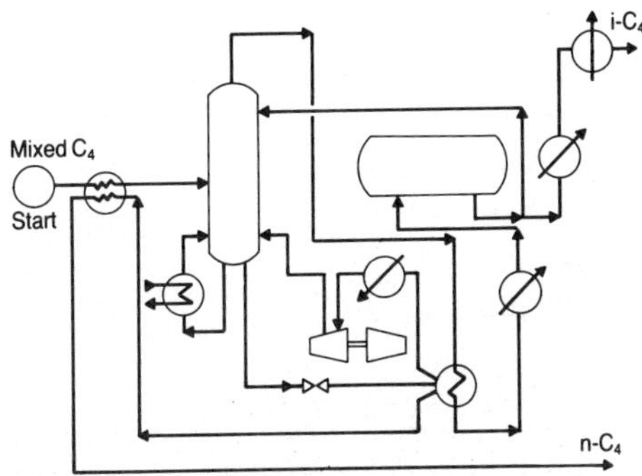

Fig.6.35 Heat Pump Configuration - Bottoms flash Compression

Table 6.18 shows the relative heat duties and power requirements of the three separation techniques discussed, and it can be seen that the heat pump schemes reduce the external energy requirements by about 50 per cent, when compared to the conventional fractionator.

The energy requirements for installations involving heat pumps depend to a great extent on the optimum discharge pressure of the compressor. For the case of over-head recompression, the approach to optimising the discharge pressure is illustrated in Fig.6.36. This shows how the discharge pressure is related to equipment sizes and capital cost differences. The detailed calculation showed that a compressor discharge pressure of approximately 230 psia would be the optimum in this case, while a similar analysis for the bottoms recompression case suggested a suction pressure of about 55 psia.

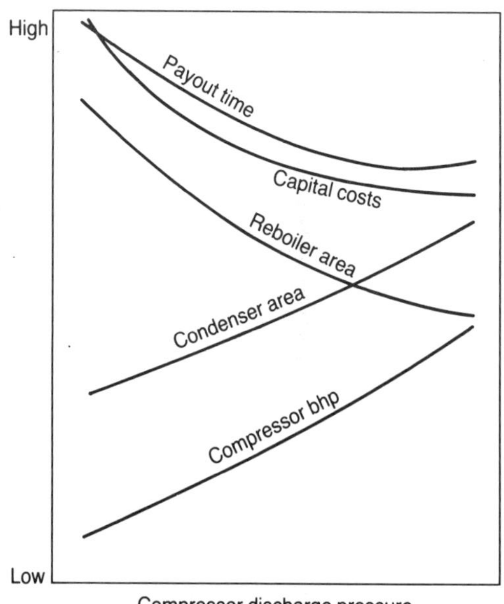

Fig.6.36 Approach to Optimizing Compressor Discharge Pressure

Following the analysis of the overall external energy requirements, which are detailed in Table 6.18, the particular scheme requirements for the heat pump schemes were detailed. In this particular project, adjacent boiler plant was sufficient only for the turbine drives, and would be insufficient to meet the requirements of the conventional fractionator reboiler. It was thus decided that a gas fired reboiler would be used, and in all the schemes the circulating pumps were driven by electric motors. The resulting steam balances are given in Table 6.19.

Based on the case of utilities for the Middle East location considered for this installation, the relative economics of the heat pumps schemes, compared with a conventional fractionator, are shown in Table 6.20. Note that the comparison is made on the basis of marginal costs, many items of plant being common to all three systems. Costing was carried out in January 1982, with no provision for contingency or escalation. The economic analysis shows that heat pumps give a significant economic advantage with paybacks of less than one year in both cases, when compared to conventional fractionation. It should also be borne in mind that the reduction in seawater cooling requirements, approaching 50 per cent, could have a significant effect on the cost of providing these services for a greenfield site project. In the overall analysis, however, economics is not the sole criterion. The relative safety, operability and reliability of the systems require evaluation.

TABLE 6.18 Heat & Power Duties of Butane Splitter Options

Heat Duty (MW)	Conventional Fractionator	Overheads Recompression	Bottoms Recompression
Reboiler	47.5	7.85	7.85
Condenser/reboiler	-	39.6	34.5
Compressor feed heater	-	2.6	1.1
Condenser/overheads Cooler	46.2	14.7	11.7
Feed/bottoms heat exchanger	5.0	5.0	2.3
Total	98.7	69.75	57.45
Power Requirements (bhp)			
Compressor	-	7430	8270
Reflux pump	300	-	300
Reboiler pump	700	-	-
Total	1000	7430	8570
Cooling Water (US gpm)	16400	5200	4200

With regard to safety, the conventional fractionator design selected makes use of an on-site forced circulation gas fired heater. The principle energy inputs in the case of the heat pump options are steam turbine driven compressors. An evaluation of the overall safety aspects of these two systems should favour the heat pump design. The only other recognisable difference between the several schemes is that the heat pump systems would contain more items of plant and a more complex instrumentation and control system.

With regard to operability, the heat pump systems are relative unknowns, because of their comparatively infrequent use. A conventional fractionator would be likely to have a greater operating stability, but in the case of the heat pump designs some back-up is provided because approximately 20 per cent of the total heat duty is provided by trim steam reboilers, leading to some enhancement of operational stability. It has been pointed out that the effect of compressor inlet conditions on overall performance and stability needs to be examined. It is possible that liquid may be formed in the compressor should the inlet fractions be at their saturation temperature, although superheating here would ensure that the fluids remained gaseous in all parts of the machine. Reliability is of course of prime importance in systems which are expected to operate continuously for 350 days/year. A detailed analysis has been carried out of the reliability of steam turbine driven compressors when compared to gas fired heaters, and it was concluded that overall the gas fired system would be more reliable. It operates at comparatively low temperatures to heat a non-fouling medium. Previous experience has shown that turbine compressor drives may be employed satisfactorily for periods of up to four years between major maintenance periods, and it should not be necessary to provide two somewhat smaller steam turbine driven systems, to ensure that one is always available as a back-up. Thus the overall economic position as given in Table 6.20 should not be changed.

TABLE 6.19 Steam Usage of Heat Pump Systems

Steam Usage (kg/h)	Overhead Recompression	Bottoms Recompression
Inlet		
Compressor Drive	34045	35818
Outlet		
Pass Out (65 psia)	17096	14636
Condensing (4 psia)	17090	21181
Pass Out		
Trim Reboiler	12818	12818
Compression Suction Heater	4272	1818
Seawater[1] (US gpm)	3500	4400

[1]Used for cooling steam turbine condenser.

TABLE 6.20 Relative Economics of Heat Pump Schemes (Middle East Location)

	Conventional Fractionator	Overheads Recompression	Bottoms Recompression
Plant (10^6\$)	Base	Base + 2.8	Base + 4.3
Utilities (10^6\$/year)			
Fuel	8.02	-	-
Steam	-	3.77	3.97
Power	0.31	-	0.09
Seawater	0.83	0.45	0.44
Total	9.16	4.22	4.50
Payback (years)[2]	Base[3]	0.6	0.8

1 350 days/year operation
2 Excludes Taxes
3 Relative economics based on marginal costs of heat pump systems

6.4 Heat Recovery from Liquid Effluents

Some data illustrating the potential of the heat pump as a device for recovering heat from liquid effluents has already been presented in Fig.6.4. Main areas of interest include the food industry, comprising also soft drinks, milk, and breweries, the textile industry, and chemicals. The food processing industry can make use of heat pumps for heat recovery from liquids in two ways. Simple recovery of heat from liquid effluents may be carried out, but in many plants a chilling, or at least a cooling, duty is required, in addition to process heat. Here the heat pump may be used in a similar manner to the refrigeration systems described in Section 6.2, but in this case the condensing and evaporating temperatures will be possibly 20°C higher, at least. In the textile industry, applications are more likely to be restricted to process heat recovery where the source of heat is finally dumped to waste. High temperature (120°C plus) may be needed in many industrial sectors.

6.4.1 Low temperature operations An example of the former where both heating and cooling can be implemented is represented by the two Templifier heat pumps at the Milk Marketing Board dairy at Bamber Bridge. This dairy processes approximately 60,000 gallons of milk per day, and the major plant comprises two bottling lines, machines for filling milk cartons, milk processing equipment, and full CIP plant.

An energy conservation study carried out at the dairy showed that the two main heat losses on the plant were at the effluent discharge outlet and via cooling towers on the chilled water circuit. Effluent was discharged from the pre-rinse on the bottle washing machine and the CIP system, while the water sent to the cooling towers originated from the cooling sections on processing plants such as pasteurizers. This heat was low grade and could only be used if heat pumps were applied. At the same time it was considered that water recovery could take place, saving significant amounts of money due to reductions in water charges and the cost of effluent disposal.

The heat pump plant at Bamber Bridge consists of two heat pumps with the condensers mounted in series. The first heat pump (No.1) is used to cool the clean water from the water recovery system. This water, which is cooled from 27°C to 24°C is supplied to the evaporator section and the condenser section of one of the heat pumps.

Heat extracted at the evaporator cools this water to 7°C, the water flow rate being 3,600 gallons per hour. This cold water is fed to the dairy's cold water storage tank, where it is distributed for final rinsing of bottles and other processing applications. The water entering the condenser, at a higher rate of 4,950 gallons per hour, is heated to 39°C before it passes to the condenser of heat pump No.2. A further rise in temperature to 60°C is effected here, the heat source being the chilled water circuit, cooled from 27°C to 24°C. The coefficient of performance of heat pump No.1 is 5.91 and the coefficient of performance of heat pump No.2 is 5.09. The flow diagram of the heat pumps is shown in Fig.6.37. The heat balance and coefficients of performance are listed in Table 6.21.

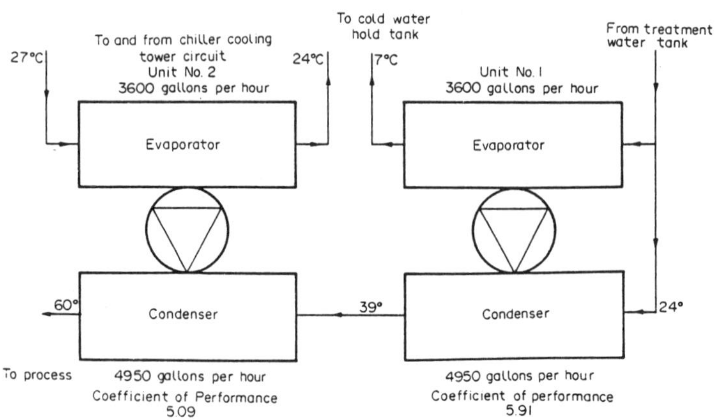

Fig.6.37 Flow Diagram of Heat Pump Circuits at Bamber Bridge

The hot water produced by the heat pumps is used for preheating boiler feed, pre-heating the domestic hot water service, bottle and crate washing, and the CIP system. The overall layout of the water and heat recovery system at the dairy is illustrated in Fig.6.38, the figures on the flow lines denoting the water temperature in degrees centrigrade.

Utilisation of the heat pumps is for 365 days per year, and any excess heat output can be used during the winter for preheating the dry goods store and the cartonning hall.

In order to ensure that data on the heat pumps' performance is regularly available, each heat pump has its own power meter, and temperature probes were fitted at all inlet and outlet points. Recordings are taken every day and checked to ensure that the system was operating to the design specification. No record was made of the water flow through the heat pumps, as this was fixed by the pump size. The plant manager did, however, point out if the flow was to fall due to pump wear an early indication would be given because the temperature would be effected, the heat rate being based on the maintenance of a fixed flow rate. It was also claimed that before the heat pumps were installed, two cooling towers had to be operated to reject the heat from the chilled water circuit. Following commissioning of the heat pumps, the use of heat pump No.2, using the cooling tower water as its heat source, enabled one cooling tower to be used only intermittently. This led to an electrical energy saving worth £3250 per annum.

The total installation cost of the heat pumps was £75,000, and a payback period of four years was estimated when the system specification had been carried out. In actual practice the payback turned out to be 2.5 years due to increased electrical savings on the cooling tower and higher energy prices. The calculation of payback presented in Table 6.22 shows a return on the investment of 36.8%. In addition, the dairy increased its running hours on production and this led to additional savings which would further reduce the payback period.

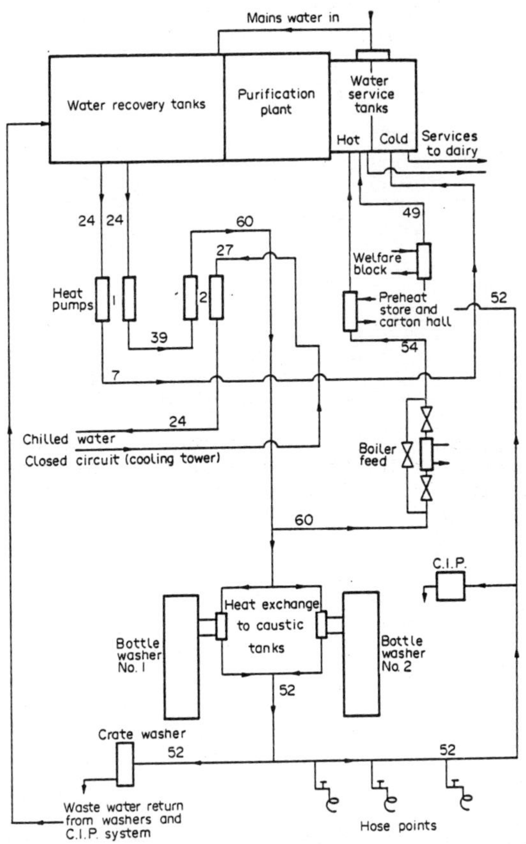

Fig.6.38 Schematic Layout of the Water and Heat Recovery System at the Bamber Bridge Dairy. The Figures on the Flow Lines Denote Water Temperature in °C

It was emphasised by the dairy manager that when considering the heat pump installation, time spent on tabulating the base information was of utmost importance. If this was incorrect and the heat pump was eventually installed, the anomalies would be indicated on commissioning.

Incidentally, the total cost of the plant including water treatment was £258,500, and this resulted in savings worth £70,000 per year. The system installed at Bamber Bridge bears a close relationship to that at the Unigate Dairy at Walsall. Here the two Templifier heat pumps also provide a cooling duty, the cooling water being used in the sterilizer, as well as hot water. The hot water duty is for preheating boiler feed water, preheating milk, the production of CIP water and space heating. As illustrated in Fig.6.39 the demand for the various heat requirements does not necessarily occur simultaneously, and therefore holding tanks are provided. Also, because the space heating is a seasonal requirement, a cooling tower has been incorporated in the system to reject surplus heat in warm weather.

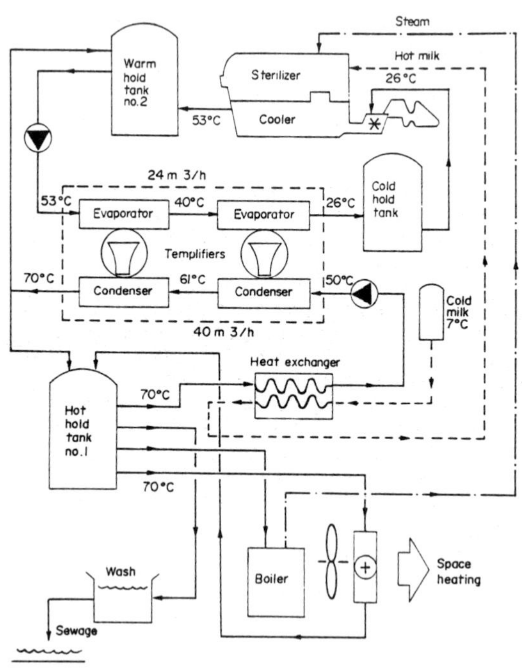

Fig.6.39 Templifier System at Unigate Dairy, Walsall

Using the two heat pumps in series, water can be cooled from 53°C to 26°C, the recovered heat being used to raise the temperature of the warm water from 50 to 70°C. The flow rates are illustrated in the figure, but the coefficient of performance of the overall system is 5.7, and the total heat output is approximately 1 MW. The installation is shown also in Fig. 6.40.

The variation of COP with source and sink temperature for Templifier heat pump of the type used in this application is shown in Fig.6.41, as derived from manufacturer's data.

Fig.6.40 Two Westinghouse Templifier Heat Pumps at Bamber Bridge Dairy

A second application of liquid-liquid heat pumps in the food industry has yielded interesting data on operating experience. Monarch Fine Foods is a company producing a variety of edible oils. The heat recovery installation, one of a number supported by the National Research Council in Canada, recovers heat from the plant cooling water system, previously served by two cooling towers.

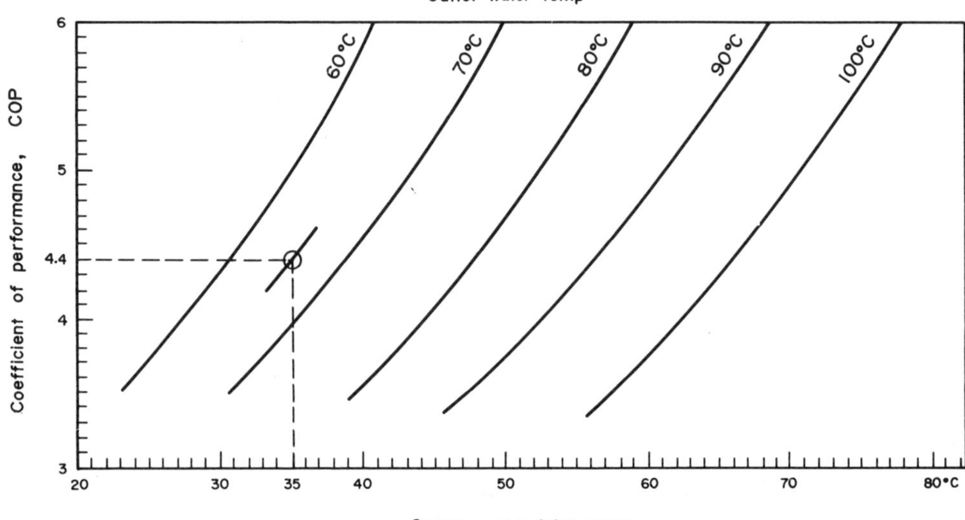

Fig.6.41 Variation of Templifier Heat Pump COP with Source and Sink Water Temperatures

TABLE 6.21 Heat Pump at Bamber Bridge - COP

Heat Pump	No.1	No.2
Evaporator inlet temperature	24°C	27°C
Evaporator outlet temperature	7°C	24°C
Flow rate	3600 gph	3600 gph
Condenser inlet temperature	24°C	39°C
Condenser outlet temperature	39°C	60°C
Flow rate	4950 gph	4950 gph
Compressor input	66.2 kW	108 kW
Heat from source	325.5 kW	443 kW
Heating capacity	391.7 kW	551 kW
COP	5.91	5.09
Overall COP	5.41	

The heat recovery system, which also incorporates a plate liquid-liquid heat exchanger, is illustrated in Fig.6.42. The water entering the heat pump evaporator at 41.2°C has its flow rate maintained at an approximately constant level by a variable speed pump. The boiler feedwater is passed through the other side of the plate heat exchanger before entering the heat pump condenser. This arrangement of course ensures that the heat pump COP is as high as possible.

TABLE 6.22 Heat Pumps at Bamber Bridge - Economics

Energy Saved:	32.18 therms/h	(gas)
	27.83 kW/h	(electricity)
Value of Gas Saved:	£45,514	
Value of Electricity Saved:	£3,250	
Total Saving:	£48,764	
Running Costs:	174.2 kW at 3.2p/kW.h = £20,346	
Annual Maintenance Cost:	£790	
Total Running Cost:	£21,136	
Net Saving:	£27,610	
Payback:	2.7 years	

The heat pump is a McQuay TPE C063 Templifier unit, based on a reciprocating compressor with R114 as the working fluid. A filter is employed on the evaporator, where the plant cooling water passes on the tube side. With a COP of 5.4, and a compressor power input (electric drive) of 93 kW, the heat pump has a payback period of 2.3 years. The original energy replaced originates from natural gas in a boiler of 85 per cent assumed efficiency.

The capital cost of the heat recovery installation is detailed in Table 6.23, while Table 6.24 shows the operating costs and the value of the energy saved.

A history of problems encountered with this "off-the-shelf" heat pump during the reported period approaching one year was compiled. These data are given in Table 6.25. A number of electrical problems were encountered, including failure of the compressor motor. As the heat pump could be bypassed, and sufficient cooling tower capacity was retained, these failures were not significant from the point of view of overall plant operation. The presence of lubricating oil in the refrigerant led to a 20 day shut-down of the heat pump. Some carry-over of oil would normally be anticipated, and while it can affect thermodynamic efficiency, the level of carry-over must have been high to warrant servicing and recharging.

Comparatively few absorption cycle heat pumps have been installed in industry, but their use in liquid-liquid duties, in this case in the textile industry, is illustrated by one such unit commissioned in October 1980 in Japan which is employed to recover heat from the waste water in dyeing. The conventional dyeing plant for woven cotton fabric is illustrated in Fig.6.43. The dyed cloth is dipped into a dyeing bath and is washed with warm water in order to fix the colours. The temperature used for washing varies in each both between 60°C and 90°C. Following dyeing the waste water becomes alkaline due to the presence of the dye stuff, and this waste water can be used for desulphurising the exhaust gases of oil fired boilers. Following the desulphurising process, the water is filtered and then discharged to waste. Feedwater preheating is effected using a heat exchanger installed in this waste stream, as the temperature of the water going to waste is at approximately 55°C. Downstream of this heat exchanger the water is still at a temperature of approximately 35°C, and is discharged at a rate of 60m³ per hour, (Ref.6.31).

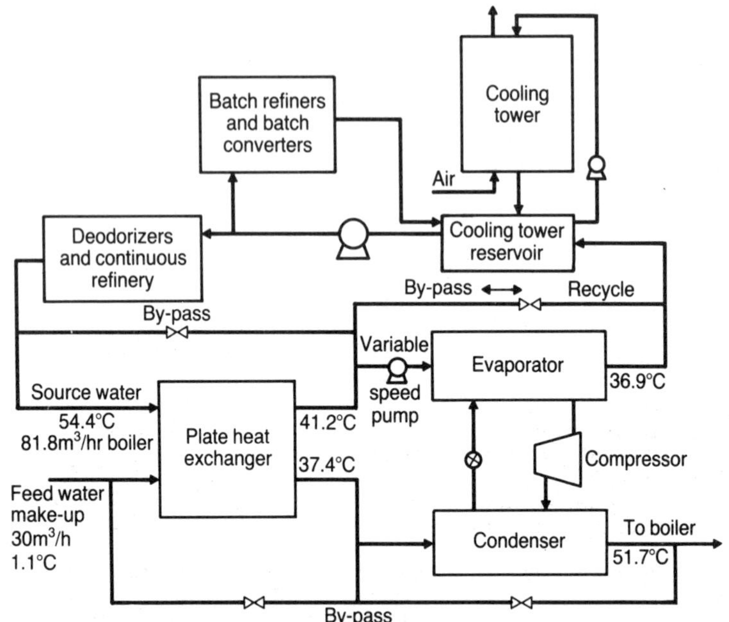

Fig.6.42 Waste Heat Recovery System and Cooling Circuit at Monarch Fine Foods, Data Average Values for 1:00-9.00 a.m., March 15, 1982

TABLE 6.23 Capital Expenditures for Waste Heat Recovery Installation (all Figures Canadian Dollars)

	Monarch
Heat Exchangers	8,200
Heat Pumps	81,090*
Water Pumps	12,664
Oil in Water Analyzer	13,800**
Miscellaneous	3,894
	$119,648
Mechanical	58,000
Electrical	14,317
Building	18,700
Total Installation	$91,017
Instrumentation	9,327
Engineering	4,104
Total Investment	$224,096

* Includes foul guard on evaporator
**Used to identify any oil which could pollute the boiler feed-water

TABLE 6.24 Savings and Operating Costs

	Monarch Fine Foods
Approximate Rate of Heat Delivered 10^9 joules/hour	
Heat Exchanger	4.5
Heat Pump	1.8
Total	6.3
Approximate Rate of Electrical Energy use kW	
Heat Pump	93
Other	4
	97
Coefficient of Performance of Heat Pump(s)	5.4
Value of Energy Replaced $/10^9 joules	4.99**
Value of Electrical Energy c/kWh	3.22
Net Value of Energy Recovered***	
Based on 6000 hr/year	$151,540
Simple Payback time (years)	1.48
(Based on Operating Experience)	
Operating hr/yr	4,021****
Net Value of Energy Recovered	$97,690
Simple Payback time years	2.29

** Replaces #6 fuel oil boiler efficiency assumed 85%
*** All monetary values in Canadian Dollars
**** Operating time of heat pump, heat exchanger operates significantly more hours per year

The absorption cycle heat pump was selected for this duty because of its comparatively insensitivity in terms of primary energy ratio to the temperature difference between the condenser and the evaporator. Also, because a comparatively high condensing temperature was required, it was felt that at that time that the life of refrigeration compressors operating with working fluids at higher temperatures would not be particularly great.

Following selection of the absorption cycle heat pump, a study was made of a number of the components. Of particular importance was the heat exchanger (the evaporator) which would be served by the waste water stream. This waste water is corrosive due to the presence of dye stuff and also the products of combustion etc. in the desulphurization process. Stainless steel type 316 was therefore used as a heat exchanger material and as a result of tests carried out over a number of months, satisfactory resistance to corrosion was demonstrated. An analysis of the waste water from the dye plant is given in Table 6.26.

The absorption cycle heat pump used in this plant is a lithium bromide/water unit and the generator is heated using natural gas. The heat pump was manufactured by Tokyo Sanyo Electric Co Ltd. The specification of the heat pump is given in Table 6.27. As described above, the warm waste water from the dyeing process is used for feedwater preheating in a conventional heat exchanger following filtration. It is then passed to the evaporator of the heat pump, where it is cooled to about 27°C prior to being discharged. The heat pump heats the warm water which has already been preheated to 38°C in the upstream heat exchanger, to 80°C for supplying the dyeing baths. The flow chart is as shown in Fig.6.44.

TABLE 6.25 Calendar History of Difficulties Encountered in Heat Pump Operation at Monarch Fine Foods

June 7 1982	Failure of Teledyne #600 oil in water analyser. This instrument must be operating to detect oil in the boiler feed water make-up. Should oil be found present, the waste heat recovery system is by-passed.
June 21 1982	Part received and installed. Operation resumed. Heat pump operating normally.
July 20 1982	Scheduled plant shut-down. No operation.
August 8 1982	Plant resumed normal operation. Heat pump operated normally.
September 13 1982	Compressor motor failed on heat pump. No operation
September 24 1982	New compressor motor installed under warranty. Operation of heat pump resumed.
October 10 1982	Lubricating oil found in refrigerant. Operation of heat pump suspended.
October 29 1982	Heat pump serviced and recharged with refrigerant. Operation of heat pump resumed.
December 1 1982	Low source water temperature due to mechanical problems with cooling towers. Operation of heat pump suspended.
December 30 1982	Problem with low source water temperature resolved. Operation of heat pump resumed.
February 22 1983	A thermister in the "foulguard" cleaning system was found faulty. Unit was shut down. During shut down an inspection indicated that overload relays were not set at design condition but at a higher value. Unit could not be operated until these were replaced.
March 14 1983	Thermister arrived and was replaced. Unit was still not operating however, because overload relays were on order.
April 14 1983	Problem encountered with anticycle relay, machine would shut down six times in one hour. Operation discontinued. During shut down an imbalance was found on the three phase winding of the compressor motor which could lead to burn-out.
April 20 1983	Problems with anticycle relay corrected by serviceman. Imbalance of electric load to motor windings corrected by serviceman. Heat pump operating resumed

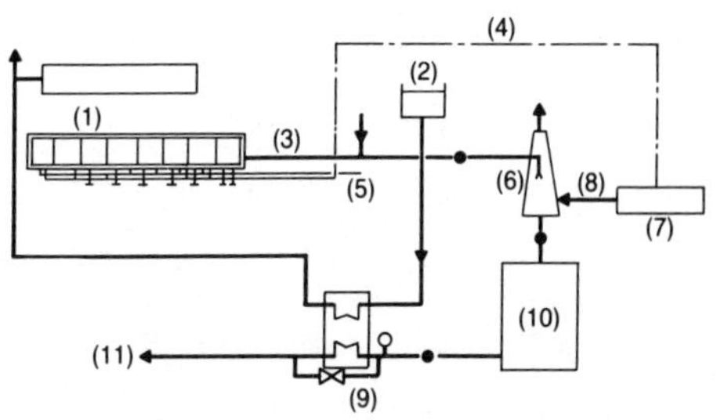

(1) Dyeing bath
(2) Elevated water tank
(3) Discharged water
(4) Steam
(5) Feed water
(6) Desulfurizing
(7) Boiler
(8) Exhaust gas
(9) Heat exchanger
(10) Waste fibre filter
(11) Waste water treating equipment

Fig.6.43 Conventional System in Dyehouse

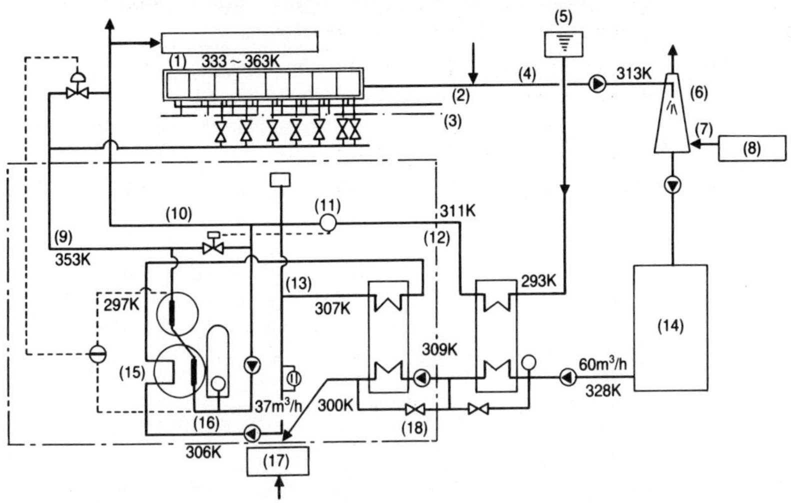

(1) Dyeing bath
(2) Feed water
(3) Steam
(4) Waste water
(5) Elevated water tank
(6) Desulfurizing
(7) Exhaust gas
(8) Oil fired boiler
(9) Hot water
(10) Automatic valve for circulation
(11) Flow switch
(12) Feed water
(13) Heat source water
(14) Filter
(15) Gas fired absorption heat pump (GT-10HP)
(16) City gas
(17) Waste water treating equipment
(18) 2nd heat exchanger

Fig.6.44 System Flowchart

The hot water in the dyeing bath must be replaced several times a day, this being necessitated by changes in the dye stuff. The gas consumption of the heat pump is controlled by the hot water outlet temperature.

A considerable amount of monitoring of this installation was made, and these included readings taken during summer and winter. The temperature of the heat source (the warm waste water) can vary by about 10°C between summer and winter, but the warm water inlet temperature is less liable to seasonal fluctuation because of the preheating undertaken by the upstream heat exchanger. It was therefore found that the hot water outlet temperature from the heat pump was stable throughout the year. It was found that the average efficiency of this system in operation was approximately the same in summer and in winter. Compared to a conventional system which would be employed in a Japanese dyeing plant, (a steam heating system using an oil fired boiler), the efficiency is approximately double, based on the primary energy input. The overall performance of the heat pump in terms of energy utilisation was calculated using data on the system efficiency and number of operating hours. These data are given in Table 6.28 and Table 6.29, for performance measured in August and December.

TABLE 6.26 Waste Water from Dye Plant - Analysis

Impurity	Proportion Present mg/l	Impurity	Proportion Present mg/l
Cd	0.02	Na	518.8
Cr	0.03	K	25.0
Cu	0.14	Ca	11.19
Fe	1.04	Mg	3.79
Mn	0.09	Al	1.23
Ni	0.13	Si	10.88
Pb	0.26	P	2.13
Zn	0.10		

Specific resistance	195 - cm
pH	90
Cl	979 ppm
SO_4	713 ppm

The fuel savings represented a financial benefit of £18,000, which when set against the installed cost of £57,000 gives a payback of about 3 years. Interestingly, the capital cost per kW of heat out seems quite low at about £60 when one compares this with quoted values of £100-£500 for gas engine driven systems. However, it is not borne out by the experience of companies outside Japan marketing absorption cycle systems, where capital costs are more on a par with competing systems.

A more conventional approach to the use of heat pumps in the textile industry is typified by the installation at Vitatex in the UK, a gas engine-driven unit used for liquid-liquid heat recovery duties. With a total heat output of 216 kW, the recovered heat is used for preheating process and boiler feed water. The heat sources are dye vats which have to be cooled. The system, illustrated in Fig.6.45, was engineered by Applied Energy Systems and cost about £50,000.

TABLE 6.27 Specification of Gas Fired Absorption Heat Pump

Warm Water System	Hot water output	3.48×10^9 J/h
	Warm water flow rate	19.8 m^3/h
	Warm water inlet temp.	311 K
	Hot water outlet temp.	353 K
Heat Source Water System	Heat source water input	1.39×10^9 J/h
	Heat source water flow rate	37 m^3/h
	Heat source water inlet temperature	306 K
	Heat source water outlet temperature	297 K
Fuel System (City gas)	Consumption	58 m^3/h (2.42×10^9 J/h)
	Low heat volume	4.16×10^7 J/Nm^3
	Specific gravity	0.65
	Design gas pressure	1961 Pa
	Effective heating surface area (generator)	12.5 m^2
Dimen-sions	Length	4.305 m
	Width	2.265 m
	Height	2.650 m
	Weight	8000 kg

6.4.2 High temperature operations All of the heat pumps described above operate at temperatures of less than 100°C. It is generally recognised that in order to make a significant impact on industrial energy conservation, industrial heat pumps must be able to operate effectively at temperatures of up to 400°C. At present commercially available units are able, on the basis of the vapour compression cycle, to achieve condensing temperatures of the order of 130°C.

TABLE 6.28 Effect on Energy Saving in August

	Absorption Heat Pump System	Conventional System (Boiler System)
Hot water output heat	5.44×10^{11} J	
Heat recovery	2.32×10^{11} J	
Energy Consumption	3.86×10^{11} J Gas consumption 9280 m^3 4.16×10^{7} J/m^3	7.56×10^{11} J
Efficiency	1.41	0.72
Energy Saving	3.7×10^{11} J Converted into fuel oil 10,000 ℓ	

TABLE 6.29 Effect of Energy Saving in December

	Absorption Heat Pump System	Conventional System (Boiler System)
Hot water output Heat	4.01 x 10 J	
Heat recovery	1.70×10^{11} J	
Energy Consumption	2.84×10^{11} J Gas consumption 6840 m^3	5.57×10^{11} J
Efficiency	1.41	0.72
Energy Saving	2.73×10^{11} J Converted into fuel oil 7,700 ℓ	

The first practically-sized heat pumps to successfully generate low pressure steam for long periods was that developed by NEI International Research & Development Company in Newcastle upon Tyne.

Fig.6.45 Gas Engine-Driven Heat Pump at the Vitatex Textiles Plant

This section describes the factors which had to be taken into account in this development, which was funded in part by the European Commission. Also described are application case studies.

Refrigerant selection: Consideration had to be given to the choice of refrigerant to be used in the heat pump before any detailed analysis of the heat cycle could be performed. There was no obvious choice of refrigerant fluid from past experience because very little work had been done with plant working at these, for heat pumps, relatively high temperatures.

In this case, every possible refrigerant fluid had to be considered for its suitability in high temperature operation. The condensing temperature in the heat pump was 120°C, and hence the first selection criterion was that the critical temperature of the refrigerant should be above 120°C.

The high condensing temperature ruled out commonly used refrigerants such as R12 and R22. Obvious selection criteria, as discussed in Chapter 3, include good thermal stability, moderate discharge pressure, low discharge temperature, high latent heat, and low viscosity.

Obviously, the final choice of refrigerant would be a compromise of all the above points. Following a decision on critical temperature the next step was to consider the operating pressures of the refrigerant at the condensing temperature of 120°C and the evaporating temperature of 60°C. This allowed the pressure ratio across the compressor to be calculated and also gave the compressor discharge pressure (Table 6.30).

For the ideal refrigerant, a moderate discharge pressure was required at the condensing temperature, otherwise it would be necessary to increase the strength of the compressor and condenser resulting in a larger and more expensive unit. The upper limit for discharge pressure was set at 2.1 MPa after consultations with the compressor manufacturers. This criterion eliminated refrigerants such as ammonia, sulphur dioxide and methyl chloride which are used widely in industrial refrigerating applications, but whose discharge pressures, for this application, are of the order of 5.0 MPa.

The remaining dozen refrigerant fluids are organic compounds made from a hydrocarbon base, with the interesting exception of water. None of them has been used widely as a refrigerant fluid on the same scale as R12, R22, ammonia etc, hence available data on their refrigerating properties was limited, and in some cases non-existent, especially at 120°C.

Explosive properties might appear to be the next most suitable criterion for selection from the remaining refrigerants. However, this may not be of prime importance for an industrial application as the heat pump may be working in an environment where flammable vapours are already present and hence a refrigerant leakage would not necessarily increase the fire risk. The amount of refrigerant leaking out to atmosphere (and hence the likelihood of explosion) can be reduced by improving the sealing of the system but this can be expensive. Ethylchloride, methyl formate and dichloroethylene would be eliminated because of explosion hazards.

Elimination by this criterion leaves the halogenated hydrocarbons and water. It was beyond the scope of the project, however, to investigate every fluid, due to the lack of published information on their properties at the temperatures concerned. Hence refrigerants 11, 21, 113, 114 and water were considered in more detail for the high temperature heat pump system.

Consideration of each refrigerant in turn led to the following observations (see Fig.6.46).

(a) Refrigerant 11

The main disadvantage of R11 was its high discharge temperature of 141°C which could cause problems with refrigerant breakdown, either thermally or chemically. The COP value of 3.9 was good

TABLE 6.30 Operating Pressures of Refrigerants

Refrigerant		Suction pressure, (at 60°C) MPa	Discharge pressure, (at 120°C) MPa	Pressure Ratio	
ASHRAE Ref. No.	Name				
717	Ammonia	2.8	6.6*	2.4	
40	Methyl chloride	1.36	4.5	3.3	High discharge pressure
764	Sulphur dioxide	1.0	4.0	4.0	
630	Methyl amine	1.02	3.85	3.8	
600a	Isobutane	0.86	2.9	3.4	
600	Butane	0.64	2.2	3.4	
114	Dichlorotetrafluorethane	0.57	2.05	3.6	
21	Dichlorofluoromethane**	0.52	1.93	3.7	
160	Ethyl chloride	0.49	1.90	3.9	
631	Ethyl amine	0.44	1.85	4.3	Acceptable discharge pressure
11	Trichlorofluoromethane	0.32	1.22	3.8	
611	Methyl formate	0.26	1.19	4.5	
30	Methylene chloride	0.20	1.05*	5.3	
610	Ethyl ether	0.23	0.99	4.3	
1130	Dichloroethylene	0.15	0.69	4.6	
113	Trichlorofluoromethane	0.15	0.60	4.6	
1120	Trichloroethylene	0.04	0.33*	8.3	
718	Water	0.02	0.2	10	

* Extrapolated values
** Now toxic and excluded from heat pump use.

and the volumetric flowrate was useful for reciprocating compressors, or centrifugal ones if a larger size unit was employed. Compression ratios and pressures presented no problems.

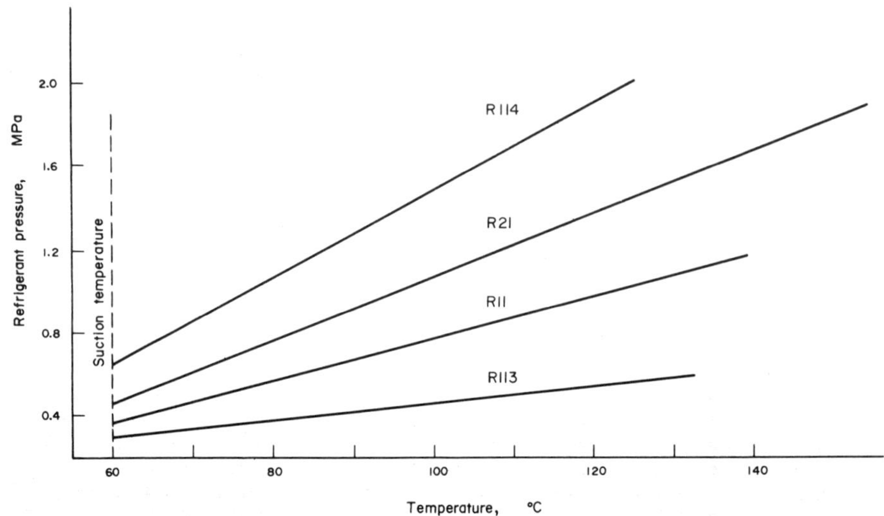

Fig.6.46 Performance of Four Candidate Refrigerants for a High Temperature Heat Pump

(b) Refrigerant 21

R21 had an even higher discharge temperature of 154°C at which refrigerant stability problems were likely to be even greater than with R11. Also, mass flowrate was rather low requiring a small compressor with a relatively large input power. Otherwise, its COP was one of the best and the compression ratio presented no problems. Subsequently, this fluid was eliminated because of its high toxicity.

(c) Refrigerant 113

The volumetric flowrate of R113 was quite high at 0.216 m^3/s, being three times as large as that for R21 and R114. As such, it may have been difficult to use R113 for reciprocating compressor, though it should be suitable for use with a centrifugal compressor with a larger power input. Its advantages were a high COP, low pressures and one of the lowest discharge temperatures which reduced the refrigerant breakdown problems.

(d) Refrigerant 114

The discharge temperature of R114 was the lowest of all the fluids under investigatiion and problems due to refrigerant breakdown were likely to be the least. Its volumetric flowrate was the same as for R21 and thus involved the same problem of compressor sizing. The suction pressure of 0.57 MPa was the largest and could have caused problems with the thrust bearings on the compressor crankshaft, depending on the design. The discharge pressure of 2.05 MPa was also the highest, approaching the limit of most compressors and in some cases exceeding the limits. While the compression ratio was the lowest, the COP at 3.19 was also the lowest.

(e) Water (R718)

Water had one of the largest COP values at 4.1 due to its high latent heat. Its other main advantage was high thermal stability in comparison to the halogenated hydrocarbons. However, the discharge temperature in this application, at 412°C was high enough to cause problems with the compressor materials. The volumetric flowrate was very large, being four times that for R113 and hence a large centrifugal compressor would have had to be used. The pressure ratio at 10 was large and would have required a multi-stage compressor and problems could have occurred with leaks into the suction pipework where the pressure was less than atmospheric.

The main conclusion from the cycle analysis was that none of the five refrigerants under consideration appeared to be ideally suited to the heat pump application under consideration. Each refrigerant posed different problems which had to be balanced against each other. Water presented an interesting case but would have only been of use in a very large heat pump utilising a large centrifugal compressor. Investigation of problems surrounding the high discharge temperature and large pressure ratio would have taken more time than was available in the project and because of this water was eliminated as a possible fluid.

In order to make a strong case for selecting one of the remaining refrigerants another criterion had to be used. The criterion chosen was the thermal and chemical stability of the refrigerants at high temperature. At a condensing temperature of 120°C breakdown is inevitable to a certain extent and the breakdown products will cause corrosion in the system as well as reducing cycle efficiency. Other factors that may cause problems are the increased solubility of the refrigerant in the oil and the increased rate of refrigerant hydrolysis, particularly important at the high discharge temperatures in this application.

Effect of dissolution in oil

Solubility of the refrigerant in the compressor oil is related to temperature and the greater the amount of dissolved refrigerant the larger will be the change in the oil properties. The most significant factor is the oil viscosity which decreases as the amount of dissolved refrigerant increases. If the viscosity falls to 10^{-2} Pa s or below, severe lubrication problems at the bearings are likely to occur. Crankcase heaters often have to be employed to reduce the amount of dissolved refrigerant.

At high temperatures, the refrigerant will react with the oil to produce carbonaceous sludges and, in some cases, acids. The sludges reduce oil performance and the acids will be corrosive.

Effect of water

Ideally, water should not be present in the system. Small amounts do, inevitably, gain access and at high temperatures hydrolyse the refrigerant with the formation of acids, which are corrosive. This problem is common to all refrigerants and, again, does not form a selection criterion. It can be overcome by inserting a refrigerant drier in the liquid line which removes both water and acid from the system. Suitable materials include activated alumina, Drierite and Silica gel.

Effect on compressor materials

Halogenated hydrocarbon refrigerants cause considerable swelling of plastics and elastomers and data are readily available. Careful material selection can overcome problems associated with a particular fluid.

The refrigerant will also react with the metals in the system, but at normal refrigerant temperatures the rate of reaction is negligible. However, condensing at 120°C the reaction rate will be much higher and breakdown problems will occur. Zinc, magnesium and aluminium containing more than 2% magnesium are not recommended for use with halogenated refrigerants at any temperature.

Refrigerant thermal stability

This was recognized as probably the most important factor in eventual refrigerant selection. Refrigerants are expensive and any breakdown will be costly. Also the products of refrigerant breakdown, such as acids, will produce corrosion problems. The problem is complex because the refrigerant can break down due to temperature alone or by reacting with the different metals or oils in the system. The breakdown rates with different metals also vary and hence it was necessary to assume some average reaction rate. Data on breakdown rates of R11, R21, R113 and R114 at high temperature is not extensive as high temperature applications have been limited. However, enough data was available to draw tentative conclusions about the relative refrigerant stabilities. Du Pont Inc give the recommended operating temperatures shown in Table 6.31.

This data was obtained from a series of sealed tube tests. A similar investigation was carried out by H.M. Parmelee in which the refrigerant was tested at different temperatures in the presence of napthenic oil, aluminium, copper and iron.

The published refrigerant stability data tend to give different values of decomposition rate according to the method of performing the test, the materials present in the test, and the length of time over which the test was performed. However, qualitative conclusions on the relative stability of the refrigerants could be made. Refrigerant 114 appeared to be consistantly much more stable than the others, probably because of the high proportion of fluorine contained in the molecule. As fluorine is a strongly electronegative element, the bonds within the refrigerant molecule become polarised and therefore much stronger, thus making the molecule more stable.

TABLE 6.31 Refrigerant Thermal Stability Data

Refrigerant Ref. No.	Maximum Temperature for Continuous Exposure, °C	Decomposition rate at 200°C with Steel, % per year
113	107	6
11	107	2
114	121	1
21	121	Not known

R113 appeared to be the next most stable refrigerant, followed by R11, and R21 was the least stable of the four. The relationship between the amount of fluorine in the refrigerant molecule and its stability is shown in Table 6.32.

TABLE 6.32 Refrigerant Fluorine Content

Refrigerant in Order of Stability	No. of Fluorine Atoms Per Molecule
114	4
113	3
11	1
21	1

Consideration of the calculated discharge temperatures for the various refrigerants reveals that in each case they were greater than the maximum continuous operating temperatures recommended by ASHRAE. It could be argued that the discharge temperature, although it is the highest in the cycle, would only be experienced by the refrigerant for a fraction of its total life in the system. However, since refrigerant breakdown was thought likely to be the biggest problem in the heat pump system it would be wisest to choose the most stable refrigerant under the given operating conditions. If the margin between operating discharge temperatures and maximum recommended temperature is now considered for the refrigerant the data in Table 6.33 is obtained.

Refrigerant 114 would be operating within 4°C of its recommended operating temperature whereas the others were at least 25°C above their recommended temperatures. Using refrigerant stability as the sole selection criterion, R114 was the obvious choice. However, thermodynamically it was the poorest fluid and it operated at the highest pressures. The next most stable fluid, R113, was one of the best thermodynamically, operated at the lowest pressures and was suited to a centrifugal compressor if a larger sized heat pump were to be used.

TABLE 6.33 Relationship Between Actual Discharge and Recommended Temperatures

Refrigerant Ref. No.	Discharge Temp. °C	Recommended temperature, °C	$t_{dis} - t_{rec}$, °C
114	125	121	4
113	133	107	26
11	141	107	34
21	154	121	33

The final choice of refrigerant was between R113 and R114. R113 would have been used with a centrifugal compressor, while R114 was selected for a 75 kW drive reciprocating machine. The relative advantages and disadvantages discussed above are summarised in Table 6.34.

Heat pump description

The heat pump The resulting heat pump system set up at IRD for laboratory testing and demonstration is shown in Fig.6.46(a) The laboratory testing comprised of a simulation of a proposed industrial application in the chemical industry in which most of the major components of the heat pump would also have been used, (Ref.6.32, 6.33, 6.34).

The main items of hardware are summarised below:

Engine

This was a gas fuelled Waukesha Type F817GU unit having a 6 cylinders and a total swept volume of 13.39 litres. Continuous shaft power rating was 75 kW at 1100 rpm, maximum intermittent power output is 115.5 kW at 1400 rpm and maximum continuous power output is 98 kW at 1400 rpm. Two phase flash evaporative engine block, cylinder head and exhaust manifold cooling was employed with mains operated electric pumps supplying and cooling water feed. Engine speed control was via a Barber-Colman manually adjusted electronic governor.

Compressor

A Grasso AC1080 wet reciprocating compressor having a swept volume of 0.071 m^3/s at 1100 rpm was used. This is an open type machine and had five banks of twin cylinders arranged around the crankshaft. Capacity control of 30 per cent, 50 per cent, 80 per cent and 100 per cent was possible through the manual operation of cylinder bypass valves.

It would have been relatively straightforward to design and construct a capacity control system for the compressor which would relate compressor loading to the heat demand so that evaporating and condensing temperatures were held constant for varying loads. However, such a system was not needed for the chemical industry application where the heat load is nominally constant under operational conditions. Therefore no attempt to design and construct an automatic load control system was made for the laboratory trial of the heat pump.

TABLE 6.34 Relative Merits of Refrigerants

	Refrigerant 114	Refrigerant 113
Advantages	(a) Much better thermal stability	(a) Low compressor pressures require lightweight pipework
		(b) Good COP
Disadvantages	(a) High discharge pressure requiring heavy pipework and compressor casing	(a) High pressure ratio requiring more expensive multistage centrifugal compression
	(b) High suction pressure could cause problems with the thrust bearing on a reciprocating compressor	(b) Lower thermal stability could cause corrosion in the system
	(c) High pressure differential could cause stress and leakage problems in the compressor	(c) Volumetric flowrate is too large for reciprocating compressors

Evaporator

This was a carbon steel shell and tube unit with refrigerant evaporation taking place in the tubes due to heat transfer from the simulated effluent stream at 60°C to 80°C on the shell side. The relative location of components is shown in Fig.6.47.

Condenser

This was also a carbon steel shell and tube unit with refrigerant condensing in the tubes supplying heat to boiling water on the shell side. Water was taken from the shell side to the engine and exhaust gas boiler for cooling/heat recovery and the 2 phase flow returned to the condenser. Steam at 100°C approximately was separated in a kettle type dome incorporated into the condenser shell.

Fig.6.46(a) High Temperature Heat Pump Showing Gas Engine, Compressor and Monitoring Equipment

Exhaust gas boiler

This was a shell and finned tube heat exchanger with water boiling in the tubes and exhaust gas passing in cross flow through the shell where it was cooled to approximately 150°C. Because of difficulties in obtaining a suitable commercial unit for reasonable cost this heat exchanger was manufactured in-house to a standard appropriate for laboratory testing only. The design standard and manufacturing quality would need to be higher on a heat exchanger for this duty for industrial use.

Heat source

In the laboratory test rig the heat source was a water stream taken at 60°C to 80°C from a thermostatically controlled storage tank. This simulated the heat source in the industrial process as far as temperature and heat flow rate were concerned.

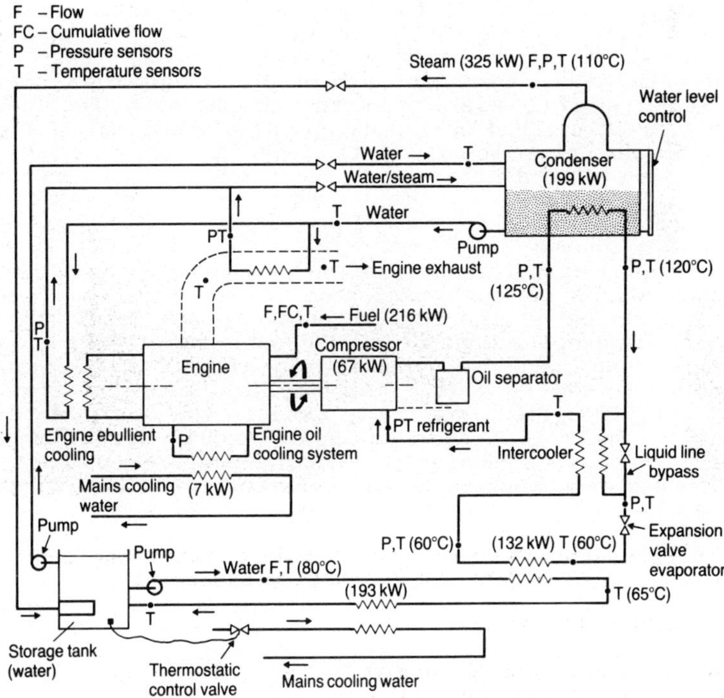

Fig.6.47 Schematic of the IRD Gas Engine Driven High Temperature Heat Pump

Heat sink

The heat sink in the laboratory test rig was water at 60°C to 80°C. The excess heat supplied over that extracted in the evaporator was dissipated through thermostatically controlled heat exchangers to the cooling water mains to maintain the system in thermal balance. Steam produced in the condenser, engine cooling jacket and the exhaust boiler was piped from the condenser into the water storage.

Refrigerant

R114 was the refrigerant used in the heat pump, based on the selection criteria discussed earlier.

Oil separator

During the research and development phase of the project, it had been observed that the refrigerant circulating in the system contained about 5 per cent by weight of engine compressor oil. Stripping the heat pump circuit down revealed that a considerable quantity of this oil had separated out in the condenser, evaporator and intercooler (approximately 2.5, 7 and 2 litres respectively) which would have inhibited heat transfer. An oil separator was, therefore, incorporated at the compressor outlet to trap oil mist from the discharging refrigerant. When sufficient oil had collected, a float operated valve opened for sufficient time to allow the oil to be returned, under the infuence of the discharge pressure, back to the compressor.

Control and instrumentation system

In the chemical industry application of the heat pump, operating conditions would be continuous and steady regarding capacity and heat sink and source temperatures. A sophisticated system of automatic capacity control giving optimised heat pump operation would not have been necessary, therefore, and for the laboratory proving trial the instrumentation and controls were only slightly modified from those used in the original research programme. Control was predominantly manual but allowed safe long term unattended running to take place and extensive instrumentation was incorporated to allow manual monitoring of process variables, i.e. temperatures, pressures and flow rates.

It was important that equipment operation under damaging conditions was avoided and to this end an automatic shut down system was incorporated. The parameters which shut down the system were as follows:

(i) Low engine oil pressure.

(ii) High engine oil temperature.

(iii) Low engine coolant flow rate.

(iv) High engine coolant pressure and temperature.

(v) High compressor discharge pressure and temperature.

(vi) Low compressor suction temperature and pressure.

(vii) Low compressor differential oil pressure.

(viii) Low gas supply pressure.

Shut down was achieved by closing the gas supply valve and the engine throttle and earthing the low tension side of the engine ignition circuit. Excessive refrigerant condensation in the compressor after shut down was prevented by a non-return valve in the discharge line and a solenoid operated isolating valve in the suction line.

Two identical thermostatic expansion valves arranged in parallel controlled the degree of superheat of the vapour leaving the evaporator to ensure that no refrigerant liquid passed into the compressor.

A manually operated valve in the steam discharge line from the condenser was used to control steam pressure.

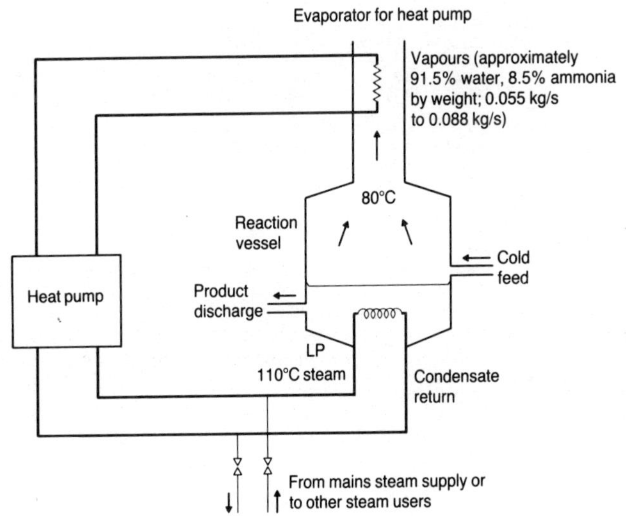

Fig.6.48 Application of High Temperature Heat Pump to a Chemical Reaction Vessel

Chemical reaction vessel heat recovery

In a particular process in the chemical industry a reaction vessel which is operating at low pressure releases a mixture of water vapour and ammonia at 80°C as a waste stream. This vapour stream has been identified as a possible heat source for a high temperature heat pump. The heating medium required for this process is low pressure steam, which also has several other uses on the site. Hence no problems are envisaged with the utilization of the steam produced by the heat pump. The vapour flowrate leaving the reaction vessel varies between 0.055 and 0.083 kg/s and is made up of approx. 91.5 per cent water vapour and 8.5 per cent ammonia (by weight). A schematic diagram of the process and the heat pump circuit is shown in Fig.6.48.

The heat recovered from source vapour varies between 125 kW and 180 kW, depending on flowrate. Heat pump economics and performance have been investigated for these limiting operating conditions and results are presented in Tables 6.35 and 6.36.

TABLE 6.35 Economic Analysis of the Application of High Temperature Heat Pumps*

Process and utilization	Year	Total Running Cost (#)	Value of Steam (#)	Annual Savings (#)	Cumulative Savings (#)	Estimated Capital Cost (#)	Payback Period (yr)	Rate of Return (%)
Chemical reaction vessel, lower Design point, 6720 h	1	12,959	25,000	12,041	12,041			
	2	14,255	27,500	13,245	25,286			
	3	15,681	30,250	14,569	39,855	60,000	4.3	29
	4	17,249	33,275	16,026	55,881			
	5	18,974	36,602	17,628	73,509			
Chemical reaction vessel, upper design point, 6720 h/yr	1	18,686	36,072	17,386	17,386			
	2	20,544	36,679	19,125	36,511			
	3	22,610	43,647	21,037	57,548	60,000	3.1	40
	4	24,871	48,012	23,141	80,689			
	5	27,358	57,813	25,455	106,144			
Maize cooking, 5760 h/yr	1	59,836	109,166	49,330	49,330			
	2	65,820	120,083	54,263	103,593			
	3	72,402	132,091	59,689	163,282	192,000	3.5	36
	4	79,641	145,300	65,659	228,841			
	5	87,606	159,830	72,224	301,165			
Whisky distillation, 5060 h/yr	1	21,220	38,238	17,018	17,018			
	2	23,342	42,062	18,720	35,738			
	3	25,676	46,268	20,592	56,330	73,000	3.8	34
	4	28,244	50,895	22,651	78,981			
	5	31,168	55,984	24,816	103,797			

* Table based on: gas price of #9.55 x 10^{-3}/kWh; maintenance cost of #0.002/kWh of engine shaft power; steam price of 8.33/tonne; total running cost = fuel cost & maintenance cost; inflation on fuel and maintenance 10% per annum; economic lifetime of 10 yr.

TABLE 6.36 Calculated Performance of the High Temperature Heat Pump in Industrial Applications

Process	COP	PER	H_S (kW)	H_C (kW)	H_E (kW)	H_F (kW)	H_R (kW)	H_T (kW)	Q_S (kg/s)	T_C (°C)	T_E (°C)
Chemical reaction vessel											
Lower design point	3.19	1.46	57	182	125	190	95	277	0.124	120	60
Upper design point	3.19	1.46	82	262	180	274	137	399	0.179	120	60
Maize cooking	3.19	1.46	307	979	672	1023	511	1490	0.632	120	60
Whisky distillation	2.86	1.36	124	355	231	413	207	562	0.252	120	55

H_S is shaft work; H_C is condenser output; H_E is process heat recovered at evaporator; H_F is fuel input; H_R is engine heat recovered; H_T is total steam output; Q_S is steam flowrate; T_C and T_E are condensing and evaporating temperatures.

Calculations based on an engine shaft efficiency of 30% and 50% heat recovery from engine.

Heat recovery from maize cooking in a distillery

The first stage consists of mixing ground maize with water at a temperature of around 65°C. The mixture is then heated rapidly to a temperature of 93°C by direct steam injection. It is important that this heating is rapid otherwise the mixture tends to gelantinise making handling difficult. After enzyme addition and holding at temperature, the mixture is finally cooled back to around 27°C. It is proposed to use the secondary cooling water from the process as the heat source for the heat pump, which would then deliver low pressure steam for the injection heating.

A schematic diagram of the proposed heat pump incorporated into the process is shown in Fig.6.49 and Tables 6.35 and 6.36 summarize system economics and operational performance.

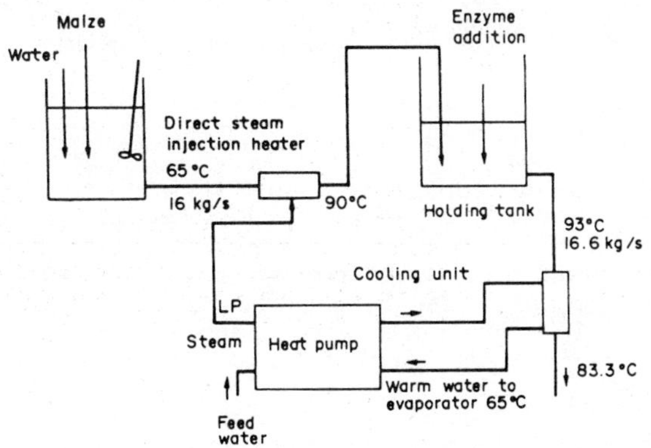

Fig.6.49 Application of a High Temperature Heat Pump to a Maize Cooking Operation

Heat recovery from distillation process cooling water

In a whisky distillery, vapours from the still are condensed in a water cooled condenser. The cooling water at a temperature of approx. 71°C is a possible source for a heat pump. The heating medium for the stills is steam at a pressure of 2.4-2.7 bar(g) which is obtained by throttling from a steam supply at 6.4-7.0 bar(g), it would not be possible to produce the required medium pressure steam directly from the heat pump, as this would require condensing temperatures of 136-140°C. However,, one solution would be to take the low pressure steam off the heat pump and compress it to the working pressure using a steam jet compressor (or thermocompressor) with the high pressure steam as the motive fluid. The sort of system envisaged is shown in Fig.6.50 together with temperatures and flow-rates. The stills are batch operated units run as individuals. The cooling water from the condensers would be available at a flow of 300 l/min (5 kg/s) for a period of 220 min a cycle. There are 30 such cycles of each still per week and a working year consists of 46 weeks. Thus the annual running time is 5060 h.

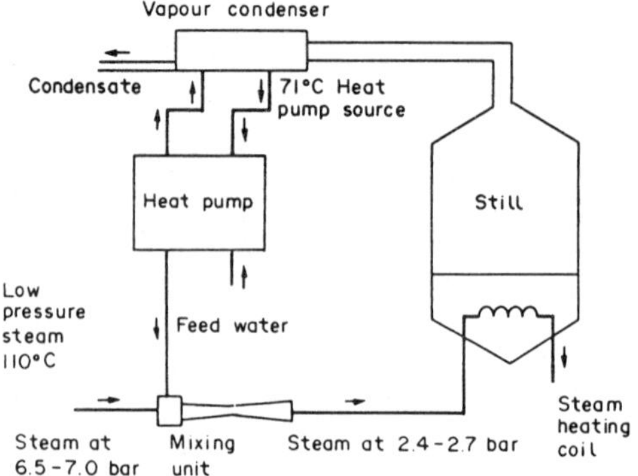

Fig.6.50 Application of Heat Pump in a Distillery

Other potential applications

Other processes where high temperature heat pumps may be applicable have been identified but insufficient process data has been made available to carry out thorough analyses. Of these the drying of textiles, carpets, paper board, pharmaceuticals and food in which air at between 100 and 200°C is required are probably the most important. Other sectors of industry where low pressure steam and pressurized hot water are used as heating media include textiles, inorganic pigments, d airies and metal finishing.

Commissioning, Operation, Measurement and Performance of Laboratory Test Rig

Initial start-up and commissioning of the heat pump was trouble free apart from a malfunction of the compressor oil separator due to a leaking float valve, and the presence of small refrigerant leaks which occurred when the system reached operating temperature. These faults were rectified.

Table 6.37 is a log for the operation of the system at IRD over the period of extended running which totalled 2059 h. Summary details of faults, method of fault rectification and maintenance efforts are given in this table. Major items of equipment in the system gave few problems and relevant comments are given as follows.

Engine

This performed well without any major problems. It was serviced at approximately 200 h intervals according to the manufacturers recommendations and this comprised of adjusting the valve clearances and spark plug and distributor points gaps, cleaning air filters and crankcase breathers and changing the oil filter. A programme of oil analysis was undertaken to gradually increase the interval between oil changes. This was increased from 200 h to over 500 h during the trial without significant signs

TABLE 6.37 Heat Pump Operation Log

Date	Faults	Fault rectification	Maintenance	Cumulative hours operation	Comments
24.9.82 to 7.12.82	Engine coolant temp. trip too low Oil separator malfunctions	Trip re-set Float leaks - float replaced	Engine oil changed to Esso HDX 30 after discussion with Esso	89	Commissioning started End of commissioning Start of testing
14.12.82 to 15.2.83	Ignition system wire on engine broken Valve in oil separator leaking Oil separator return pipe ruptured	Renewed wire Renewed float/valve essembly Line replaced	Engine serviced engine oil changed 23 kg refrigerant added	289	Included Christmas holiday period Analysis of engine oil showed it to be satisfactory after 200 h
16.2.83 to 3.3.83	Compressor oil differential pressure low (1.35 to 1.6 bar). Specifed as 2 to 2.5 bar	Big ends checked - no wear Pressure regulator set beyond range - gave 1.8 bar	Engine serviced Engine oil changed Engine and compresor oils analysed	570	Restarted test. 0.5% oil in refrigerant - satisfactory Engine oil satis factory after 271 h Comp. oil satis-factory after 570 h
7.3.83 to 8.4.83	Refrigerant low	Charged with 30 kg refrigerant	Engine serviced Engine oil changed and analysed	920	Engine oil satisfactory after 350 h
19.4.83 to 11.11.83	Refrigerant low on several occasions (vapour bubbles) in liquid line site glass) Small refrigerant leaks detected at isolating valves Comp. oil dif. press. still low (1.2 to 1.3 bar) and erratic Oil migration from comp. to oil separator on occasions during shut down due to faulty seating of valve Regular tripping of system due to high suction temp. Fault eventually traced to undersizing of effluent/cooling mains heat exchangers when temp. difference reduced in summer weather Loss of water from system during operation eventually traced to leaking tube in exhaust gas boiler	Additional refrigerant added. New seals fited New press. regulator fitted but partially successful only No remedial action but non return valve in connecting line needed Additional heat exchanger fitted Weld repair to tube plus cleaning of heat transfer surfaces.	Check by comp. manu-facturer's Engineer - comp. in good condition Engine and comp. oil analysed. Engine serviced.	1510	Engine oil satis-factory after 500 h Comp. oil satis-factory after 1500 h
3.1.84 to 24.2.84	Bypass valves on comp. cylinders not operating on occasions due to sticking solenoids. Migration of oil from comp. after shutdown (see above). Refrigerant leaks from spindles of thermostatic expansion valves. consistant effluent loss of about 50 l/day towards and of test period. No obvious leaks apparent, could be exhaust gas boiler again (see above)	Solenoids cleaned No remedial action New seals fitted, 26.5 kg refrigerant added Not yet rectified	Comp. oil analysed	2027	Comp. oil satis-factory after 2027 h
24.2.84 to 2.5.84				2059	Part load and off-design. Point operation of performance testing

of oil deterioration. A small but persistent oil leak from the sump gasket accounted for an estimated 1/3 to 1/2 of the oil consumption of 130 h/l.

Compressor

The compressor performed well without sign of mechanical or performance deterioration despite running at lower than specified oil differential pressure (see Table 6.37). The cause of this was probably the relatively high level of refrigerant dissolved in the oil compared to conventional lower temperature operation. Modifications to the pressure regulator as suggested by the compressor manufacturers to increase the differential pressure only resulted in pressure fluctuation. Oil analyses performed over the test period showed no deterioration up to 2027 h. There was occasional sticking of the bypass valve solenoids towards the end of the test period but cleaning eliminated this problem.

Evaporator and condenser

No problems occurred and there were no signs of deteriorating performance.

Exhaust gas boiler

Thermally this item performed satisfactorily but it did develop a leak at a cracked weld which required repair. A unit manufactured to industrial standards would be required for an industrial application of the heat pump.

Oil separator

This component worked effectively in separating oil from the compressor discharge and reduced the oil level in the refrigerant to about 0.5 per cent compared to the 5 per cent to 6 per cent that had been present during the previous development work without the oil separator. However, it did not functiion properly on numerous occasions due to the float valve not sealing properly and allowing oil to migrate from the compressor during shut down. An improvement in this aspect of mechanical performance would be needed for an industrial application.

Heat pump circuit

Problems in the heat pump circuit which arose were due entirely to leaks but as these were often difficult to locate because of their small size a significant amount of time was spent in rectification. Leaks from valve stems in particular were a problem to locate as they seemed to be erratic. This experience has indicated that higher specifications for performance should be applied to the valves used in an industrial application of the high temperature heat pump and care should be taken to ensure they can operate at the higher temperature.

Water circuits

A seal in one of the water pumps failed during the trial but this was easily repaired. The only other major problem to occur was due to initial undersizing of the effluent/cooling mains heat exchangers. This fault manifested itself as a high compressor suction temperature due to high evaporating and effluent temperatures in the warmer summer weather. The addition of another cooler in series eliminated the problem.

Controls and instrumentation

These items worked quite satisfactorily. Safety shut down procedures were adequate and coped with all malfunctions which occurred. Operational control was mainly manual but once operational conditions were set, the system ran steadily. No faults with instrumentation occurred and sufficient information on operational parameters was produced to satisfactorily analyse thermal performance.

Thermal performance

The system was run for periods at 50 per cent, 80 per cent and 100 per cent compressor load during the long term trial and at engine speeds between 1100 and 1350 rpm. However, the evaporating and condensing temperatures were maintained at about the design values, 60°C and 120°C respectively, apart from at the end of the trial when some off-design tests were carried out. Checks showed performance to have been consistent over the test period.

Concluding Remarks

In the 1st Edition of this book, the final application discussed was a proposal for a high temperature absorption cycle heat pump - high temperature in this context being in the range 320°C to 560°C. This remains an application area, in terms of temperature, some way from commercial application, and it may be argued that heat pump technology has not progressed sufficiently, in spite of the resources which have been expended in research, development and demonstration.

This would, however, be unreasonable when, in most domestic and commercial building applications the heat pump is a proven technology acceptable where conditions favour it economically. The development of absorption cycle units which have good efficiencies and are cost-effective will have a major impact on future heating systems, and the variety of industrial heat pump types will benefit from improved system engineering, while in some cases awaiting a return to attractiveness brought about by an upturn in energy costs at some future date.

REFERENCES

6.1 Lyle, O. The Efficient Use of Steam, HMSO, London, 1947.

6.2 Heaton, A.V. and Benstead, R. Steam recompression drying. Paper E3 in Proc. 2nd Int. Symposium on Large Scale Applications of Heat Pumps, York, 1984. BHRA, Cranfield, 1984.

6.3 Reay, D.A. Industrial Energy Conservation. 2nd Edition, Pergamon Press, Oxford, 1979.

6.4 Reay, D.A. Heat Recovery Systems - A Directory of Equipment E & FN Spon, London, 1979.

6.5 Neill, D.T. and Jensen, W.P. Geothermal powered heat pumps to produce process heat. Proc. 11th Intersoc. Energy Conv. Engng. Conf., Nevada, pp 802-807, Sept. 1976.

6.6 Alefeld, G. A high temperature absorption heat pump as topping process for power generation. Energy, Vol.3, pp 649-656, 1978.

6.7 Anon. A century of refrigeration by Sulzer. Sulzer Technical Review, No.1, pp 41-43, 1978.

6.8 Charity, L.F. et al. Heating water with a milk cooler using heat pump principle. Agric. Engng., pp 216, 219, April 1952.

6.9 Anon. Waste heat recovery saves £15000 a year. The Engineer, p 24,25, 5 Feb. 1976.

6.10 Witt, J.A. and Whelan, A. Economics of industrial heat pumps: A case study of plastics processing. Proc. 9th Congress of Union Internationale D'Electrothermie, Cannes, 20-24 Oct. 1980.

6.11 Anon. Heat pump maintains plastic injection moulding output. Case History, Electricity Council Note EC3822, 1978.

6.12 Berghmans, J. Report for the International Energy Agency Implementing Agreement on Advanced Heat Pumps. Annex III: Heat Pump Systems Applied in Industry, Leuven University, Belgium, Nov. 1981.

6.13 Braham, G.D. Energy management looks after lost heat. Electrical Review, Vol. 203, No. 22, Dec. 8, 1978.

6.14 Anon. Heat pumps. A new application for screw compressors. Howden Journal, Issue No. 14, Jan. 1976.

6.15 Reay, D.A. Industrial Energy Conservation. A Hand book for Engineers and Managers. 2nd Edition. Pergamon Press, Oxford, 1979.

6.16 Hodgett, D.L. Efficient drying using heat pumps. The Chemical Engineer, pp 510-512, July/August 1976.

6.17 Oliver, T.N. Process drying with a dehumidifying heat pump. Paper C2. Proc. Conf. on Ind. Applications of Heat Pumps, Coventry, 24-26 March 1982. BHRA Cranfield 1982.

6.18 Flikke, A.M. et al. Grain drying by heat pump. Agricultural Engng, Vol. 38, No. 8, pp 592-597, 1957.

6.19 Davis, C.P. A study of the adaptability of the heat pump to drying shelled corn. M.Sc. Thesis, Purdue University, USA, 1949.

6.20 Laroche, M. and Solignac, M. Heat pump application to drying in agricultural and industrial fields. Rev. Gen. Therm., Vol. 15, No. 179, pp 989-995, 1976.

6.21 Anon. EEC Demonstration Project Profile No.17. Heat pump using the steam generated by a thermomechanical process. CEC Brussels, 1984.

6.22 Mezon, J. Theoretical study of an industrial absorption dryer. Proc. Absorption Heat Pump Congress, Paris, 20-22 March 1985. European Commission Report EUR10007 EN, Luxembourg, 1985.

6.23 Harris, P.S. The dairy industry. Energy Audit Series No.3, Dept. Energy and Dept. Industry, Harwell, 1978.

6.24 Bucher, F. Making savings in thin film evaporators. Processing, pp 73-75, June 1978.

6.25 Westbrook, N.J. Evaporation advances in the chemical industry. Processing, pp 39, 41, September 1978.

6.26 Frazen, P. Heat transformers for the reactional utilization of waste heat. Krupp Technical Information Report, Krupp Industrietechnik, Greven-broich, West Germany, 1984.

6.27 Null, H.R. Heat pumps reduce distillation energy requirements, Oil and Gas Journal, pp 96-98, 9 Feb. 1976.

6.28 Anon. Heat pump refinements. New Scientist, 22 Jan. 1976.

6.29 Watanabe, H. Field experience with large absorption heat pump and heat transformers. Proc. IEA Int. Heat Pump Congress, Graz, Austria, 1984, pp 357-367. IEA, Paris, 1984.

6.30 Barnwell, J. and Morris, C.P. Heat pump cuts energy use. Hydrocarbon Processing, Vol. 61, Pt. 7, pp 117-119, July 1982.

6.31 Kannoh, S. Heat recovery from warm waste water at dyeing process by absorption heat pump. J. Heat Recovery Systems, Vol. 2, No. 5/6, pp 443-451, 1982.

6.32 Smith, D.B., Eustace, V.A., and Reay, D.A. Operating performance of a heat pump for the production of process steam. Paper E5. Proc. 2nd Int. Conf. on Large Scale Applications of Heat Pumps, York, 25-27 Sept. 1984, BHRA, Cranfield, 1984.

6.33 Eustace, V.A. and Smith, D.B. High temperature heat pump applied in the chemical industry. CEC Report, Demonstration Projects, DG17, Brussels (to be Published).

6.34 Eustace, V.A. Testing and application of a high temperature gas engine driven heat pump. Heat Recovery Systems, Vol. 4, No. 4, pp 257-263, 1984.

APPENDIX 1
Refrigerant Properties

This Appendix starts with a list of ASHRAE designated refrigerants for general reference, classified as described in Chapter 3. Only a very few are in common use, and this is reflected by the fact that only eight pressure-enthalpy diagrams are included here (reproduced by permission of ICI Limited, Mond Division).

ASHRAE Standard Designation of Refrigerants (ASHRAE Standard 34-67, ANSI B79.1)

Refrigerant Number	Chemical Name	Chemical Formula
Halocarbon Compounds		
10	Carbontetrachloride	CCl_4
11	Trichlorofluoromethane	CCl_3F
12	Dichlorodifluoromethane	CCl_2F_2
13	Chlorotrifluoromethane	$CClF_3$
13B1	Bromotrifluoromethane	$CBrF_3$
14	Carbontetrafluoride	CF_4
20	Chloroform	$CHCl_3$
21	Dichlorofluoromethane	$CHCl_2F$
22	Chlorodifluoromethane	$CHClF_2$
23	Trifluoromethane	CHF_3
30	Methylene Chloride	CH_2Cl_2
31	Chlorofluoromethane	CH_2ClF
32	Methylene Fluoride	CH_2F_2
40	Methyl Chloride	CH_3Cl
41	Methyl Fluoride	CH_3F
50*	Methane	CH_4
110	Hexachloroethane	CCl_3CCl_3
111	Pentachlorofluoroethane	CCl_3CCl_2F
112	Tetrachlorodifluorcethane	CCl_2FCCl_2F
112a	Tetrachlorodifluoroethane	CCl_3CClF_2
113	Trichlorotrifluoroethane	CCl_2FCClF_2
113a	Trichlorotrifluoroethane	CCl_3CF_3
114	Dichlorotetrafluoroethane	$CClF_2CClF_2$
114a	Dichlorotetrafluoroethane	CCl_2FCF_3
114B2	Dibromotetrafluoroethane	$CBrF_2CBrF_2$
115	Chloropentafluoroethane	$CClF_2CF_3$
116	Hexafluoroethane	CF_3CF_3
120	Pentachloroethane	$CHCl_2CCl_3$
123	Dichlorotrifluoroethane	$CHCl_2CF_3$
124	Chlorotetrafluoroethane	$CHClFCF_3$
124a	Chlorotetrafluoroethane	CHF_2CClF_2
125	Pentafluoroethane	CHF_2CF_3
133a	Chlorotrifluoroethane	CH_2ClCF_3
140	Trichloroethane	CH_2CCl_3
142b	Chlorodifluoroethane	CH_3CClF_2
143a	Trifluoroethane	CH_3CF_3
150a	Dichloroethane	CH_3CHCl_2
152a	Difluoroethane	CH_3CHF_2
160	Ethyl Chloride	CH_2CH_2Cl
170*	Ethane	CH_3CH_3
218	Octafluoropropane	$CF_3CF_2CF_3$
290	Propane	$CH_3CH_2CH_3$
Cyclic Organic Compounds		
C316	Dichlorohexafluorocyclo-butane	$C_4Cl_2F_6$
C317	Chloroheptafluorocyclo-butane	C_4ClF_7
C318	Octafluorocyclobutane	C_4F_8

Refrigerant Number	Chemical Name	Chemical Formula
Azeotropes		
500	Refrigerants 12/152a(73.8/26.2)	CCl_2F_2/CH_3CHF_2
501	Refrigerants 22/12/(75/25)	$CHClF_2/CCl_2F_2$
502	Refrigerants 22/115(48.8/51.2)	$CHClF_2CClF_2CF_3$
503	Refrigerants 23/13(40.1/59.9)	$CHF_3/CClF_3$
504	Refrigerants 32/115(48.2/51.8	$CH_2F_2/CClF_2CF_3$
505	Refrigerants 12/31(78.0/22.0)	CCl_2F_2/CH_2ClF
506	Refrigerants 31/114(55.1/44.9)	$CH_2ClF/CClF_2CClF_2$
Miscellaneous Organic Compounds		
Hydrocarcarbons		
50	Methane	CH_4
170	Ethane	CH_3CH_3
290	Propane	$CH_3CH_2CH_3$
600	Butane	$CH_3CH_2CH_2CH_3$
600a	Isobutane (2 methyl propane)	$CH(CH_3)_3$
1150**	Ethylene	$CH_2 = CH_2$
1270**	Propylene	$CH_3CH = CH_2$
Oxygen Compounds		
610	Ethyl Ether	$C_2H_5OC_2H_5$
611	Methyl Formate	$HCOOCH_3$
Nitrogen Compounds		
630	Methyl Amine	CH_3NH_2
631	Ethyl Amine	$C_2H_5NH_2$
Inorganic Compounds		
702	Hydrogen (Normal and Para)	H_2
704	Helium	He
717	Ammonia	NH_3
718	Water	H_2O
720	Neon	Ne
728	Nitrogen	N_2
729	Air	$.210_2, .78N_2, .01A$
732	Oxygen	0_2
740	Argon	A
744	Carbon Dioxide	$C0_2$
744A	Nitrous Oxide	N_20
764	Sulfur Dioxide	$S0_2$
Unsaturated Organic Compounds		
1112a	Dichlorodifluoroethylene	$CCl_2 = CF_2$
1113	Chlorotrifluoroethylene	$CClF = CF_2$
1114	Tetrafluoroethylene	$CF_2 = CF_2$
1120	Trichloroethylene	$CHCl = CCl_2$
1130	Dichloroethylene	$CHCl = CHCl$
11132a	Vinyliden Fluoride	$CH_2 = CF_2$
1140	Vinyl Chloride	$CH?_2 = CHCl$
1141	Vinyl Fluoride	$CH_2 = CHF$
1150	Ethylene	$CH_2 = CH_2$
1270	Propylene	$CH_3CH=CH_2$

* Methane, ethane, and propane appear in the Halocarbon section in their numberical order, but these compounds are not halocarbons.

** Ethylene and propylene appear in the Hydrocarbon section to indicate that these compounds are hydrocarbons, but are properly identified in the section Unsaturated Organic Compounds.

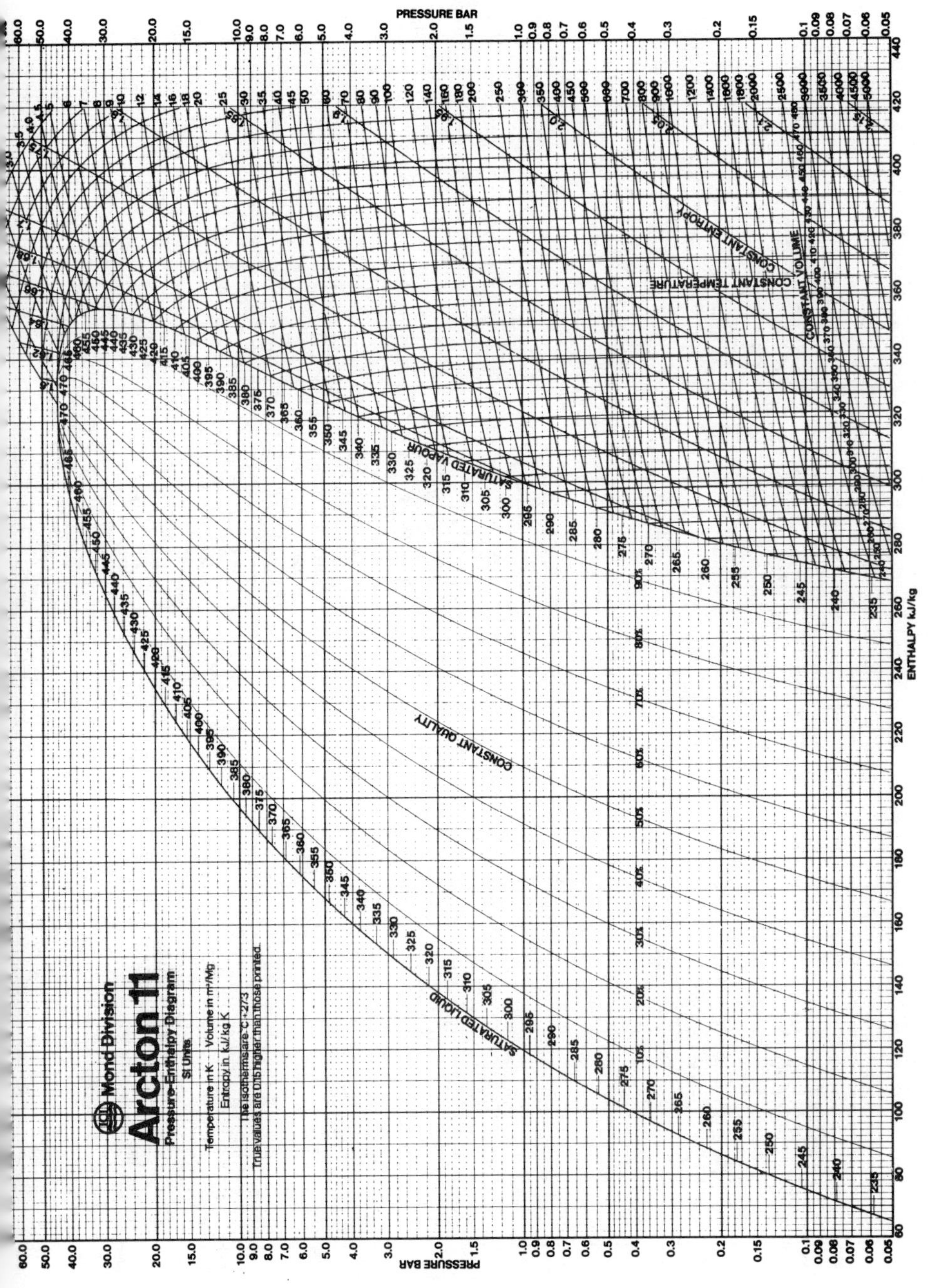
Mond Division
Arcton 11
Pressure-Enthalpy Diagram
SI Units
Temperature in K   Volume in m³/Mg
Entropy in kJ/kg K
The isotherms are C+273
True values are 0.15 higher than those printed.
PRESSURE BAR
ENTHALPY kJ/kg
SATURATED LIQUID
SATURATED VAPOUR
CONSTANT QUALITY
CONSTANT ENTROPY
CONSTANT TEMPERATURE
CONSTANT VOLUME

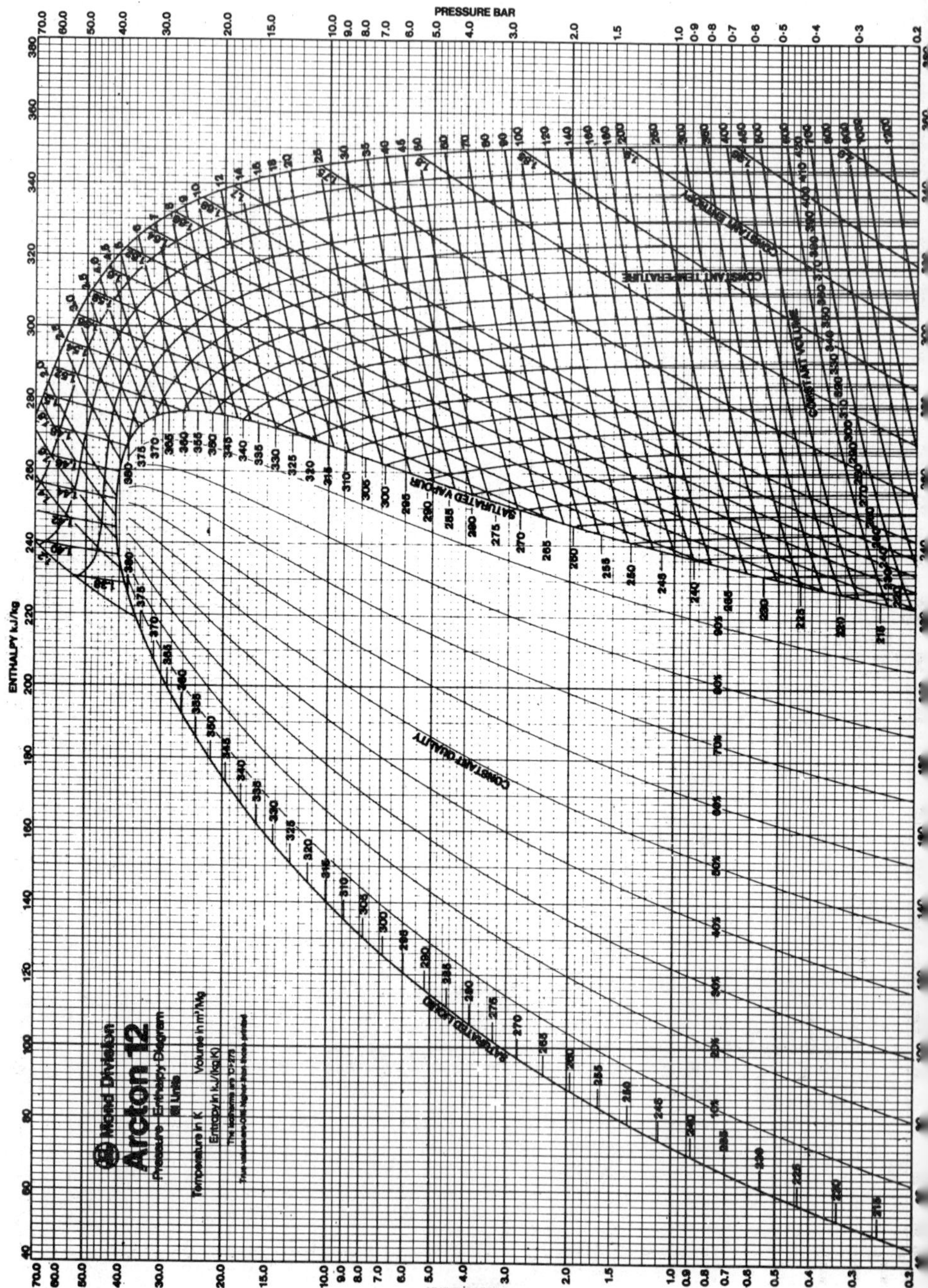
PRESSURE BAR
ENTHALPY kJ/kg
Mond Division
Arcton 12
Pressure–Enthalpy Diagram
SI Units
Temperature in K
Volume in m³/kg
Entropy in kJ/(kg K)
SATURATED VAPOUR
SATURATED LIQUID
CONSTANT QUALITY
CONSTANT ENTROPY
CONSTANT TEMPERATURE
CONSTANT VOLUME

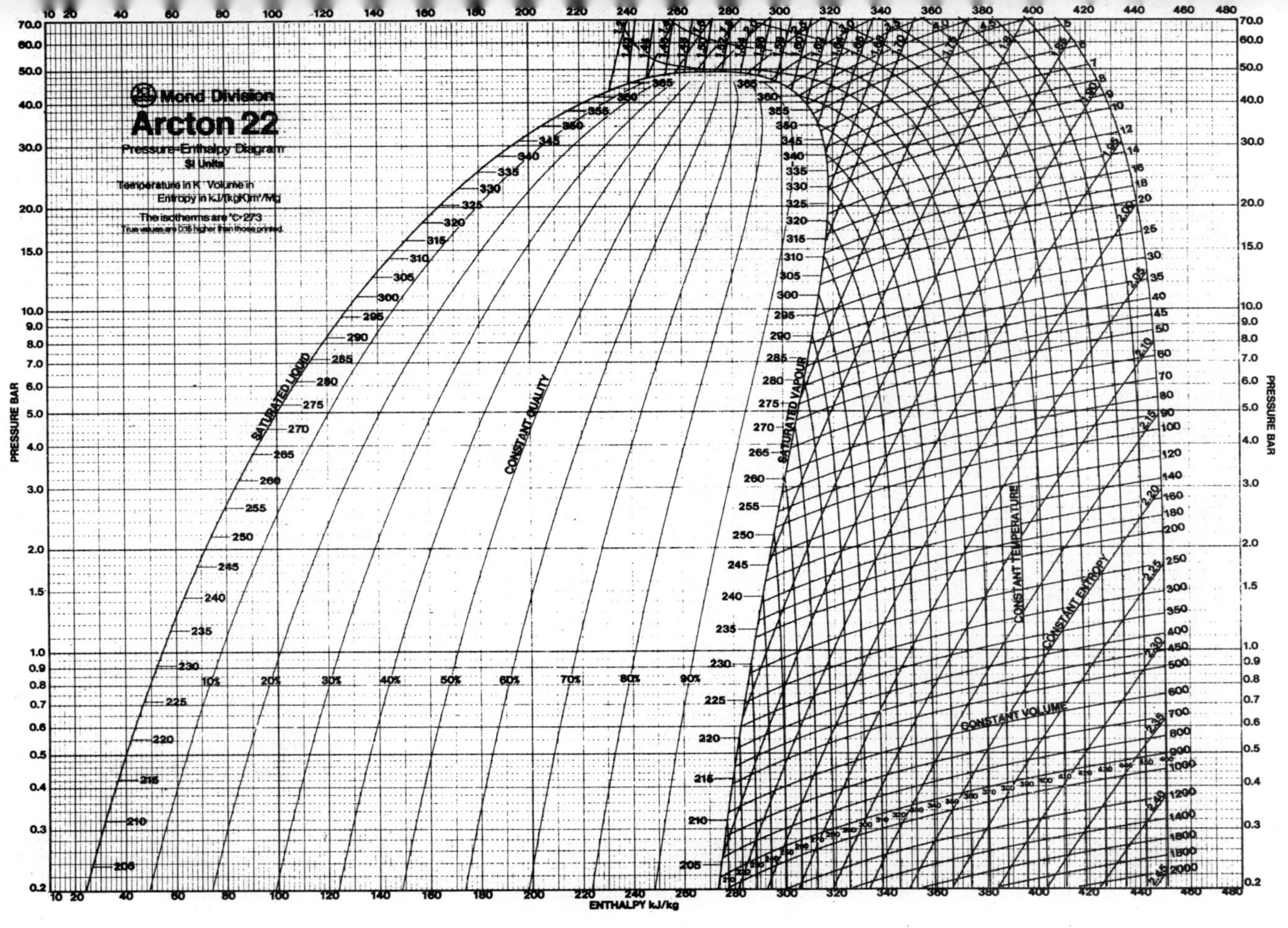
Mond Division
Arcton 22
Pressure-Enthalpy Diagram
SI Units
Temperature in K Volume in
Entropy in kJ/(kgK) m³/Mg
The isotherms are °C+273
PRESSURE BAR
ENTHALPY kJ/kg
SATURATED LIQUID
CONSTANT QUALITY
SATURATED VAPOUR
CONSTANT TEMPERATURE
CONSTANT ENTROPY
CONSTANT VOLUME

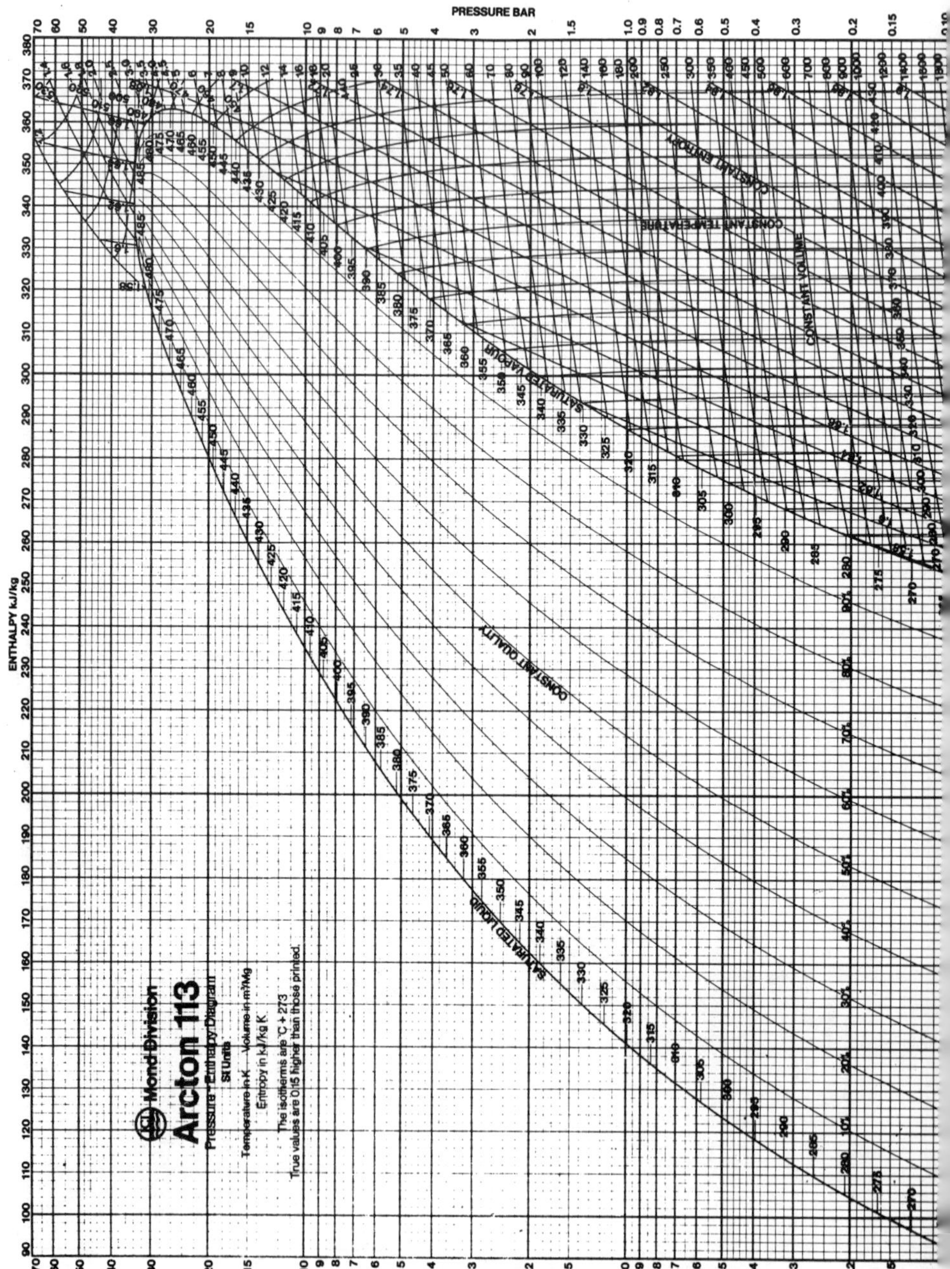
PRESSURE BAR
ENTHALPY kJ/kg
CONSTANT ENTROPY
CONSTANT TEMPERATURE
CONSTANT VOLUME
SATURATED VAPOUR
CONSTANT QUALITY
SATURATED LIQUID
ICI Mond Division
Arcton 113
Pressure–Enthalpy Diagram
SI Units
Temperature in K   Volume in m³/kg
Entropy in kJ/kg K
The isotherms are °C + 273
True values are 0.15 higher than those printed.

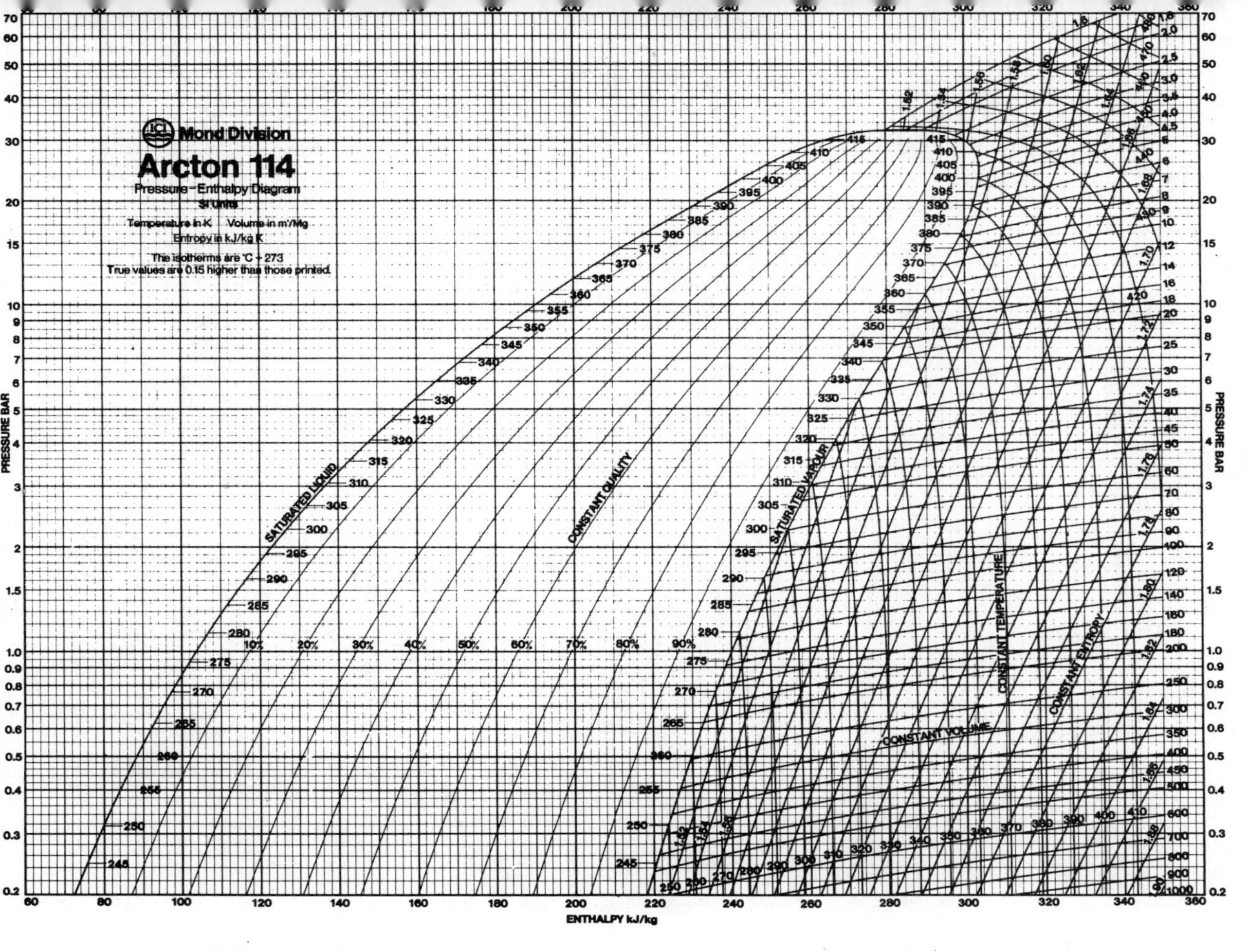
ICI Mond Division
Arcton 114
Pressure–Enthalpy Diagram
SI Units
Temperature in K   Volume in m³/Mg
Entropy in kJ/kg K
The isotherms are °C + 273
True values are 0.15 higher than those printed.
PRESSURE BAR
ENTHALPY kJ/kg
SATURATED LIQUID
SATURATED VAPOUR
CONSTANT QUALITY
CONSTANT TEMPERATURE
CONSTANT ENTROPY
CONSTANT VOLUME

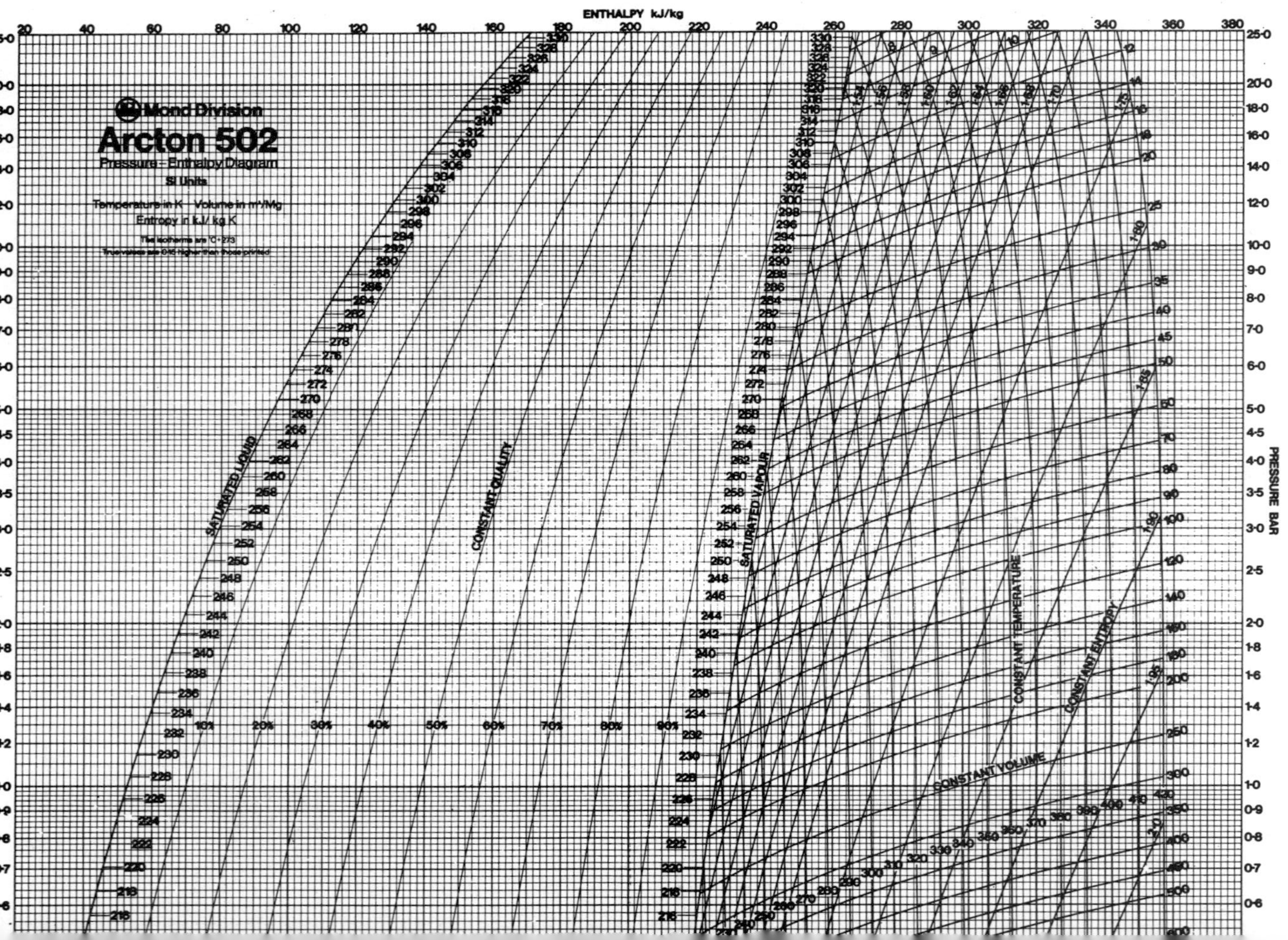
ENTHALPY kJ/kg
PRESSURE BAR
Mond Division
Arcton 502
Pressure - Enthalpy Diagram
SI Units
Temperature in K Volume in m³/Mg
Entropy in kJ/kg K
SATURATED LIQUID
CONSTANT QUALITY
SATURATED VAPOUR
CONSTANT TEMPERATURE
CONSTANT ENTROPY
CONSTANT VOLUME

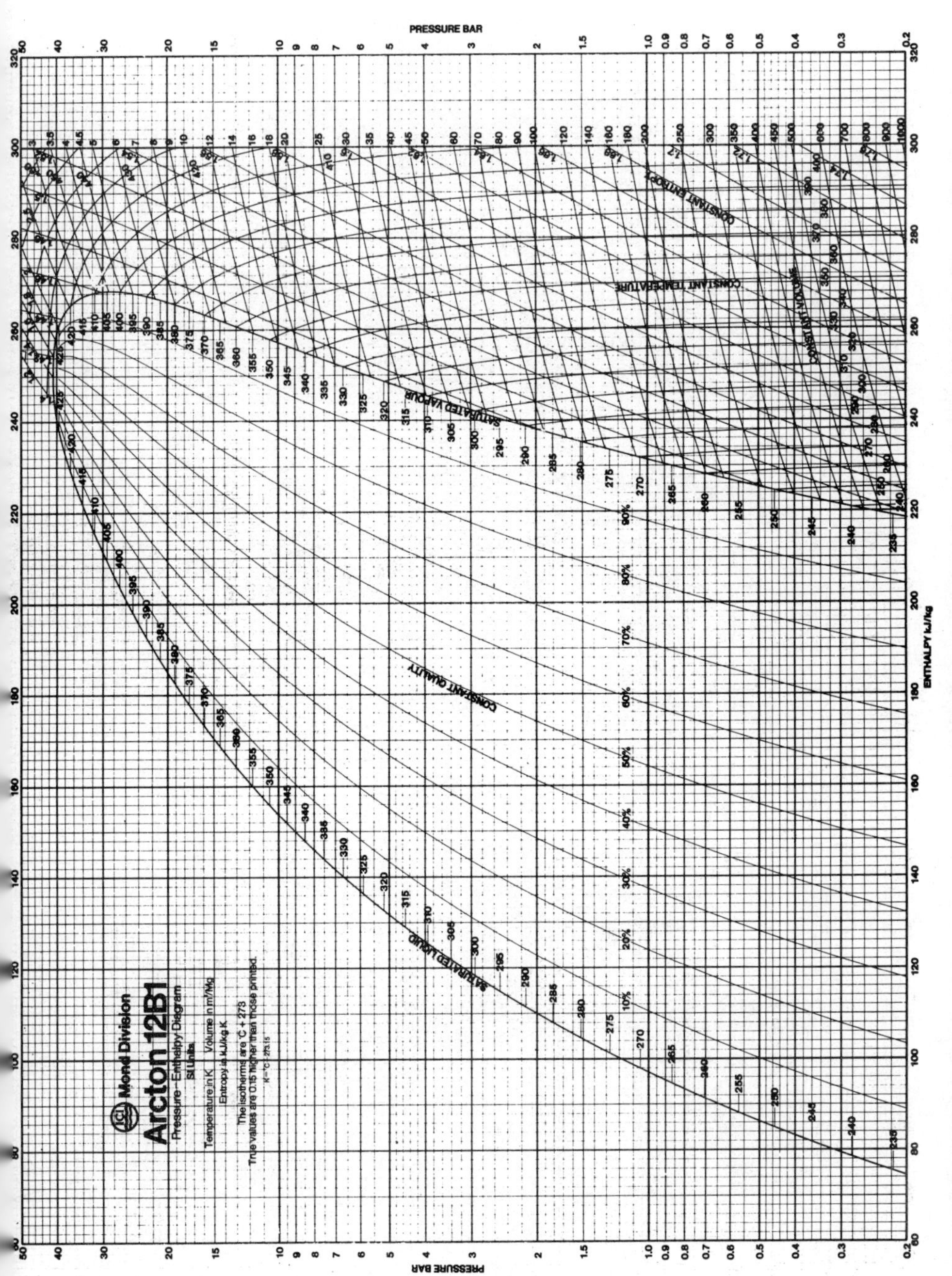
ICI Mond Division
Arcton 12B1
Pressure–Enthalpy Diagram
SI Units
Temperature in K   Volume in m³/kg
Entropy in kJ/kg K
The isotherms are °C + 273
True values are 0.15 higher than those printed.
K=°C 273.15
PRESSURE BAR
ENTHALPY kJ/kg
SATURATED VAPOUR
SATURATED LIQUID
CONSTANT QUALITY
CONSTANT ENTROPY
CONSTANT TEMPERATURE
CONSTANT VOLUME

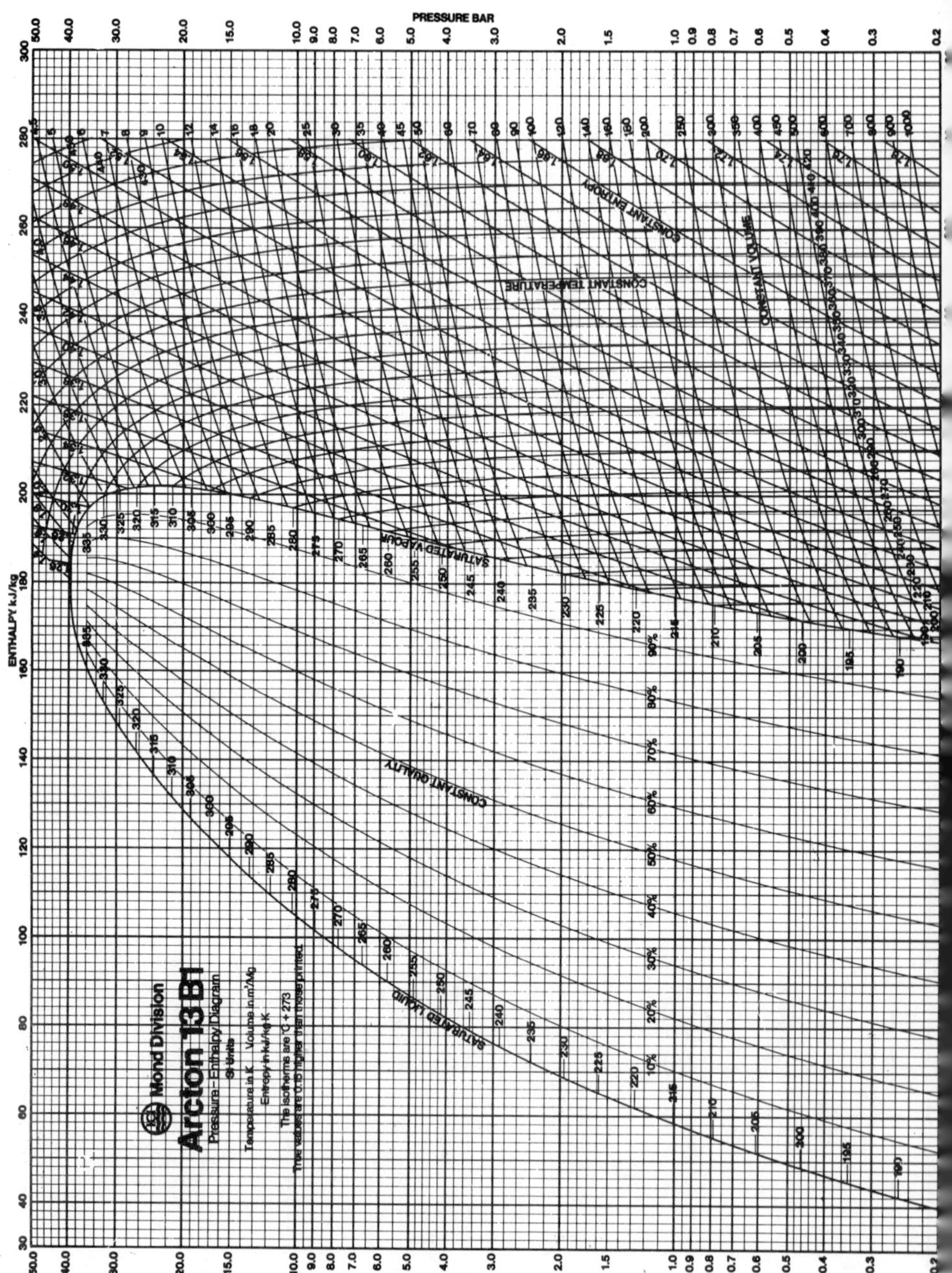
PRESSURE BAR
ENTHALPY kJ/kg
Mond Division
Arcton 13 B1
Pressure–Enthalpy Diagram
SI Units
Temperature in K Volume in m³/Mg
Entropy in kJ/kg K
The isotherms are °C + 273
CONSTANT ENTROPY
CONSTANT TEMPERATURE
CONSTANT VOLUME
SATURATED VAPOUR
CONSTANT QUALITY
SATURATED LIQUID

APPENDIX 2
Bibliography

This Bibliography lists a number of selected references not directly mentioned in the main text. The following categories are covered:

- Papers of Historical Interest
- General Papers and Heat Pump Types
- Domestic Heat Pumps
- Heat Pumps in Large Buildings
- Industrial Heat Pumps

Papers of Historical Interest

Kell, J.R. and Martin, P.L. The Nuffield College Heat Pump. J.Inst. Heating & Ventilating Engnrs., pp 333-356, Jan. 1963.

Montagnon, P.E. The economics of heat pumps. Ibid, pp 233-239, Dec. 1958.

Vestal, D.M. and Fluker, B.J. Earth as a heat source for heat pumps. Heating, Piping & Air Condit., pp 117-123, Aug. 1956.

Pietsch, J.A. The unitary heat pump industry - 25 years of progress. ASHRAE Jnl., Vol. 19, Pt. 7, pp 15-18, July 1977.

Macadam, J.A. Heat pumps - the British experience. Building Research Establishment Note N117/74, UK Dept. of Environment, Dec. 1974.

Johnson, W.E. and Bloomfield, N.J. Economic and technical aspects of the heat pump. ASHVE Transactions, Vol. 54, pp 201-220, 1948.

Kroeker, J.D. and Chewning, R.C. A heat pump in an office building (the first in the USA). Ibid.

Evans, J.H. Heat pump principles and applications. J.Inst. Heating & Ventilating Engnrs., pp 217-251, Aug. 1948.

Sumner, J.A. A description of the Norwich heat pump. Norwich Corpn. Electricity Dept. Publication, Jan. 1946.

Thomson, W. (later Lord Kelvin). Heating and refrigeration of air. Cambridge Mathematical Jnl., 1852.

Sumner, J.A. Domestic heating by the heat pump. J.Inst. Heating & Ventilating Engnrs., July 1955.

Thomson, W. On the economy of heating or cooling of buildings bymeans of currents of air. Proc. Glasgow Phil Soc., Vol.III, pp 269-272, Dec. 1852.

Haldane, T.G.N. The heat pump - an economical method of producing low grade heat from electricity. I.E.E. Jnl., Vol. 68, pp 666-675, June 1930.

Egle, M. The heating of the Zurich town hall by the heat pump. S.E.V. Bulletin, Vol. 29, pp 261-273, 27 May 1938.

Kemler, E.N. and Oglesby, S. Heat Pump Applications. McGraw-Hill, New York, 1950.

Ambrose, E.R. and Sporn, P. The sun as a source of heat for heat pumps. Refrigeration Engng., Vol. 63, No. 11, pp 39-42, 138-148, Nov. 1955.

Achenbach, P.R. and Davis, J.C. Analysis of electric energy usage in Air Force houses equipped with air-to-air heat pumps. US National Bureau of Standards Monograph No. 51, 1962.

Anon. Festival Hall heat pumps. Industrial Heating Eng., Vol.13, Pt. 69, pp 198-203, 206, July 1951.

General Papers & Heat Pump Types

Pannkoke, E. The heat pump. Heating, Piping & Air Conditining, Vol. 47, Pt. 2, pp 23-27, Feb. 1975.

Leonard, L.H. High temperature heat recovery in refrigeration. US Patent 3922873, Dec. 2, 1975.

Vickers, H.H. and Sage, R.W. Studies in silencing a diesel driven heat pump. Heating, Piping & Air Conditiioning, Vol. 29, Pt. 5, pp 163-167, May 1957.

Nye, L.B. Heating and cooling with a gas engine heat pump. Gas Age, Vol. 130, pp 25-30, June 1965.

Crow, R. and Pepper, J. The heat pump - reviewing an option. EPRI Journal, Vol.1, No. 8, pp 20-25, Oct. 1976.

Farrell, T. Thermoelectric heat pumping. Electricity Council Research Centre Report ECRC/R844, Sept. 1976.

McMullen, J.T. et al. Energy Resources and Supply. John Wiley, London, 1976.

Schindelhauer, G. The Wankel compressor and its possible use in heat pumps. Elektrowarme International, Vol. 34, No. A3, pp 133-135, May 1976.

Werden, R.G. Improving heat pump reliability. ASHRAE Transactions, Vol. 82, Pt. 1, pp 372-386, 1976.

Thielbahr, W.H. Heat exchanger technology needs for conservation research and technology. Naval Weapons Center Tech. Memo. No. 2930, California, Dec. 1976.

Loyd, S. and Starling, C. The Heat Pump. BSRIA Bibliography 103, Building Services Research & Information Assn. Bracknell, UK, 1975.

Ambrose, E.R. Heat Pumps and Electric Heating. Wiley, New York, 1966.

Anon. Directory of Certified Unitary Air Conditioners and Heat Pumps. Air Conditioning Refrigeration Institute, USA, 1973.

Andreiff, G. Heat pumps. Revue Pratique Froid, Vol. 25, No. 322, pp 71-78, Nov. 1972.

Cole, M.H. and Pietsch, J.A. Qualification of heat pump design. ASHRAE Jnl., Vol. 15, No.7, pp 43-47, July 1973.

Griffith, M.V. Heat pump operation in Great Britain. Electrical Research Assn. Tech. Report Y/T22, 1958.

Wilcutt, K.E. Heat pump reliability in Alabama. Proc. ASHRAE Symposium on Applications & Reliability, 1972.

Griffith, M.V. Power station heat pumps. Heating, pp 307-312, Sept. 1960.

Greiner, P.C. Marketing heat pumps in the US means energy conservation. Elektrowarme International, Vol. 31, pp A78-A83, March 1973.

Rittner, E.S. On the theory of the Peltier heat pump. J.Appl. Phys., Vol. 30, No. 5, pp 702-707, May 1959.

Matsuda, T. et al. A new air-source heat pump system. ASHRAE Jnl., pp 32-35, August 1978.

Anon. GRI research and development programme. Gas Research Institute Digest, Vol. 6, No. 2, March/April 1983.

Anon. Heat Pump Systems. A Technology Review. IEA/OECD, Paris, 1982.

Anon. US Heat Pump Research and Development Projects. US Dept. of Energy Report DOE/CE-0035, August 1982.

Reay, D.A. Heat pump research and development in the USA. J.Heat Recovery Systems, Vol. 3, No. 3, pp 165-176, 1983.

Anon. Proceedings of SERC Heat Pump Research Grant Holders Meeting, Abingdon, 19-20 June 1983. SERC, 1983.

Bokelmann, H. et al. Working fluids for sorption heat pump. Paper A 1.18, Proc. Int. Seminar, Energy Saving in Building, The Hague, Nov. 1983.

Knoche, K.F. et al. Periodically operating absorption heat pump. Paper A.1.15 (Ibid).

Cheron, J. and Rojey, A. Absorption-Resorption heat pump for space heating. Investigation of solute-solvent pairs. Paper A1.14 (Ibid).

Smith, I.E. High temperature lift absorption heat pumps. Paper A.1.12, (Ibid).

Domestic Heat Pumps

McClory, P. (Editor). How to Use Natural Energy. The Natural Energy Centre, London, 1978.

Sumner, J.A. Domestic Heat Pumps. Prism Press, Dorchester, UK, 1976.

Seymour-Walker, K.J. Experimental low energy houses. Energy Digest, pp 11-16, Oct. 1976.

Anon. Heat pumps: a major residential role? Domestic Heating News, Vol. 15, No. 5, pp 10-13, May 1975.

Blundell, C.J. Domestic heat pumps - time for reappraisal. Electronics & Power, Vol. 22, No. 10, pp 686-688, Oct. 1976.

Heap, R.D. A preliminary report on the domestic heat pump at Alton. Electricity Council Research Centre Report ECRC/N819, May 1975.

Nilsson, O. and Nyberg, H. Development of heat pumps for space heating in Sweden and experiences of pilot plant. VIII Congress of Union Internationale d' Electrothermie, Liege, 11-15 Oct. 1976.

Anon. Solar design eliminates collectors. Heating, Piping & Air Conditioning, Vol. 47, No. 7, July 1975.

Anon. Heat pump reliability shows big gains. Electrical World, pp 78-80, 1 August 1973.

Jones, E.C. Unitary heat pumps in multi-family dwellings. Proc. ASHRAE Symposium on Heat Pumps - Application & Reliability, 1972.

Waterkotte, K. Water-to-water heat pump for a single-family home. Elektrowarme International, pp A39-A43, Jan. 1972.

Heap, R.D. Domestic heat pumping in Britain. Proc. 6th Int. Conf. of Climatistics, Paper II.04.01, Milan, March 1975.

Krumme, W. Air-to-water heat pump in a single family house. Elektrowarme International, Vol. 32, pp 35-41, Jan. 1974.

Brockmeyer, H. Packaged heat pumps. Elektrowarme International, A2. Vol. 31, pp A65-A71, March 1973.

Seymour-Walker, K.J. and Freund, P. Future developments for buildings (heat pumps). Electrical Review, Vol. 198, No. 19, pp 33-34. 14 May 1976.

Brundrett, G.W. et al. Energy conservation in buildings. Coal Energy Quarterly, No. 15, pp 14-28, Winter 1977.

Blundell, C.J. Optimisating heat exchangers for air-to-air space heating heat pumps in the UK. Int. J. Energy Research, Vol. SMC-7, No. 5, pp 340-149, May 1977.

McMullan, J.T. et al. A new approach to domestic heat pumps. Electrical Review, Vol. 197, No. 21, pp 673-674, 21 Nov. 1975.

Buick, T.R. et al. Ice detection in heat pumps and coolers. Int. J. Energy Research, Vol. 2, No. 1, pp 85-98, Jan-March 1978.

Yaneske, P.P. and Forrest, I.D. Thermal response of rooms with intermittent forced convective heating. Build.Serv.Eng. Vol. 46, No. 1, April 1978.

Sumner, J.A. History of a home built heat pump. Electrical Review, Vol. 196, No. 12/13, 28 March/4 April, 1975.

Kuhlenschmidt, D. and Merrick, R.H. An ammonia-water absorption heat pump cycle. Symposium AC-83-05, Proc. ASHRAE Semi-Annual Meeting, Atlantic City, Jan. 23-26, 1983.

Murphy, K.P. and Phillips, B.A. Development of a residential gas absorption heat pump. Ibid.

O'Dell, M.P. et al. Solar heat pump systems with refrigerant-filled collectors. Symposium AC-83-11, Ibid.

C. Ramet et al. Compression heat pumps operating with non-azeotropic pairs. In: Energy Saving in Buildings. Proceedings of International Seminar, The Hague, 14-16 Nov. 1983. Results of the European Communities Energy R&D and Demonstration Programmes, (1979-1983), CEC, Luxembourg, 1984.

Strong, D.T.G. The development of a directly fired domestic heat pump for domestic and light commercial application. Paper 4.8, Proc. Heat Pump Contractors Meeting, Brussels. Commission Report EUR8077EN, CEC Luxembourg, 1982.

Paulick, W. I.C. Engine Driven Heat Pumps with a Heat Power upto 20 kW. In: Energy Saving in Buildings, Proc. CEC Seminar, The Hague, 14-16 Nov. 1983, pp 281-287. D. Reidel, Dordrecht, 1984.

Lloyd, A.S. Heat pump water heating systems. Heating, Piping & Air Conditioning, Vol. 55, Pt. 5, pp 83-86, 91-94, May 1983.

Heat Pumps in Large Buildings

Anon. London's first privately-developed IED offices. Energy Digest, Vol. 3, Pt. 4, pp 25-29, Aug. 1974.

Hardy, A.C. and Mitchell, H.G. Building a climate - the Wallsend project. Jnl. Inst. Heating & Ventilating Engineers. Vol. 38, pp 71-84, July 1970.

Biehl, R.A. and Werden, R.G. The energy bank. Heating, Piping, & Air Conditioning, Vol. 49, Pt. 7, pp 53-60, Jan. 1977.

Anon. Air Conditioning for Victory House. Heating & Air Cond. Jnl., Vol. 46, Pt. 537, pp 30-31, Oct. 1976.

Shepherd, L. and Smith, A. Air conditioning and heat recovery in the headquarters of the London Electricity Board.Electrowarme International, Vol. 31, Pt. A4, July 1973.

Bridgers, F.H. Energy conservation: pay now, save later. ASHRAE Jnl., Vol. 15, Pt. 10, pp 47-52, Oct. 1973.

Kent. H.S. Engineering Building designed with heat relcaim in mind. Heating, Piping & Air Conditioning, Vol. 42, Pt. 11, pp 79-82, Nov. 1970.

Quinn, G.C. Heat pumps for a million-sq-ft training centre. Elec. Constr. & Maintenance, Vol. 74, Pt. 1, pp 67-69, Jan. 1975.

Anon. Waste heat utilisation in Electricity Board office block. Steam & Heating Eng., Vol. 37, Pt. 441, pp 39-40, Aug. 1968.

Bridgers, F.H. Application and design of electric heat pump systems. Heating, Piping & Air Conditioning, pp 98-104, Feb. 1967.

Anon. Air Conditioning the Liverpool Daily Post and Liverpool Echo Building. Heating & Air Cond. Jnl., Vol. 43, No. 511, June 1974.

Perry, E.J. The role of refrigeration in energy conservation. ASHRAE Jnl., Vol. 19, Pt. 12, pp 17-21, Dec. 1977.

Anon. Gas motor heat pump with additional heat recovery. Sulzer Review, No. 3, pp 136-137, 1977.

Juttemann, H. Heat pumps in large buildings. Heiz. Luft. Haustechn., Vol. 25, No. 4, pp 124-130, April 1974.

Hegedus, T. Air conditioning and heat recovery in pharmaceutical laboratories. Building System Design, Vol. 68, Pt. 2, pp 23-31, 1971.

Schlitt, R. and Schlosser, W. Air conditioning systems with small heat pump units in four large all electric buildings. Electrowarme Internat., Vol. 31, pp A213-218, July 1973.

Bazzoni, J.P. Newspaper plant heating is hot off the press. Heating, Piping & Air Conditioning, Vol. 38, pp 122-125, Oct. 1966.

Dodson, C. Commercial heat pump systems. Building Services Eng., Vol. 44, No. 8, pp A14-17, Nov. 1976.

Bennett, J.H. Packaged terminal air conditioners and closed loop water-to-air heat pump systems. ASHRAE Jnl., Vol. 15, Pt. 12, pp 57-62, Dec. 1973.

Ratai, W.R. Heat pump saves school energy. Building Systems Design, pp 6-9, March 1972.

Birk, D. and Fusenig, R. Heat pumps for dehumdification and heat recovery in a school's swimming bath at Irrel. Elektrowarme International, Vol. 32, Pt. A3, pp A143-A146, May 1974.

Brady, J. Analysis of the impact of heat pump technology on theIrish energy system to the year 2000. Energy Case Study Series, No. 2, National Science Council, Ireland, 1977.

Anon. Energy Saving in Buildings. Proceedings of International Seminar, The Hague, 14-16 Nov. 1983. Results of the European Communities Energy R&D and Demonstration Programmes, (1979-1983), CEC, (or D. Reidel, Dordrecht),

Atkinson, G.S. et al. The development of heat engine driven heat pump systems. In: Proc. SERC Heat Pumps Research Grant Holders Meeting, Abingdon, 19-20 June 1983. SERC, 1982.

Martin, D.J. ETSU seminar on Rankine-Rankine heat pumps for space heating in non-domestic buildings. Background Paper. ETSU, Jan. 1983.

Cole. Heat pumps - a users view. Proc. Heat Sense 1982 Conference, The Energy Manager, Cafe Royal, London, MCM Publications London.

Hoggarth, M.L. and Pickup, G.A. The role of gas fuelled heat pumps for space heating. Proc. Heat Pumps for Buildings Conference, 14-15 April 1983, Publ. Construction Industry Conference Centre, Welwyn, 1983.

Currie, W.M. and Martin, D.J. Building energy conservation - the role for heat pumps. Proc. Heat Pumps for Buildings Conference, 14-15 April 1983, Nottingham CICC Bookshop, Welwyn, 1983.

King, R. Rankine engine driven Rankine cycle heat pumps in the USA. A review of work undertaken by Mechanical Technology Incorporated, New York. BHRA Fluid Engineering, TN1944, 1983.

Sharma, V. and Dann, B. The gas engine heat pump installation at the Purley Way Service Centre. Paper 5. In: Heat Pumps for Buildings - 2, Proc. Conference 24 and 25 October 1984, (Editor, A.F.C. Sherratt), Construction Industry Conference Centre, 1984.

Industrial Heat Pumps

L'Hermitte, J. and Donay, D. The heat pump in industrial applications. EDF Journal of Electrical Industry Information, Nov. 1974, Paris, (In French).

Picciotti, M. and Kaiser, V. Multilevel optimisation in ethylene technology, Chemistry & Industry, pp 928-934, 16 Dec. 1978.

Miller, W. Energy conservation in timber drying kilns by vapour recompression. Forestry Prodn. Jnl., Vol. 27, pp 54-58, 1977.

Witt, J.A. Light industrial applications of heat pumps. Electrical Review, pp 12-22, 1976.

Anon. Heat pump can recycle industrial heat. Electrical World, Vol. 183, No. 10, pp 68-69, 15 May 1975.

Anon. Waste heat utilisation in glass processing. Elektrowarme International, Vol. 34, No. A5, pp 244-247, Sept. 1976.

Thornaby, D.L. Heat recovery techniques. Proc. Conf. on Energy - The economics & techniques of efficient use. Electricity Council, London, 14 April 1976.

Solignac, M. Applications of the heat pump for drying purposes. Paper III 5, Proc. VIII Conference of Union Internationale d' Electrothermie, Liege, 11-15 Oct. 1976.

Lewis, G.P. et al. Sulphuric acid plant Rankine cycle waste heat recovery. Proc. 11th Intersoc Energy Conv. Engng. Conf., Vol. 2, pp 1182-1186, Nevada, Sept. 1976.

Witt, G.A. and Aylott, G.W. Heat pump blows hot and cold. Process Engng., pp 81-83, Nov. 1976.

Stork, R. Timber drying kilns with heat pumps - fundamentals and experience. Elektrowarme International, Vol. 32, Pt. A4, pp 184-187, July 1974.

Martin, R.J. Design factors for industrial heat pump installations. Mechanical Engng., Vol. 74, Pt. 4, pp 280-284, April 1952.

Eickenhorst, H. Use of heat pumps in waterworks. Elektrowarme International, Vol. 29, pp 147-151, March 1971.

Baxter, W.R.S. et al. Drying by heat pumps. IEE Conf. Publ. (London), No. 149, 1977.

Bauder, H.J. High temperature heat pump applications and their limitations. Sulzer Technical Review, No. 3, pp 22-28, 1982.

Heppenstall, T. Heat pumps for industrial energy saving as part of the United Kingdom energy strategy. Proc. M.I.T.E.A., Conference, Bolzano, Italy, 1983.

Kannoh. Heat recovery from warm waste water at dyeing process by absorption cycle heat pump. Inst. Symposium on Industrial Application of Heat Pumps. University of Warwick, 24-26 March 1982.

Zegers, P. and Knobbout, J. An overview of work on industrial and domestic heat pumps in the energy R&D programme of the European Community. Int. Symposium on Industrial Applications of Heat Pumps, University of Warwick, 24-26 March 1982.

Boland, D. and Hill, R. Heat Pumps - Application in an energy integrated process. Paper F1, Ibid, 1982.

Burton, D.C. New York State thermally activated industrial heat pump applications study. MTI Final Report 81TR53, to NYS Energy Research & Development Authority, June 1981.

APPENDIX 3
Heat Pump Manufacturers

This Appendix lists manufacturers and suppliers of heat pumps.

Names and addresses are given, with key letters and numbers which indicate the product covered.

Key to products:

A	Equipment manufacturer	K	Swimming pools
B	Agent	M	Power generation in remote areas
C	Turnkey contractor	N	Fossil fuel-fired power plant
D	Consultant	O	District heating
E	Designer & contractor	P	Municipal waste disposal
F	Distributor	Q	Incinerators
G	Industrial processes	R	Offshore
H	Commercial buildings	S	Agricultural
I	Domestic	T	Solar
J	Marine		

1. Electric drive: air-air
2. Electric drive: air-water
3. Electric drive: water-water (liquid-liquid)
4. Electric drive: ground-air
5. Electric drive: ground-water
6. Engine drive: air-air
7. Engine drive: air-water
8. Engine drive: water-water (liquid-liquid)
9. Engine drive: ground-air
10. Engine drive: ground-water
11. Turbine drive
12. Heat pump dehumidifier
13. Absorption cycle: air-air
14. Absorption cycle: air-water
15. Absorption cycle: water-water (liquid-liquid)
16. Absorption cycle: ground-air
17. Absorption cycle: ground-water
18. Heat transformer
19. Thermoelectric heat pump

20. Steam-producing heat pump*
21. Thermocompressor (jet pump)
22. Reverse cycle water-air unit*
23. Heat pump with integral recuperator*
24. Chemical heat pump
25. Brayton cycle heat pump
26. Solar absorption cycle heat pump
27. Heat pump - unspecified

* Description based on that submitted by equipment supplier.

AAF Ltd 2 G H
Bassington Lane
Cramlington
Northumberland
NE23 8AF
UK

ACFT 27 H
122 av des Pyrēnēes
33140 Le Point-de-la-Maye
France

ACR Heat Transfer Manufacturing 27 H
Rollesby Road
Kings Lynn
Norfolk
PE30 4LN
UK

AiAX (UK) Ltd 1,2,6,7 G,H
Veneto House
Park Drive
Rayners Lane
Middlesex
MA2 7LT
UK

AL-KO Polar GmbH 1,2 H,I
BAB-Ausfaht Burgau
Postfach 51
8876 Jettingen-Scheppach
West Germany

APV Hall Commercial Ltd 2,3 G,H
10 Holdom Avenue
Saxon Park Industrial Estate
Bletchley
Milton Keynes
MK1 1QU
UK

APV Hall Products Ltd 2,3,7,8 G,H
Hythe Street
Dartford
Kent
DA1 1BU
UK

ASCU Hickson Ltd 12 H,I
7A Elgin Road
Calcutta 700020
India

ASEA-STAL Ltd 27, G,H
Wells House
229-231 High Street
Sutton
Surrey
SM1 1LD
UK

AWAK-Warmepumpen 27 H,I
Postfach 674
D-8630 Coburg
West Germany

Aero-Plast 1 H,I
Azay-le-Brulẽ
79400 Saint-Maixent-'Ecole
France

Air Comfort Ltd 1 H,I
Unit 3
Green lane Industrial Estate
Bordesley Green
Birmingham
B9 5QP
UK

Airaqua Engineering Ltd 1,2,3 H,J,K
PO Box 7
Wilmslow
Cheshire
SK9 2LG
UK

Aircalo 27 H
Rue Jules-Massenet
33160 Saint-Medard-en-Jales
France

Airedale International Air Conditioning 13 G,H
Clayton Wood Rise
West Park
Leeds
LS16 6RF
UK

Airoheat (UK) Ltd 1,2,3,4,5 G,H,K
20 Broomgrove Road
Sheffield
S10 2LR
UK

Aldridge Air Control Ltd 6,7 H
Middlemore Lane
Aldridge
Walsall
West Midlands
WS9 8SP
UK

Allbrook & Hashfield Ltd 6,7,8 G,H
153 Huntingdon Street
Nottingham
NG1 3NG

Alsthom Atlantique Rateau 1,2,3 G
141 rue Rateau
F-93123 La Courneauve Cedex
France

American Air Filter Co Inc (AAF) 1,2 H
215 Central Avenue
Louisville
Kentucky 40201
USA

Anco Products UK Ltd 2,3 I,K
Daish Way
Dodnor Industrial Estate
Newport
Isle of Wight
PO30 5XJ
UK

Andrew Engineering 2,3,13 G,H
Warwick Street Industrial Estate
Storforth Lane
Chesterfield
Derbyshire
S40 2TT
UK

Andrews Industrial Equipment Ltd 1,2,3,4,5 G,H,I,K
Springhead Enterprise Park
Springhead Road
Northfleet
Gravesend
Kent
DA11 8HD
UK

Anglo Nordic Thermal Holdings Ltd 27 H,I
74 London Road
Kingston upon Thames
Surrey
KT2 6FZ
UK

Apples 1, I
12 Route de Chainteaville
77140 Nemours
France

Applied Energy Systems 6,7,8,13,14,15 G,H
1 Whippendell Road
Watford
Herts
WD1 7LZ
UK

Aquatech Marketing Ltd 27 H
Unit 1
Hambridge Lane
Newbury
Berks
RG14 5UF
UK

Atlas Copco (Great Britain) Ltd 27 G
PO Box 79
Swallowdale Lane
Hemel Hempstead
Herts
HP2 7HA

Atmospheric Control Engineers Ltd 1,2,3,6,7,8
St. Anne's House
North Street
Radcliffe
Manchester
M26 9RN
UK

BBC York 2,3,4,27 G,H
Cottl Daimler St 6
Postfach 5180
6800 Mannheim 1
West Germany

BOC-Linde Refrigeration Ltd 27 G
Stonefield Way
South Ruislip
Middlesex
HA4 ONT
UK

Bauknecht GmbH 27 H,I
Heidenklinge 22
7000 Stuttgart 1
West Germany

Bero-Energie GmbH 6,7,8,27 G,H
D-3016 Seelze/Hannover
West Germany

F.H. Biddle Ltd St. Mary's Road Nuneaton Warwickshire CV11 5AU UK	1,2,3 G,H,I
Boamanco 8 Woolley Street Bradford on Avon Wiltshire BA15 1AE UK	1,
Bock GmbH Postfach 1129 D-7440 Nurtingen West Germany	1,27, H,I
Borsig GmbH Berliner Strasse 27-33 D-1000 Berlin 27 West Germany	14,15,18,27 G
Robert Bosch GmbH Geschaftsbereich Junkers D 7314 Wernau West Germany	1,2,27, H,I
Briau SA Avenue du Prieure La Riche 37000 Tours (BP0903-37009 Tours Cedex) France	1,2,3, G,H,I
CEM 12 rue Portalis 75383 Paris Cedex 08 France	1,2,3, G
CIAT Rue de Rhone 01350 Culoz France	1,2,3,6,7,8, G,H,I
CIAT UK 5 The Byfleet Tech Centre Canada Road Byfleet KT14 7JN UK	1,2,3,27, G,H
CLIREF ZI Les Meurieres 69780 Mions France	1,2,3, G,H,I

COMELA ZI Sain-Bel 69210 L'Arbresle France	1,2,327, H
COMETH 60 rue du Fg. Poissonniere 75010 Paris France	27, H
Calorex Heat Pumps Ltd The Causeway Maldon Essex CM9 7PU UK	1,2,3,12, G,H,K
Cannon Air Engineering Broadway House The Broadway Wimbledon London SW19 1RL UK	14, G,H
Carlyle Air Conditioning Co Ltd 197 Knightsbridge London SW7 1RB UK	1,2,3,27, H
Carrier (UK) Distribution Ltd Priory House Marsh Road Alperton Lane Wembley Middsx HA0 1ES UK	1,2,3,6,7,8,11,15, G,H,J
Carrier/Drysys 90 rue Rouget de Lisle BP 82-92153 Suresnes France	1,6 G
Church Hill Systems Frolesworth Lutterworth Leicestershire LE17 5EE UK	2,3,5, G,H,I,J
Climate Equipment Ltd Highlands Road Shirley Solihull West Midlands BN90 4NL UK	1, G,H,I,J 2,3,6,7,8,27

Commac Ltd 15 Dryden Court Renfrew Road London SE11 4NH UK	27, H
Command-Aire Corporation 3221 Speight Avenue Box 7916 Waco Texas USA	2,27, G,H,I
Conder M & E Products Enviroware Division Abbotts Barton House Worthy Road Winchester Hants SO23 7SJ UK	1, H
Crepelle & Cie 2 Place Guy de Dampierre BP29-59011 Lille Cedex France	1,2,3, G
Creusot-Loire Division Energie BP 31 Cedex F-71208 Le Creusot France	1,2,3,6,7,8, G,H
Croll-Reynolds Co Inc 751 Central Avenue Westfield New Jersey 07091 USA	22, G
Cuenod Thermotechnique SA Rue des Buchillons BP193-Ville-la-Grand 74102 Annemasse Cedex France	1, H,I
Dantherm Ltd Hither Green Clevedon Avon BS21 6XT UK	1, G,H
De Dietrich & Cie 67110 Niederbronn-les Bains France	1,3, H,I

Delchi SpA 27, G,H,I
via R Sanzio 9
200 58 Villasanta
Italy

Delrac Ltd 1,2,3,19, G,H
128 Malden Road
New Malden
Surrey
KT3 6DD
UK

Drycool Equipment Ltd 1,12
64 Solent Road
Havant
Hants
PO9 1JH
UK

Dunham-Bush International 1,2,3,27, G,H
175 South Street
West Hartford
Conn. 06110
USA

Dunham-Bush Ltd 3, H,J
Fitzherbert Road
Farlington
Portsmouth
PO6 1RR
UK

ECE Environmental Control Equipment Ltd 1, G,H
Coronation Works
Eynsford
Kent
DA4 0AB
UK

Eaton-Williams Group 1,12, H,I
Dry-Aire Division
Station Road
Edenbridge
Kent
TN8 6EG
UK

Electra Air Conditioning Services Ltd 1, H
Unit 20
Kingsbury Trading Estate
Church Lane
London
NW9 8AU
UK

Envirosystems Ltd 1,4,5, G,H,I
Hampsfell Road
Grange over Sands
Cumbria
LA11 6BE
UK

Fabdec Ltd 1, H
Grange Road
Ellesmere
Shropshire
SY12 9DG
UK

Falcon Howard Heat Pumps Ltd 2, H
50/3 Palace Road
Bangalore 560 052
India

Flakt Industrie 1, H,I
26 quai Carnot
92212 Saint-Cloud
France

Flaktfabriken 1,2, M,I
Fach
104 60 Stockholm
Sweden

Flebu Ltd 2,14, I
Cotswold Cottage
Common Road
Northleigh
Oxfordshire
OX8 6RA
UK

Freidrich Air Conditioning & Refrigeration Co 1,2, H
4200 N PanAm Expressway
Post Office Box 1540
San Antonio
Texas 78295
USA

Frimair 1,2,3, H
Z.1. Sud de Longvic
21600 Longvic
France

Fuji Electric Co 1,2,3, H,I
Energy Technology Dept
New Yurakucho Building
12-1 Yurakucho 1-chome
Chiyoda-ku 1-chome
Japan

GEA Spirogills Ltd 1,2,3,6,7,8,19, G
Greencoat House
Francis Street
London
SW1P 1DH
UK

GEA Warmetauscher Happel KG 27, G,H
A-4673 Gaspoltschofen 00
Austria

GEC Energy Systems 1,2, G,H
Cambridge Road
Whetstone
Leicester
LE8 3LH
UK

GEWE MbH 27, H
Austrasse 19
7315 Weilheim-Teck
West Germany

Gaswarmepumpen Gesellschaft mbH 6,7,8, G,H
Postfach 103252
4300 Essen 1
West Germany

Generale de Fonderie 1,2, H,I
8 Place d'Iena
75783 Paris Cedex 16
France

Girdwood Halton (Air Conditioning) Ltd 22, H
The Street
Hockering
Dereham
Norfolk
NR20 3HL
UK

Gotaverken Energy Systems AB 3,8,20, G,I
POB 8734
S-40275 Gothenburg
Sweden

Grenco, Bedrijfskoeling VB 1,2,3,4,5,6,7,8,9,10, G,H
Postbus 205
S'Hertogenbosch
Netherlands

Herzog Klimatechnik 1,2,3, H,I
Halverstrasse 67
D-5885 Schalksmuhle
West Germany

Hindustan Brown Boveri Ltd 6,7,8, H
Brown Boveri House
Race Course Circle
Baroda-390 007
India

Hiross Ltd 1,2,3, G
Totman Crescent
Weir Industrial Estate
Rayleigh
Essex
SS6 7UY
UK

Hitachi Europe GmbH 1,2,27, H,I
Jagerhofstrasse 32
4000 Dusseldorf
West Germany

Hitachi Ltd 1,2, G,H
Industrial Sales Division
6 Kinda-Surgadai 4-Chome
Chiyoda-ku
Tokyo
Japan

Hitachi Zosen 1, H
Hitachi Shipbuilding & Engineering Co Ltd
6-14 Edobori
1-Chome, Nishi-ku
Osaka 550
Japan

Hubbard Commercial Products Ltd 27, H
26 Perivale Industrial Estate
Horsenden Lane South
Greenford
Middlesex
UB6 7RJ
UK

IC Enterprise 1,2,3, G,H
68 Avenue J.B. Clement
BP 404-92103 Boulogne-Billancourt Cedex
France

IMI Marstair 1, H,I
Bradford Road
Brighouse
West Yorks
HD6 1PT
UK

N.V. Jaga 1,2,3, H,I
Verbindingsweg S/N
B-3610 Diepenbeek
Belgium

R.G. Jolliffe & Co Ltd 6, G
6 Amerley Station Road
London
SE20 8PT
UK

Jyoti Ltd 27, H
Energy Division
Tandalja
Baroda 391 410
India

KKK Ltd 1,2,3, G,J
POB 23
Oxford House
Wellingborough
NorthantsNN8 4JY
UK

KVS-Klimatechnik Geratebau GmbH
7000 Stuttgart 30
Leoberner Strasse 73a
West Germany

Kestner 27, G
7 rue de Toul
BP 44-59003 Lille Cedex
France

Knudsen Koling A/S 1,2, G,H
Rugvanget 10-14
Taastrup 2630
Denmark

Kobe Steel Ltd 27, G
Aluminium & Copper Division
Tekko Building
No. 8-2, 1-Chome
Marunouchi, Chiyoda-ku
Tokyo
Japan

Komatsu Ltd 6,7,8, G,H
Thermal Engineering Dept
2597 Chinomiya
Hiratsuka
Kanagawa-Ken
Japan

H. Krantz GmbH & Co 6, G,H
D 5100 Aachen
West Germany

LOOS 27, H,I
8820 Gunzenhausen
West Germany

Lahmeyer Aktiengesellschaft 27, H
D-5353 Mechernich/Eifel
West Germany

Lebrun 1,2,3, G,H
220 rue de General de Gaulle
59110 la Madeleine
France

Lennox Industries 1,2, H
PO Box 43
Lister Road
Basingstoke
Hampshire
RG22 4AR
UK

Leroy-Somer 1,2, H,I
Bd. Marcellin Leroy
16015 Angouleme Cedex
France

Linde Aktiengesellschaft 1,2,3,6,7,8,11, G,H
Werksqruppe Industriekalte
Postfach 501610
5000 Koln 50 (Surth)
West Germany

Luwa GmbH 1,2,27, H,I
Hanauer Landstr. 200
6000 Frankfurt /M.1
West Germany

MAN Untemehmensbereich 6,7,8,27, G,H
GHH Sterkrade, Bahnhofstrasse 66
Postfach 11 02 40
4200 Oberhausen 11
West Germany

MJN Ltd - Newcastle 27, H
Norris House
Crawhall Road
Newcastle upon Tyne
NE1 2BZ

MacWhirter Heating 1,2,13, G,H,K
Service House
North Road
Cardiff
CF4 3XN
UK

Metro 2, I
Bymoseveij 1-3
DK-3200 Helsinge
Denmark

Mitsubishi 1,2,27, H,I
Ratinger Str. 45
4000 Dusseldorf 1
West Germany

Mitsubishi Electric Corporation 1,2, H,I
Home Appliance & Heating & Air
Conditioning Division
Mitsubishi Electric Building
2-2-3 Marunouchi
Chiyoda-ku
Tokyo
Japan

Mitsubishi Electric UK Ltd 1,2,27, H,I
Hertford Place
Maple Cross
Rickmansworth
Herts
WD3 2BJ
UK

Moducel 1, G,H,K
165 King Street
Fenton
Stoke on Trent
ST4 3ES
UK

Myson Copperad Ltd 1,2,27, H,K
Old Wolverton Road
Old Wolverton
Milton Keynes
Bucks
MK12 5PT
UK

Myson Group plc 1,2, H,I
The Industrial Estate
Ongar
Essex
CM5 9RE
UK

Oakwood Thermo Products 1,2,3,6,7,8, G,H,I
Oak House
Market Place
Macclesfield
Cheshire
SK10 1ER
UK

Officine de Seveso Spa 1,2,27, H,I
Via Orobia 3
20139 Milano
Italy

Pfluger Apparatebau GmbH & Co KG 27, H,I
Postfach 3056
4690 Herne 1
West Germany

Precision Air Control Ltd 1, G
Station Road
Edenbridge
Kent
TN8 6EG
UK

Qualitair (Air Conditioning) ltd 1, H
Castle Road
Eurolink
Sittingbourne
Kent
ME10 3RH
UK

Quiri & Cie 1,2,6,7,8,13,14,15, G
46 route de Bischwiller
67300 Schiltigheim BP190
67042 Strasbourg Cedex
France

RA (Air Conditioning) Company 1, G,H
Hollands Road
Haverhill
Suffolk
CB9 8BT
UK

Recotherm Ltd 1,2, H,K
224 Station Road
Kings Heath
Birmingham
B14 7TE
UK

Rekord 27, G,H
Ruckelshausen GmbH & Co KG
Pf. 1180
6102 Pfungstadt
West Germany

Renault Techniques Nouvelles Appliquees
8-10 avenue Emile Zola
92109 Boulogne-Billancourt
France

Rendamax BV 13,14,15,16,17, G
PO Box 1035
NL-6460 Kerkrade
The Netherlands

Rose Group 6,7,8, G,H
Al North Orbital Trading Estate
Napsbury Lane
St. Albans
Herts
AL1 1XB
UK

SAGEM 1,2,3, G,H
6 avenue d'Iena
75785 Paris Cedex 16
France

Sabroe A/S 6,7,8,9,10, G,H
PO Box 1810
DK 8270 Hoj Bjerg
Denmark

Sanyo Electric Co Ltd 1,2,3,13,14,15,21, G,H
Solar Energy Systems
100 Dainichi Higashimachi
Moriguchi-shi
Osaka
Japan

Sanyo Machine Works Ltd 22, G,H
Solar Division
Okimura
Nishiharu-cho
Nioschikasugai-gun
Aichi
Japan

Saphair Ltd 1,2,3, G,H,J
12 Market Hill
Southam
Warwickshire
CV33 0HF
UK

Saunier Duval 2, I
6 rue Lavoisier
93107 Montreuil-sous-Bois
France

Schafer Werke GmbH 27, H
Geschaftsbereich Heiztechnik
Postfach 1120
5908 Neunkirchen/Pfannenberg
West Germany

Schrieber & Cie 27, H
Krautmuhlenweg 5
5100 Aachen
West Germany

Searle Manufacturing Co Ltd Newgate Lane Fareham Hampshire PO14 1AR UK	2, G
Sekisui Chemical Co Ltd Energy Equipment Division 2-4-4 Kinugasa-cho Kita-ku Osaka Japan	27, G
Siemens AG Postfach 3240 D-8520 Erlangen West Germany	1,2,3,4,5,6,7,8,9,10,12,27, G,H,O
Stal-Levin Ltd River Pinn Works High Street Yiewsley Middlesex UB7 7TA UK	2,3, G,J
Steefane Ltd Reed House The Hill Blunham Bedford MK44 3JE UK	1,2,3, G,H
Stiebel Eltron 25 Lyreden Road Brackmills Northampton NN4 0ED UK	1,2,3, H,I
Stiebel-Eltron 3450 Holzminden West Germany	1,2,3,13,14,15, H,I
Sulzer Bros. (UK) Ltd Westmead Farnborough Hampshire GU14 7LP UK	1,2,3,6,7,8, G,H
Sulzer Brothers Ltd CH-8401 Winterthur Switzerland	3,7,8,11,20, G,H,O

Sulzer Escher Wyss GmbH — 3,7,8,11,20, G,H
D-8990 Lindau-Bodensee
Postfach 1380
West Germany

Sumak GmbH — 27, G,H
7250 Leonberg
Postfach 1565
West Germany

TRW Energy Engineering Division — 25, G
8301 Greensboro Drive
McLean
Virginia 22102
USA

Technibel — 1,2,3, G,H,I
Route Departmentale 28
BP262-01600 Trevoux
France

Tekno Term Systems AB — 27, H,I
Fack
S-201 10 Malmo
Sweden

Teknokyl OY — 24,27, I
PO Box 13
SF-03101 Nummela
Finland

Temperature Ltd — 1,4, G,H,I
Newport Road
Sandown
Isle of Wight
PO36 9PH
UK

Thermal Engineering Systems Ltd — 1,2,3, G,H
Clay Lane
Uffculme
Nr. Cullompton
Devon
EX15 8AJ
UK

Thermal-Werke — 27, H
6832 Hockenheim
Postfach 1680
West Germany

Thermat — 1,2,3, G,H
27 avenue de Penhoet
44608 Saint-Nazaire
France

Thermo Electron Corporation 20, G
101 First Avenue
Waltham
Massachusetts 02154
USA

Toshiba UK Ltd 27, H,I
Toshiba House
Finley Road
Camberley
Surrey
GU16 5JJ
UK

Trace Heat Pumps ltd 1,2,3,4,5,22, G,H,I
Trace House
Eastways Industrial Park
Witham
Essex
CM8 3YJ
UK

Trane (United Kingdom) Ltd 1,2,3, G,H,I,K
1 Gastons Wood
Reading Road
Basingstoke
Hants
RG24 0TW
UK

Trembath Refrigeration Ltd 12, H
414 Purley Way
Croydon
CR9 4BT
UK

Trendpam Engineering Ltd 1,2, H
Unit 17, Barwell Trading Estate
Leatherhead Road
Chessington
Surrey
KT9 2NY
UK

Tri-Therm (Temperature Controls) Ltd 1,2,3,23, G,H
Unit 4
96 Brook Street
Colchester
Essex
CO1 2UZ
UK

Turbo Heat Pumps ltd 1, G,H
6 Hornsby Square
Southfields Industrial Park
Basildon
Essex
SS15 6SD
UK

Uranus SA BP 52 69890 La Tour de Salvagny France	1,12, G,H,I,K
Wartsila Vasa Diesel PO Box 244 SF-65101 Finland	6,7,8, G,H
Wiegand (London) Ltd 34 York Way London N1 9AB UK	21, G
York International Gardiners Lane South Basildon Essex SS14 3HE UK	1,2,3, G,H
Zaegel Held 35 rue du General-Leclerc 67210 Obernai France	2,3, H,I
Zohar Engineering Co Ltd 13 Nakhalat-Benjamin Street 65-161 Tel Aviv Israel	27, H,I

The above data were taken from Heat Recovery Systems & CHP, Vol.7, No.1, 1987.

APPENDIX 4
Conversion Factors

Length	1 ft	= 0.3048 m
Area	1 ft^2	= 0.0929 m^2
	1 in^2	= 6.451 cm^2
Volume	1 ft^3	= 0.0283 m^3
	1 gallon	= 4.546 litres
	1 US gallon	= 0.833 Imperial gallon
Volume Rate of Flow	1 gal/min	= 0.0758 litres/s
Mass	1 lb	= 0.4536 kg
	1 ton	= 1.016 tonnes
	1 tonne	= 1000 kg
Mass Flow Rate	1 lb/hr	= 1.259×10^{-4} kg/s
Density	1 lb/ft^3	= 16.019 kg/m^3
Force	1 lbf	= 4.448 N
Pressure	1 lbf/in^2	= 6.894 kN/m^2
	1 bar	= 10^5 N/m^2
	1 atm.	= 101.325 kN/m^2
Dynamic Viscosity	1 Poise	= 0.1 Ns/m^2
	1 $lbf.s/ft^2$	= 0.047 Ns/m^2
Energy	1 kW h	= 3.6×10^6 J
	1 hp h	= 2.684×10^6 J
	1 Btu	= 1.055 kJ
	1 Btu	= 0.251 k cal
Power	1 hp	= 0.745 kW
	1 hp	= 1.013 metric hp
Temperature	(°F-32) x 5/9	= °C

Quantity of Heat	1 Btu	= 1.055 kJ
	1 k cal	= 4.186 kJ
Heat Flow Rate	1 Btu/h	= 0.293 W
	1 k cal/h	= 1.163 W
Density of Heat Flow	1 Btu/ft²h	= 3.154 W/m²
Thermal Conductivity	1 Btu/ft h °F	= 1.730 W/m°C
Coefficient of Heat Transfer	1 Btu/ft² h °F	= 5.678 W/m²°C
Specific Heat Capacity	1 Btu/lb °F	= 4.186 x 10^3 J/kg °C
Enthalpy	1 Btu/lb	= 2.326 J/g
	1 k cal/kg	= 4.186 J/g
Calorific Value (Volume Basis)	1 Btu/ft³	= 0.037 J/cm³
	1 therm/gal	= 2.32 x 10^4 J/cm³

Energy Equivalents

Calorific Value of fuel oil:	177000 Btu per gallon 186 MJ per gallon
Average efficiency of steam raising:	125 pounds of steam per gallon fuel oil 12.47 kg per litre fuel oil.
Heating value of steam:	1050 Btu per pound 1108 MJ per pound 2443 MJ per kg
One tonne coal equivalent:	27.3 GJ (heat supplied)

Primary fuel equivalent conversion factors:

Fuel oil	46 MJ/litre
Electricity (UK)	14.40 MJ/kWh
Gas	146 MJ/therm
Derv	43 MJ/litre
Other liquid fuels	46 MJ/litre
Coal	29.8 GJ/tonne

APPENDIX 5
Nomenclature

A	Area
COP	Coefficient of performance
h	Enthalpy
I	Current
k	Thermal conductivity
L	Length
ℓ	Electrical resistivity
M	Mass flow rate
M_{He}	Mass flow rate of helium
PER	Primary energy ratio
p	Pressure
Q	Heat flux
R	Ideal gas constant
	also electrical resistance
S	Seebeck coefficient
s	Entropy
T	Temperature
W	Work
X	Concentration
Δt	Temperature difference
η	Efficiency

Index

Absorbers, 29, 97
Absorption cycle, 25, 97
  absorber, 27, 97
  buildings, 80, 195
  domestic, 143
  drying, 236
  evaporation, 102
  fluids, 30
  industry, 261
  solution pumps, 102
Air
  heat sink, 105
  heat source, 109
Air-air heat pumps,
  domestic, 105, 147
Air conditioning, 77, 160
  heat reclaim, 163, 176
  packages, 149
American Air Filter Co., 172
  Enercon, 172
Ammonia/water, 97
Applications, 1
  boiling, 203, 217
  butane splitter, 251
  chemical plant, 279
  commercial buildings, 160
  dairy industry, 253
  decentralised, 172
  dehumidification, 141, 222
  distillation, 242
  district heating, 117, 176
  domestic, 46, 105
  drying, 203, 217
  dyeing, 259
  evaporation, 203, 217
  industrial, 202
  injection moulding, 209
  maize cooking, 281
  refrigeration, 108, 163, 202
  schools, 186
  sports complexes, 189
  steam raising, 268
  textile drying, 228
  timber kilning, 227
  tumble dryers, 140
ASHRAE, 49
Atkins, W. S., 209

Bamber Bridge, 253
Beghin-Say, 234
Boilers, 276
Boiling, 203, 217
Bottle washing, 254
Brayton cycle, 40
British Gas, 195
Building Research Establishment, 129
Buildings
  cooling, 70, 160, 165
  heating, 70, 160
  refrigeration, 163

Capital costs, 8, 79, 105, 189
Carnot, 3
  cycle, 14
Carrier, 144
Case histories
  Bamber Bridge, 253
  British Gas, 195
  Fichtel & Sachs, 153
  Hawaii, 149

IEA, 212
IRD, 266
Lindenberg, 192
MANWEB, 168
Monarch Fine Foods, 257
National Research Council, 257
Northgate Arena, 191
Philips, 123
Revell, 211
Sala, 180
Sulzer, 170
Warendorf, 181
CEM CERCEM, 43
Centrifugal compressors, 62
Climasol, 109
Coefficient of performance, 15, 20, 50
calculation of, 15, 22, 27
Collectors, solar, 120
Commercial buildings, 160
Compressors
centrifugal, 62
oil, 57, 61
reciprocating, 55
reliability - see Case histories
rotary vane, 53
screw, 60
selection, 53, 267
slugging, 57
Condensors, 'double bundle', 166
Control
defrost, 76, 112
performance, 70, 93, 278
thermostats, 94
Conversion factors, 329
Cooling, 70, 108, 162
buildings, 160, 165
Cooling towers, 166
Costs
capital, 79, 105, 189
installation, 80, 105
operating, 5, 79, 189
refrigerant, 52
Cycles
absorption, 25
Brayton, 40
Carnot, 14
comparison, 33
Rankine, 19
Rankine-Rankine, 54
Stirling, 37
thermoelectic, 44

Dairy industry, 253
Decentralised air conditioning, 172
Defrosting, 76, 112
methods, 112
Dehumidification, 223
Denmark, 148, 186
Diesel engine, 186
Distillation, 242
District heating, 117, 176
Domestic
applications, 46, 105
heat storage, 130
space heating, 106
water heating, 148
Dortmund University, 94
'Double bundle' condensor, 167
Drives
electric, 66
gas engine, 34, 67, 96, 154
gas turbine, 41
Merlin, 5
noise, 78
steam turbine, 216
Stirling, 37, 81
Drying
industrial, 203, 217
Dyeing, 259
Dunham-Bush, 89, 166

Economics, 79, 105, 189, 209
Efficiency, isentropic, 58, 65
Electric drives, 66
Electricity Council Research Centre, 92, 204
Electrolux, 1, 25
'Enercon', 172
Engines
gas, 34, 67, 96, 154, 274
Stirling, 37, 81
European Commission, 12
ESDU, 92
Evaporation, 18
applications, 202, 217
Evaporators, 88
flooded, 83
frosting, 112
heat pump, 102
multiple-effect, 237
Expansion, 17

Ferranti, 7
Fichtel & Sachs, 153
Flooded evaporators, 83
Frosting, 112

Gas engines
drives, 34, 67, 96, 154
efficiency, 153
heat recovery, 275
Gas turbines, 41

General Electric, 111
Grain drying, 224
Ground heat, 117, 180

Heat exchangers, 82
  condenser, 83
  'double bundle', 166
  evaporator, 83
  ground, 119
  heat transfer, 82
  intercooler, 21, 33, 91
  run-around coils, 191
  shell and coil, 91
  shell and tube, 89
  sizing, 92
Heating
  district, 116, 176
  space, 106, 160
  supplementary, 111
Heat multipliers, 3
Heat pumps
  absorption cycle, 25, 81, 97
  air-air, 107, 110
  applications, 105, 160, 202
  Brayton, 40
  COP, 22, 50, 111
  costs, 7, 79, 105, 189
  decentralised, 172
  design, 49
  electric, 66
  gas engine-driven, 67
  heat sinks, 108
  heat sources, 109, 124
  history, 3
  maintenance, 76, 262
  Rankine-Rankine, 54
  solar, 120
  Stirling, 37
  thermoelectric, 44
  theory, 14
Heat reclaim
  buildings, 163
Heat recovery
  buildings, 108
  engines, 276
  industry, 202
  swimming pools, 189
Heat sinks
  air, 105, 147
  water, 107
Heat sources
  air, 105, 109
  ground, 117, 180
  industrial effluent, 205
  relative merits, 105
  sewage, 179
  solar, 112, 120, 135
  water, 113, 147, 253
  well water, 113
Heat storage
  domestic, 130
  techniques, 130
Heat transformers, 36, 240
Hermetic compressors, 55
History, 3
HTFS, 92

Industrial applications, 202
  dehumidification, 222
  refrigeration, 202
Ingersol Rand, 59
Injection moulding, 209
Installation costs, 79, 105
Intercooler, 21, 33
Internal combustion engines - see Gas engines
International Energy Agency, 10, 212
Isentropic efficiency, 58, 65

Kelvin, Lord, 1, 3
Kolbusz, 116
Komatsu, 67

Laboratoires de Marcoussis, 46
Lennox, 136
Link, 209
Liquid effluent as heat source, 205
'Liquifrigor', 194
Liquid pumps, 102, 105
Lithium bromide, 31, 99
Lubrication, 59
Lucas Industries, 7, 110

Maintenance, 79, 262
Manufacturers, 308
MAN New Technology, 36
MANWEB building, 168
Materials selection, 93
McQuay, 259
Mechanical efficiency, 20
Mechanical vapour recompression, 205, 237
Metro, 148
Monsanto Chemicals, 242
Multiple effect evaporators, 240

NEI International Research & Development Co., 35, 96, 266
Noise, 78, 94
Nomenclature, 331

Non-azeotropic mixtures, 85
Northgate Arena, 190
Norwich, 5
Nuffield College, Oxford, 7

Office buildings
  air conditioning, 160
  heat pumps, 160
Oil
  compressors, 57
  free, 59
  mixing, 84, 271, 278
  selection, 267
Ontario Hydro, 126
Operating costs, 5, 79, 189, 209

Pay-back period - see Case histories
Peltier effect, 44
Pennsylvania Power & Light Co., 134
PER, 24, 43
Philips NV, 123
Pressure, 218, 269
Prestcold Ltd, 210
Prime movers, 66
Production, 10
Protection of system, 73
Pumps solution, 102, 105

Rankine cycle, 19
Rankine-Rankine cycle, 54
Raoult's law, 31
Reciprocating compressors, 55
Refrigerants
  costs, 52
  and oil, 83, 278
  properties, 49, 218, 289
  selection, 29, 49
  stability, 52, 272, 273
Refrigeration equipment
  heat reclaim, 163, 207
  buildings, 108
  industrial, 207
Revell, 211
Reversibility, 15
  defrosting, 76, 112
Rotary vane compressors, 53
Royal Festival Hall, 5
Run-around coils, 191

Safety, 73
Sanyo, 99, 101
Schools, 186
Screw compressors, 60
Seebeck coefficients, 45
Sewage, 179
Shell and tube heat exchangers, 89
Sizing of heat exchangers, 92
Slugging, 57
Soil temperature, 117
Solar heat, 112, 120
  collectors, 120, 135
Solvents, 29
Solution pumps, 102
Space heating, 105, 106, 161
Split systems, 109, 147
Sports complexes, 189
Start-up, 74
Steam
  as working fluid, 204
  compression, 204, 235
Stiebel Eltron, 103
Stirling engines, 37
Storage
  of heat, 130, 185
Subcooling, 21
Sulzer, 170
Sumner, J. A., 7
Superheating, 17
Supplementary heating, 110
Swimming pools, 189
Switzerland, 3

Temperature Ltd
  Versa Temp, 174
'Templifier', 205, 256
Theory, 14
Thermal stability
  refrigerants, 52, 272, 273
Thermoelectric effect, 44
Thomson, W. - see Kelvin, Lord
Timber kilning, 225, 227
Tokyo Gas Co., 67
Tumbler dryers, 141
Turbines, 40, 215
Turbocompressors, 44

Unigate, 256
Union Carbide, 89
'Unitop', 170

Valves, 57
Vapour recompression
  mechanical, 203, 235
Variable speed compressors, 64
'Versa Temp', 174
Vitatex, 264, 267
Volumetric efficiency, 21

Warendorf, 181
Water
  heat sink, 147
  heat source, 113
  in homes, 147
  in industry, 253
  types, 113, 116
  from wells, 113
Weathertron, 111
  Well water, 113
Westinghouse Templifier, 205, 256
Working fluids
  compatibility, 29
  non-azeotropic mixtures, 85
  selection, 29, 32, 49

Zimmern screws, 61
Zurich, 3